Studies in Systems, Decision and Control

Volume 38

Series editor

Janusz Kacprzyk, Polish Academy of Sciences, Warsaw, Poland
e-mail: kacprzyk@ibspan.waw.pl

About this Series

The series "Studies in Systems, Decision and Control" (SSDC) covers both new developments and advances, as well as the state of the art, in the various areas of broadly perceived systems, decision making and control- quickly, up to date and with a high quality. The intent is to cover the theory, applications, and perspectives on the state of the art and future developments relevant to systems, decision making, control, complex processes and related areas, as embedded in the fields of engineering, computer science, physics, economics, social and life sciences, as well as the paradigms and methodologies behind them. The series contains monographs, textbooks, lecture notes and edited volumes in systems, decision making and control spanning the areas of Cyber-Physical Systems, Autonomous Systems, Sensor Networks, Control Systems, Energy Systems, Automotive Systems, Biological Systems, Vehicular Networking and Connected Vehicles, Aerospace Systems, Automation, Manufacturing, Smart Grids, Nonlinear Systems, Power Systems, Robotics, Social Systems, Economic Systems and other. Of particular value to both the contributors and the readership are the short publication timeframe and the world-wide distribution and exposure which enable both a wide and rapid dissemination of research output.

More information about this series at http://www.springer.com/series/13304

Alexander Barkalov · Larysa Titarenko
Malgorzata Kolopienczyk
Kamil Mielcarek · Grzegorz Bazydlo

Logic Synthesis for FPGA-Based Finite State Machines

Springer

Alexander Barkalov
Institute of Informatics and Electronics
University of Zielona Góra
Zielona Góra
Poland

Kamil Mielcarek
Institute of Informatics and Electronics
University of Zielona Góra
Zielona Góra
Poland

Larysa Titarenko
Institute of Informatics and Electronics
University of Zielona Góra
Zielona Góra
Poland

Grzegorz Bazydlo
Institute of Informatics and Electronics
University of Zielona Góra
Zielona Góra
Poland

Malgorzata Kolopienczyk
Institute of Informatics and Electronics
University of Zielona Góra
Zielona Góra
Poland

ISSN 2198-4182 ISSN 2198-4190 (electronic)
Studies in Systems, Decision and Control
ISBN 978-3-319-24200-2 ISBN 978-3-319-24202-6 (eBook)
DOI 10.1007/978-3-319-24202-6

Library of Congress Control Number: 2015949406

Springer Cham Heidelberg New York Dordrecht London

Printed on acid-free paper

Springer International Publishing AG Switzerland is part of Springer Science+Business Media (www.springer.com)

Preface

The up-to-day state of the art in the computer science is characterized by three major factors. The first factor is a development of ultra complex VLSI such as "system-on-programmable chips" having billions of transistors and hundreds of millions of equivalent gates. The second factor is a development and widespreading hardware description languages (HDL) such as VHDL and Verilog that permits to capture a design with tremendous complexness. The third factor is an existence of different computer-aided design (CAD) tools. It allows designing very complex projects in the satisfactory time-to-market. These three factors affected significantly the modern process of hardware design. Now the hardware design is very similar to the development of computer programs. An application of HDLs together with CAD-tools allows concentrating the designer's energy on the basic problems of design, whereas a routine work remains the prerogative of computers.

Tremendous achievements in the area of semiconductor electronics has turned microelectronics into nanoelectronics. Actually, we observe a real technical boom connected with achievements in nanoelectronics. It results in development of very complex integrated circuits, particularly in the area of field programmable logic devices. Our book targets field-programmable gate arrays (FPGA). First FPGA chips were introduced by Xilinx in 1985, i.e. exactly 30 years ago. They were used mostly as devices implementing simple random and glue logic. Up-to-day FPGAs have up to 7 billion of transistors. So, they are so huge, that it is enough only one chip to implement a rather complex digital system including a datapath and a control unit. Because of the extreme complexity of FPGA chips, it is very important to develop effective design methods targeting their particular properties. It means that the design methods should be technology-depended.

As it is known, any digital system can be represented as a composition of a datapath and a control unit. Logic circuits of operational blocks forming a datapath have regular structures. It allows using standard library elements of CAD tools (such as counters, multibit adders, multipliers, multiplexers, decoders and so on) for their design. A control unit coordinates interplay of other system blocks producing a sequence of control signals (microoperations). These control signals cause executing required actions by a datapath. As a rule, control units have irregular structures. It makes process of its design very sophisticated. Many important features of a digital system, such as hardware amount, performance, power consumption and so on, depend to a large extent on characteristics of its control unit. Therefore, to design competitive digital systems with FPGAs, a designer should have fundamental knowledge in the area of logic synthesis and optimization of logic circuits of control units. As experience of many scientists shows, design methods used by standard industrial packages are far from optimal. Especially it is true in the case of designing complex control units. It means that a designer could be forced to develop his own design methods, next to program them and at last to combine them with standard packages to get a result with desired characteristics. To help such a designer, this book is devoted to solution of the problems of logic synthesis and reduction of hardware amount in control units. We discuss the case when a control unit is represented by the model of finite state machine (FSM). The book contains many original synthesis and optimization methods based on the taking into account the peculiarities of FPGA chips and an FSM model in use. One of the peculiarities of FPGA chips is existence of embedded memory blocks (EMB). We try to implement with EMBs as much of control unit's circuits as it is possible. It results in reducing the irregular part of the control units described by means of Boolean functions. It permits decreasing for the total number of look-up table (LUT) elements in comparison with logic circuits based on known models of FSM. Also, it decreases the number of interconnections in the resulting circuits. In turn, it makes the problem of place-and-routing much simpler. The third benefit is the reducing power dissipation in comparison with FSM circuits implemented only with LUTs. In our book, control algorithms are represented by graph-schemes of algorithms (GSA). This choice is based on obvious fact that this specification provides the simple explanation of the methods proposed by the authors.

To replace LUT-based subcircuits in FSM logic circuits by EMB-based parts, it is necessary to diminish the number of arguments in corresponding systems of Boolean functions. To do it, we propose using the methods of structural decomposition, such as: (1) the replacement of logical conditions; (2) the encoding of the collections of microoperations; (3) the transformation of FSM objects, when the internal states are represented as functions of microoperations and vice versa. Also, we use the existence of the classes of pseudoequivalent states of Moore FSMs to compress structure tables and diminish the number of state variables. It simplifies the system of input memory functions and, therefore, decreases the number of EMBs in the resulting FSM circuit. We combine this approach with using EMBs for implementing the system of output functions (microoperations). It allows a significant decreasing for the number of LUTs, as well as eliminating a lot of interconnections in the FSM logic circuit. It saves area occupied by the circuit and diminishes the resulting power dissipation. Of course, it leads to more sophisticated synthesis process than the one targeting only LUT-based circuits.

The process of FSM logic synthesis is reduced to a transformation of a control algorithm into some tables describing the behaviour of FSM blocks. These tables are used to find the systems of Boolean functions, which can be used to implement logic circuits of particular FSM blocks. In order to implement corresponding circuits, this information should be transformed using data formats of particular industrial CAD systems. We do not discuss this step is in our book. Our book contains a lot of examples showing design of FSMs with using the proposed methods. Some examples are illustrated by logic circuits.

The book contains Introduction, eight chapters and Conclusion. It is written by a research group from University of Zielona Góra (Poland). Professors Alexander Barkalov and Larysa Titarenko wrote Introduction, two first chapters and Conclusion. Chapters 3 and 4 are written by Ph.D. Kamil Mielcarek. Chapters 5 and 6 are written by Ph.D. Grzegorz Bazydlo. Chapters 7 and 8 are written by Ph.D. Malgorzata Kolopienczyk.

Chapter 1 provides some basic issues connected with finite state machines and field programmable logic devices. The basic models of Mealy and Moore FSMs are presented. The classical design methods are presented for three kinds of control units: microprogrammed automata, microprogram control units and compositional microprogram control units. Main methods of hardware reduction are given for finite state machines implemented with field programmable logic devices, such as the replacement of logical conditions, encoding of collections of microoperations and encoding the fields of compatible microoperations.

Chapter 2 is devoted to application of field programmable gate arrays in design of logic circuits of FSMs. The basic features of FPGA are analysed. It is shown that embedded memory blocks allow effective implementing systems of regular Boolean functions. Next, the basic problems of FSM design are considered. Different state assignment methods are analysed in details, as well as the methods of functional decomposition. At last, there are discussed the methods of hardware reduction for FPGA-based FSMs.

Chapter 3 presents the original methods of hardware reduction based on the transformation of object codes of Mealy FSMs. The principle of object code transformation (OCT) is introduced. Two types of objects are introduced: internal states and collections of microoperations. Two types of basic models of Mealy FSMs with OCT are described. The design methods are proposed for the EMB-based FSMs with transformation of the states into the collections of microoperations. Next, the design methods are shown allowing the transformation of the collections of microoperations into the states. The models of FSMs with the replacement of logical conditions and OCT are discussed. At last, the analysis of the proposed methods is executed giving conditions of their application.

Chapter 4 deals with the original methods of hardware reduction based on the transformation of object codes of Moore FSMs. Two types of basic models of Moore FSMs with OCT are described, as well as EMB-based structures corresponding to these models. The design methods are proposed for the EMB-based FSMs with transformation of states into the collections of microoperations. Next, the design methods are shown allowing the transformation of the collections of microoperations into the states. The models of FSMs with the replacement of logical conditions and OCT are discussed. The additional hardware reduction is achieved due to using the classes of pseudoequivalent states.

Chapter 5 deals with optimization of logic circuits of Moore FSMs based on using two and three sources of codes of classes of pseudoequivalent states (PES). First of all, the application of this method for CPLD-based FSMs is discussed. Next, the models with two sources of class codes are discussed and corresponding design methods are proposed. This approach requires the usage of a multiplexor to choose a particular source. Also, the models with three sources of class codes are discussed and corresponding design methods are proposed. It is shown how the replacement of logical conditions can be used in multisource models of FSMs. At last, it is shown that the hardware reduction can be obtained due to increasing the number of class variables.

Chapter 6 is devoted to hardware reduction based on using many directions of input memory functions in Moore FSMs. First, the hardware reduction methods are proposed for the two-directional Moore FSMs. They are based on the special state assignment allowing the decreasing for the number of literals in sum-of-products representing input memory functions. Next, the design methods are proposed for the three-directional Moore FSMs. It is shown that the number of directions can be increased. It leads to simplifying the input memory functions in comparison with the single-directional models. The last part of the chapter is devoted to combining the replacement of logical conditions with many directions of state codes.

Chapter 7 deals with design of Mealy FSMs based on using embedded memory blocks. The methods of trivial EMB-based implementation of logic circuits of Mealy FSMs are discussed. In this case, only one EMB is enough for implementing the circuit. Next, the optimization methods are discussed based on encoding of the collections of microoperations and replacement of logical conditions. Also, the methods are discussed based on encoding of the rows of FSM structure table. All these methods lead to two-level models of Mealy FSMs. Next, these methods are combined together for further optimizing the hardware amount in FSM logic circuits. The last section considers different methods proposed for diminishing the hardware amount in LUTer implementing the block of replacement of logical conditions. The chapter includes a lot of tables with results of investigations of proposed methods for the standard benchmarks.

Chapter 8 deals with design of Moore FSMs based on using embedded memory blocks. The methods of trivial EMB-based implementation of logic circuits of Moore FSMs are discussed. In this case, only a single EMB is enough for implementing the logic circuit. Next, the optimization methods are discussed based on the structural decomposition leading to two-level models of FSMs. It is shown how to use the classes of PES for decreasing the number of EMBs in the final circuit. The last section considers different methods proposed for diminishing the hardware amount in LUTer implementing the block of replacement of logical conditions. It is shown that at least 17 different models can be used for optimizing the LUTer.

We hope that our book will be interesting and useful for students and Ph.D. students in the area of Computer Science, as well as for designers of modern digital systems. We think that proposed FSM models enlarge the class of models applied for implementation of control units with modern FPGA chips.

Zielona Góra
May 2015

Alexander Barkalov
Larysa Titarenko

Contents

1 Background of Finite State Machines and Programmable Logic . . . 1
1.1 Basic Models of FSMs . . . 1
1.2 Logic Synthesis for Microprogram Automata . . . 4
1.3 Logic Synthesis for Microprogram Control Units . . . 9
1.4 Logic Synthesis for Compositional MCUs . . . 15
1.5 Hardware Reduction for FPLD-Based FSMs. . . . 21
References . . . 29

2 Field Programmable Gate Arrays in FSM Design . . . 33
2.1 General Characteristic of FPGAs. . . . 33
2.2 Trivial Implementing FPGA-Based FSMs. . . . 38
2.3 Methods of State Assignment . . . 45
2.4 Hardware Reduction for FPGA-Based FSMs . . . 50
References . . . 60

3 Object Codes Transformation for Mealy FSMs . . . 65
3.1 Principle of OCT for Mealy FSMs . . . 65
3.2 Synthesis of FPGA-Based Mealy FSMs with Transformation of States . . . 73
3.3 Synthesis of FPGA-Based Mealy FSMs with Transformation of CMOs . . . 80
3.4 Replacement of Logical Conditions in Mealy FSMs with OCT . . . 87
3.5 Analysis of Proposed Methods . . . 90
References . . . 95

4 Object Codes Transformation for Moore FSMs . . . 97
4.1 Principle of OCT for Moore FSMs . . . 97
4.2 Synthesis of FPGA-Based Moore FSMs with Transformations of States . . . 104
4.3 Synthesis of FPGA-Based Moore FSMs with Transformation of CMOs . . . 112

4.4 Replacement of Logical Conditions in Moore FSMs with OCT 120
References 126

5 Distribution of Class Codes in Moore FSMs 129
5.1 The Case of CPLD-Based FSMs. 129
5.2 Two Sources of Class Codes in FPGA-Based Moore FSMs 133
5.3 Three Sources of Class Codes in FPGA-Based Moore FSMs 141
5.4 Replacement of Logical Conditions and Distribution of Class Codes 143
5.5 Increasing the Number of Class Variables 147
References 158

6 Hardware Reduction in Multidirectional Moore FSMs 161
6.1 Hardware Reduction in Two-Directional Moore FSMs. 161
6.2 Hardware Reduction in 3-Directional Moore FSMs 171
6.3 Replacement of Logical Conditions for K-Directional Moore FSMs 186
References 191

7 Design of EMB-Based Mealy FSMs 193
7.1 Trivial Implementing Mealy FSMs 193
7.2 Encoding of Objects in Mealy FSMs. 200
7.3 Replacement of Logical Conditions for Mealy FSMs. 211
7.4 Hardware Reduction for BRLC. 223
References 236

8 Design of EMB-Based Moore FSMs 239
8.1 Trivial Implementing Moore FSMs 239
8.2 Structural Decomposition for Moore FSMs. 243
8.3 Optimization of BIMF Based on Pseudoequivalent States. 251
8.4 Optimizing LUTer in Replacement of Logical Conditions 257
References 276

Conclusion 277

Index 279

Abbreviations

BF	Block of functions
BGSA	Block graph-scheme of algorithm
BIMF	Block of input memory functions
BM	Benchmark
BMO	Block of microoperations
BRAM	Block of random access memory
BRLC	Block of replacement of logical conditions
BSCT	Block of state codes transformer
CAD	Computer-aided design
CLB	Configurable logic block
CMCU	Compositional microprogram control unit
CMO	Collection of microoperations
CPLD	Complex programmable gate arrays
CT	Counter
DMO	Distribution of microoperations
DRAM	Distributed RAM
EMB	Embedded memory block
EMBer	Logic block consisting from EMBs
FCO	Fields of compatible microoperations
FPGA	Field-programmable gate arrays
FPLD	Field-programmable logic devices
FSM	Finite state machine
FT	Formula of transitions
GAL	Generic array logic
GFT	Generalized formula of transitions
GSA	Graph-scheme of algorithm
IOB	Input–output block
LUT	Look-up table
LUTer	Logic block consisting from LUTs
MCU	Microprogram control unit
MI	Microinstruction

MO	Microoperation
MPA	Microprogrammed automaton
MPI	Matrix of programmable interconnections
MX	Multiplexer
OCT	Object codes transformation
OLC	Operational linear chain
PAL	Programmable array logic
PEO	Pseudoequivalent outputs
PES	Pseudoequivalent states
PIA	Programmable interconnection array
PLA	Programmable logic array
PLS	Programmable logic sequencer
PROM	Programmable read-only memory
RAM	Random access memory
RG	Register
RLC	Replacement of logical conditions
ROM	Read-only memory
SBF	System of Boolean functions
SFT	System of formulae of transitions
SOP	Sum of products
ST	Structure table

Chapter 1
Background of Finite State Machines and Programmable Logic

1.1 Basic Models of FSMs

Finite state machines (FSM) are sequential devices whose outputs depend on prehistory of operation [1, 5]. As any digital circuit, an FSM can be represented as a "black box" having inputs $x_e \in X$ and outputs $y_n \in Y$ (Fig. 1.1a).

Let the symbol $X(t)$ stand for the value of inputs in the instant of time $t(t = 0, 1, 2, \ldots)$. Let the symbol $Y(t)$ stand for the value of outputs in the instant of time $t(t = 0, 1, 2, \ldots)$. An FSM starts its operation in the instant $t = 0$. This instant is determined by the pulse *Start*. Each instant of time t is determined by the pulse *Clock*.

The prehistory of operation is determined by the vector $\langle X(0), X(1), \ldots, X(t-1)\rangle$. It means that each output function $y_n (n \in 1, \ldots, N)$ is represented as:

$$y_n(t) = f_n(X(0), X(1), .., X(t-1), X(t)). \tag{1.1}$$

To be able to implement FSM circuits, it is necessary to simplify functions (1.1) [6]. To do it, the internal states are used representing the prehistory of FSM's operation [32].

Internal states form the set $A = \{a_1, \ldots, a_M\}$. They are represented by binary strings called state codes [7]. It is enough R binary variables $T_r \in T$ to encode the states of FSM, where:

$$\lceil log_2 M \rceil \leq R \leq M. \tag{1.2}$$

State codes are named "binary" if $R = \lceil log_2 M \rceil$; they are named "one-hot" if $R = M$ [13, 46].

To keep state codes, a state memory is used. If an FSM state is changed, then the content of state memory should be changed, too. To do it, they use input memory functions creating the set $\Phi = \{\phi_1, \ldots, \phi_R\}$.

A. Barkalov et al., *Logic Synthesis for FPGA-Based Finite State Machines*, Studies in Systems, Decision and Control 38, DOI 10.1007/978-3-319-24202-6_1

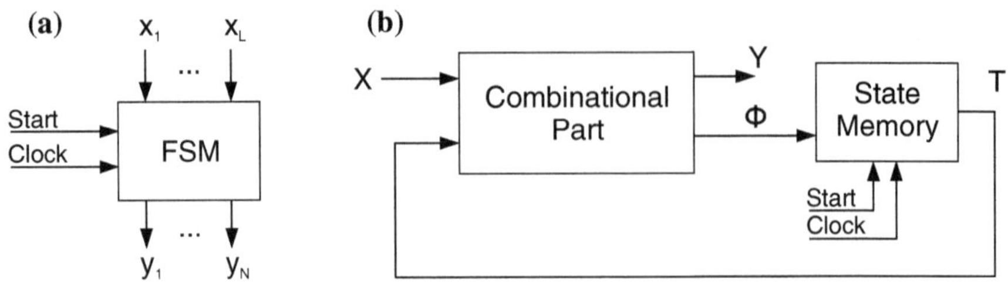

Fig. 1.1 Simplest modes of FSMs

As a rule, an FSM is represented as a composition of a combinational part and the state memory (Fig. 1.1b) [6]. The combinational part implements systems:

$$Y = Y(T, X), \tag{1.3}$$

$$\Phi = \Phi(T, X). \tag{1.4}$$

There are two main models of FSMs. In the Mealy FSM, Eq. (1.3) determines output functions. In the case of Moore FSM, output functions depend only on state variables:

$$Y = Y(T). \tag{1.5}$$

In the practice of digital design, they use D flip-flops to implement the state memory [16, 37]. In this case, the state memory is represented as a register (RG) having R of flip-flops. It gives the set $\Phi = \{D_1, \ldots, D_R\}$. We use this approach in our book.

To construct functions (1.1)–(1.5), it is necessary to describe the behaviour of FSM. It can be done in many ways [6, 43]. In this book, we use the language of graph-schemes of algorithm (GSA) to represent an FSM [6, 7]. Let us name FSM inputs as "logical conditions" and FSM output functions as "microoperations".

An GSA Γ is a directed connected graph having finite set of vertices [6]. There are 4 different types of vertices shown in Fig. 1.2: start, end, operator and conditional.

There is only output in the *Start* vertex (Fig. 1.2a). This vertex corresponds to the beginning of operation. There is only input in the *End* vertex (Fig. 1.2b). This vertex corresponds to the finishing of operation. An operator vertex contains a collection of

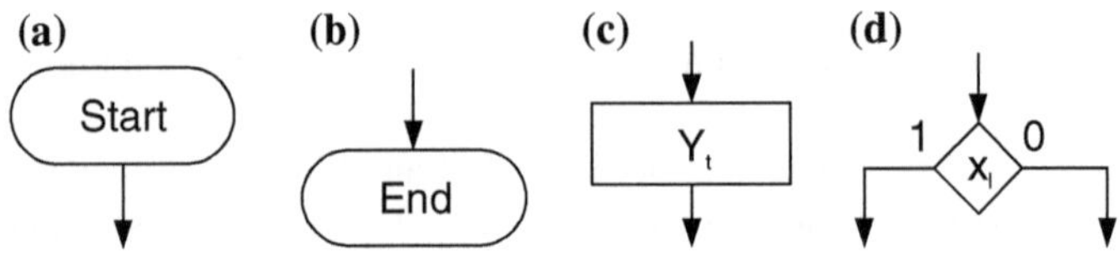

Fig. 1.2 Types of vertices

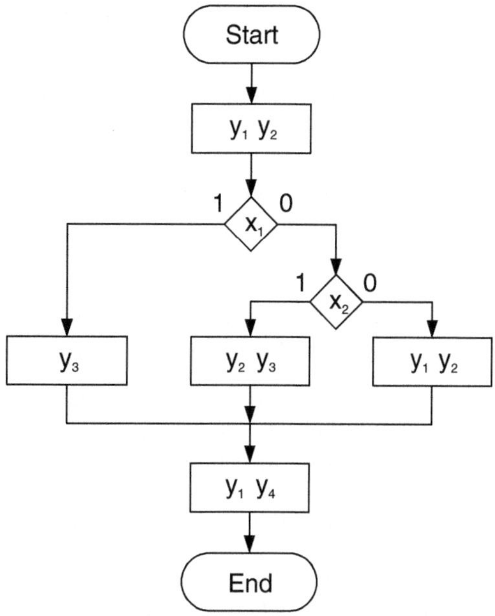

Fig. 1.3 Graph-scheme of algorithm Γ_1

microoperations (CMO) $Y_t \in Y$. These microoperations are produced and executed during one cycle of FSM's operation. A conditional vertex contains an element of the set X used for branching. There are two different values of Boolean variable $x_e \in X$. An example of GSA is shown in Fig. 1.3.

Two sets can be derived from GSA Γ_1: the set of logical conditions $X = \{x_1, x_2\}$ and the set of microoperations $Y = \{y_1, \ldots, y_4\}$. There are the following collections of microoperations in the operator vertices of GSA Γ_1: $Y_1 = \{y_1, y_2\}$, $Y_2 = \{y_3\}$, $Y_3 = \{y_2, y_3\}$ and $Y_4 = \{y_1, y_4\}$. There is no implicit states visible on GSA Γ_1. This set should be constructed.

FSMs are used for implementing logic circuits of control units [33, 34]. The methods of logic synthesis depends strongly on the following factors:

1. Type of a control unit.
2. Technology (devices family) used for implementing logic circuits.
3. Peculiarities of an FSM model.

There are three main types of control units used in digital systems [12, 37]:

1. Microprogram automata (MPA).
2. Microprogram control units (MCU).
3. Compositional microprogram control units (CMCU).

Let us discuss the peculiarities of these control units.

1.2 Logic Synthesis for Microprogram Automata

Let us start from the simplest Mealy MPA. It is named P Mealy MPA [16] and consists from two main blocks (Fig. 1.4). The block of functions (BF) implements the systems (1.3) and (1.4). It is a combinational part of MPA. The register RG implements the state memory having R_0 of flip-flops:

$$R_0 = \lceil log_2 M_0 \rceil. \quad (1.6)$$

In (1.6), the symbol M_0 stands for the number of states of Mealy MPA.

There are the following steps in the synthesis method of P Mealy MPA [6]:

1. Constructing a marked GSA and finding the set of internal states A.
2. Executing the state assignment.
3. Constructing the structure table of MPA.
4. Deriving functions Y and Φ from the structure table.
5. Implementing MPA logic circuit using particular logic elements.

Let the symbol $P(\Gamma_j)$ stand for the case when the P Mealy MPA is synthesized starting from a GSA Γ_j. Let us discuss an example of synthesis for the $P(\Gamma_1)$ Mealy MPA.

To construct the marked GSA, the following rules are used [6, 7]:

- state a_1 marks the output of *Start* vertex and the input of *End* vertex;
- there is a mark of state $a_m (m \in \{2, \ldots, M_0\})$ on the input of any vertex following an operator vertex;
- each input is marked only once.

Using these rules, the marked GSA Γ_1 is obtained (Fig. 1.5a). It is easy to find the set $A = \{a_1, a_2, a_3\}$ for the Mealy MPA $P(\Gamma_1)$ MPA. It gives the value of $R_0 = 2$ and the sets $T = \{T_1, T_2\}$ and $\Phi = \{D_1, D_2\}$.

During the second step, a binary code $K(a_m)$ is assigned for each state $a_m \in A$. This step is named the state assignment [32]. There are many variants for executing the state assignment. As a rule, the zero code is assigned for the initial state $a_1 \in A$. So, there are $2^{R_0} - 1$ different ways for encoding the state $a_2 \in A$. Also, there are

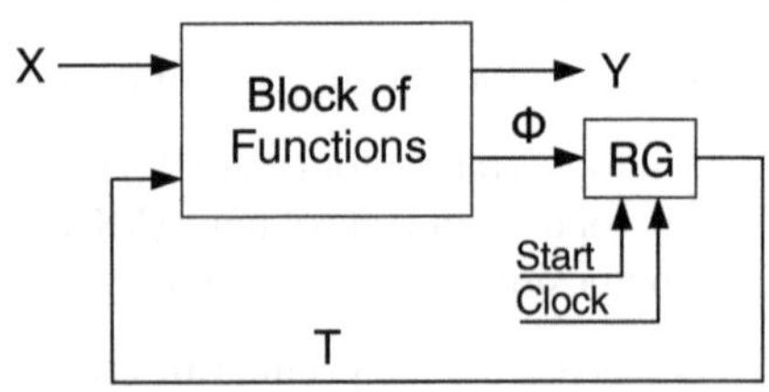

Fig. 1.4 Structural diagram of P Mealy MPA

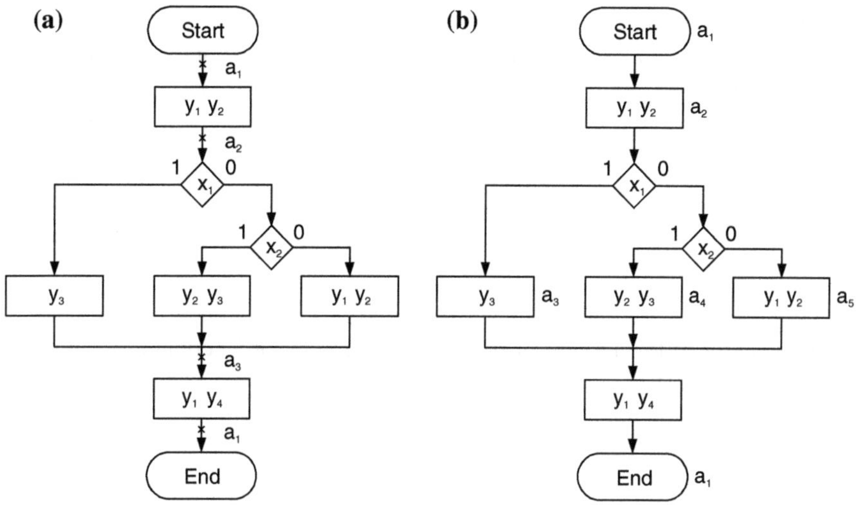

Fig. 1.5 Marked GSA Γ_1 with Mealy (**a**) and Moore (**b**) states

$2^{R_0} - 2$ different ways for encoding the state $a_3 \in A$. The following formula shows the quantity of different variants of the state assignment (n_{SA}):

$$n_{SA} = \frac{(2^{R_0} - 1)!}{(2^{R_0} - (M_0 - 1))!}. \tag{1.7}$$

Let us use the simplest way of state assignment in the discussed case. It is reduced to assignment of a binary equivalent of the number $(m - 1)$ for the state $a_m \in A$. It gives the following codes: $K(a_1) = 00$, $K(a_2) = 01$, $K(a_3) = 10$.

A structure table (ST) of MPA represents transitions between states $a_m \in A$. It is constructed using the rules [6, 7]. The ST includes the following columns: a_m is a current state; $K(a_m)$ is a code of the state $a_m \in A$; a_s is a next state (a state of transition); $K(a_s)$ is a code of the state $a_s \in A$; X_h is an input signal determining the transition $\langle a_m, a_s \rangle$ and it is equal to the conjunction of some logical conditions (or their complements); Y_h is a CMO generated during the transition $\langle a_m, a_s \rangle$; Φ_h is a set of input memory functions equal to 1 to load the code $K(a_s)$ into RG; h is a number of transition ($h \in \{1, \ldots, H_0\}$). Table 1.1 is an ST for Mealy MPA $P(\Gamma_1)$. It includes $H_0 = 5$ rows.

Each row h of ST corresponds to a function $F_h (h = \overline{1, H_0})$:

$$F_h = A_{mh} X_h. \tag{1.8}$$

In (1.8), the symbol A_{mh} stands for the conjunction of state variables $T_r \in T$ corresponding to the code $K(a_m)$ from the h-th row of $ST(h = \overline{1, H_0})$. Functions F_h are used as product terms of functions (1.3) and (1.4). Because of it, we use the name "term" for any function $F_h (h = \overline{1, H_0})$.

Table 1.1 Structure table of Mealy MPA $P(\Gamma_1)$

a_m	$K(a_m)$	a_s	$K(a_s)$	X_h	Y_h	Φ_h	h
a_1	00	a_2	01	1	y_1y_2	D_2	1
a_2	01	a_3	10	x_1	y_3	D_1	2
		a_3	10	$\bar{x_1}x_2$	y_2y_3	D_1	3
		a_3	10	$\bar{x_1}\bar{x_2}$	y_1y_2	D_1	4
a_3	10	a_1	00	1	y_1y_4	–	5

There are the following sum-of-products (SOP) forms of functions (1.3) and (1.4) derived from an ST:

$$y_n = \bigvee_{h=1}^{H_0} C_{nh}F_h(n = \overline{1, N}); \tag{1.9}$$

$$D_r = \bigvee_{h=1}^{H_0} C_{rh}F_h(r = \overline{1, R_0}). \tag{1.10}$$

In these equations, $C_{nh}(C_{rh})$ is a Boolean variable equal to 1 if and only if (iff) the symbol $y_n(D_r)$ is written in the h-th row of $ST(h = \overline{1, H_0})$.

The following expressions can be derived from Table 1.1 for the terms:

$$\begin{aligned} F_1 = \bar{T_1}\bar{T_2};\ \ F_2 = \bar{T_1}T_2x_1;\ \ F_3 = \bar{T_1}\bar{T_2}\bar{x_1}x_2; \\ F_4 = \bar{T_1}T_2\bar{x_1}\bar{x_2};\ \ F_5 = T_1\bar{T_2}. \end{aligned} \tag{1.11}$$

There is no state having the code 11. So, this code can be treated as "don't care" [31]. Using this code, it is possible to get the following system of terms:

$$\begin{aligned} F_1 = \bar{T_1}\bar{T_2};\ \ F_2 = T_2x_1;\ \ F_3 = T_2\bar{x_1}x_2; \\ F_4 = T_2\bar{x_1}\bar{x_2};\ \ F_5 = T_1. \end{aligned} \tag{1.12}$$

Using rules of Boolean algebra [31], it is possible to get the minimized SOP of functions (1.9) and (1.10):

$$\begin{aligned} &y_1 = F_1 \vee F_4 = \bar{T_1}\bar{T_2} \vee T_2\bar{x_1}\bar{x_2} \vee T_1; \\ &y_2 = F_1 \vee F_3 \vee F_4 = \bar{T_1}\bar{T_2} \vee T_2x_1; \\ &y_3 = F_2 \vee F_3 = T_2x_1 \vee T_2x_2; \\ &y_4 = F_5 = T_1; \\ &D_1 = F_2 \vee F_3 \vee F_4 = T_2; \\ &D_2 = F_1 = \bar{T_1}\bar{T_2}. \end{aligned} \tag{1.13}$$

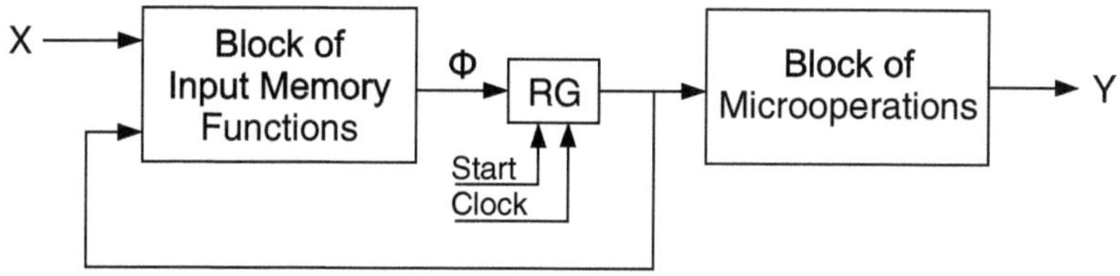

Fig. 1.6 Structural diagram of P Moore MPA

These expressions are used for implementing the logic circuit of Mealy FSM $P(\Gamma_1)$. The circuit of BF is implemented using the system (1.13). The circuit of RG is implemented in a trivial way. It is a library element for computer-aided design (CAD) system [21]. The last step of discussed method depends on logic elements in use. We will discuss it a bit later.

Now, let us discuss the structural diagram of P Moore MPA (Fig. 1.6). It includes the register RG and two combinational blocks. The block of input memory functions (BIMF) implements the system (1.4). The block of microoperations (BMO) implements the system (1.5).

There are M_1 of states in the set A for a Moore MPA. To encode them, it is required at least R_1 of state variables:

$$R_1 = \lceil log_2 M_1 \rceil. \tag{1.14}$$

There are the same steps in synthesis of Mealy and Moore MPAs. But they are executed in a different way. Let us discuss an example of synthesis for Moore MPA $P(\Gamma_1)$.

To construct the set A, the following rules are used [6]:

- both *Start* and *End* vertices are marked by state a_1;
- each operator vertex is marked by a unique state.

Applying these rules to GSA Γ_1 (Fig. 1.3) leads to the marked GSA shown in Fig. 1.5b.

There are $M_1 = 6$ elements in the set A. It gives us $R_1 = 3$, $T = \{T_1, T_2, T_3\}$ and $\Phi = \{D_1, D_2, D_3\}$. Let us encode the states $a_m \in A$ in the trivial way: $K(a_1) = 000, \ldots, K(a_6) = 101$.

There are the following columns in ST of Moore MPA: a_m, $K(a_m)$, a_s, $K(a_s)$, X_h, Φ_h, h. There is an CMO $Y(a_m) \subseteq Y$ in the column a_m. The CMO $Y(a_m)$ includes microoperations written in the operator vertex of a GSA marked by the state $a_m \in A$. In the discussed case, there are $H_1 = 8$ rows in the ST (Table 1.2).

This table is used for deriving systems (1.4) and (1.5). The terms F_h of (1.4) are determined by (1.8). For example, there are the following terms: $F_1 = \bar{T}_1\bar{T}_2\bar{T}_3$, $F_2 = \bar{T}_1\bar{T}_2T_3x_1$, $F_5 = T_2\bar{T}_3$. To get the term F_5, the “don't' care” assignment 110 is used.

Table 1.2 Structure table of Moore MPA $P(\Gamma_1)$

a_m	$K(a_m)$	a_S	$K(a_S)$	X_h	Φ_h	h
$a_1(-)$	000	a_2	001	1	D_3	1
$a_2(y_1y_2)$	001	a_3	010	x_1	D_2	2
		a_4	011	$\bar{x}_1x_2$	D_2D_3	3
		a_5	100	$\bar{x}_1\bar{x}_2$	D_1	4
$a_3(y_3)$	010	a_6	101	1	D_1D_3	5
$a_4(y_2y_3)$	011	a_6	101	1	D_1D_3	6
$a_5(y_1y_2)$	100	a_6	101	1	D_1D_3	7
$a_6(y_1y_4)$	101	a_1	000	1	–	8

Functions $y_n \in Y$ are determined by the following expression:

$$y_n = \bigvee_{r=1}^{R_1} C_{nm}A_m (n = \overline{1, N}). \tag{1.15}$$

In (1.15), the Boolean variable C_{nm} is equal to 1 iff there is the relation $y_n \in Y(a_m)$. The conjunction A_m is determined as the following:

$$A_m = \bigvee_{r=1}^{R_1} T_r^{l_{mr}} (m = \overline{1, M_1}). \tag{1.16}$$

In (1.6), the symbol l_{mr} stands for the value of r-th bit of the code $K(a_m)$, where $l_{mr} \in \{0, 1, *\}$ and $T_r^0 = \bar{T}_r$, $T_r^1 = T_r$, $T_r^* = 1$ $(r = \overline{1, R_1})$. The symbol "*" corresponds to the "don't care" value of a particular bit.

Let us point out that the system (1.15) can be represented by a truth table [31] with inputs $T_r \in T$ and outputs $y_n \in Y$. In the discussed case, it is Table 1.3.

There is the obvious connection between Tables 1.2 and 1.3. Functions $y_n \in Y$ are determined for more than 50 % of possible state assignments. It takes place for any Moore FSM if they use exactly R_1 of state variables. It means that functions (1.5) are regular [13]. In the next chapters, we show how to use this property.

Table 1.3 Truth table of system Y

$K(a_m)$	$Y(a_m)$	m	$K(a_m)$	$Y(a_m)$	m
$T_1T_2T_3$	$y_1y_2y_3y_4$		$T_1T_2T_3$	$y_1y_2y_3y_4$	
000	0000	1	100	1100	5
001	1100	2	101	1001	6
010	0010	3	110	****	*
011	0110	4	111	****	*

Using rules of expansion, the following equations can be extracted from Table 1.3:

$$y_1 = \bar{T}_1\bar{T}_2T_3 \vee T_1; \quad y_2 = \bar{T}_1T_3 \vee T_1\bar{T}_3; \quad y_3 = T_2; \quad y_4 = T_1T_3. \tag{1.17}$$

These equations include 6 terms having 11 literals. There are 8 terms having 15 literals in functions $y_n \in Y$ from (1.13). So, functions (1.5) are simpler than functions (1.3) for equivalent Mealy and Moore MPAs.

MPAs are equivalent if they are synthesized for the same GSA Γ. There are the following relations for equivalent Mealy and Moore MPAs:

$$R_0 \leq R_1; \tag{1.18}$$

$$H_0 \leq H_1. \tag{1.19}$$

It seems that using models of Mealy MPAs always leads to circuits with less amount of hardware. But it is not true due to two properties of Moore FSMs [10, 13]:

1. System Y is a system of regular functions.
2. Set A can be divided by classes of pseudoequivalent states.

States $a_m, a_i \in A$ are pseudoequivalent states (PES) if they mark operator vertices whose outputs are connected with the input of the same vertex of a GSA Γ [10]. Using this definition, it could be found a partition $\Pi_A = \{B_1, \ldots, B_I\}$ where B_i is a class of PES.

In the discussed case, there is the partition $\Pi_A = \{B_1, .., B_4\}$ with the classes $B_1 = \{a_1\}, B_2 = \{a_2\}, B_3 = \{a_3, a_4, a_5\}$ and $B_4 = \{a_6\}$. It follows from the analysis of GSA Γ_1 (Fig. 1.5b).

Let us assume that each class $B_i \in \Pi_A$ can be represented by a single block of a GSA having inputs $a_m \in B_i$ and only one output. It leads to a block GSA (BGSA). In the discussed case, we can get the BGSA $B(\Gamma_1)$ (Fig. 1.7).

In the next chapters, we show how to use BGSAs for reducing hardware amount in circuits of Moore MPAs.

1.3 Logic Synthesis for Microprogram Control Units

Microprogram control units were proposed by Wilkes [44, 45]. In 1951, he proposed to use the intermediate level between the computer software and hardware. This level is now named "firmware" [2, 3]. According to the principle of microprogrammed control, each computer instruction is represented as a microprogram. A microprogram is a sequence of microinstructions (MI) kept into a control memory of MCU. There are the following steps [1, 16] in the typical method of MCU's design:

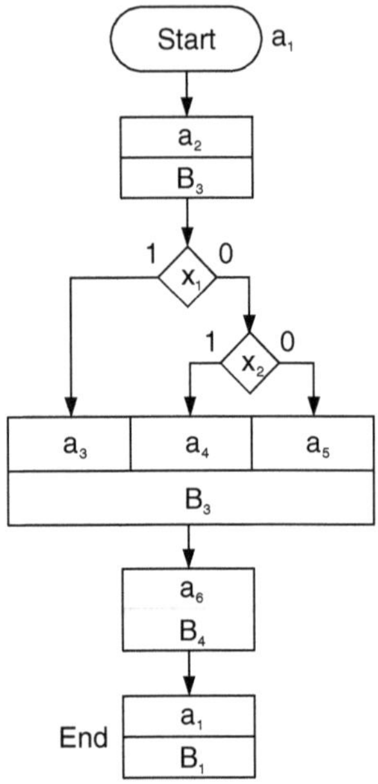

Fig. 1.7 Block graph-scheme of algorithm $B(\Gamma_1)$

1. Transformation of initial GSA Γ.
2. Constructing the set of microinstructions with a given format.
3. Executing the microinstruction addressing.
4. Executing the encoding of the fields of MIs.
5. Constructing the MCU logic circuit with particular logic elements.

The peculiarities of a design method depend tremendously on the mode of microinstruction addressing [42]. There are three main methods of microinstruction addressing [36].

1. Compulsory addressing.
2. Natural addressing.
3. Combined addressing.

A microinstruction is divided by fields such as: FY, FX, FA_0, FA_1. The field FY is an operational part of MI. It contains a code $C(Y_t)$ of a CMO $Y_t \subseteq Y$ evoked by a microinstruction. The field FX contains a code $C(x_l)$ of a logical condition $x_e \in X$ checked to find an address of the next MI to be executed. The field FA_0 contains an address of transition for two cases: (a) unconditional transition and (b) transition if $x_e = 0$. The field FA_1 contains an address of transition for the case when $x_e = 1$. In the case of compulsory addressing, there are all these fields in the microinstruction format (Fig. 1.8).

Let us use the following denotations: x_e^t is a logical condition checked in the instant of time t; $[FX]^t$, $[FA_0]^t$, $[FA_1]^t$ is a content of the corresponding field in the instant of time t; A^t is an address of microinstruction executed in the instant of time

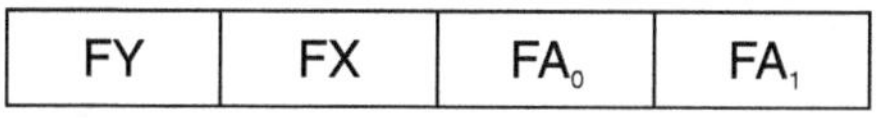

Fig. 1.8 Format of microinstruction with compulsory addressing

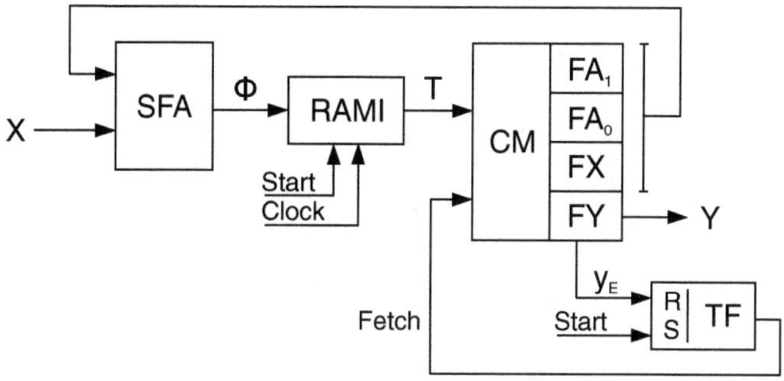

Fig. 1.9 Structural diagram of MCU with compulsory addressing of microinstructions

$t(t = 0, 1, \ldots)$. Let us presume that the following law is used to determine the address of transition A^{t+1}:

$$A^{t+1} = \begin{cases} [FA_0]^t \text{ if } x_e^t = 0; \\ [FA_0]^t \text{ if } [FX]^t = \emptyset; \\ [FA_1]^t \text{ if } x_e^t = 1. \end{cases} \quad (1.20)$$

The second line in (1.20) corresponds to the unconditional transition.

There are four main blocks in MCU with compulsory addressing (Fig. 1.9): address sequencer SFA; register of microinstruction address RAMI; control memory CM and fetch flip-flop TF.

There is the following mode of operation of MCU from Fig. 1.9. A single pulse *Start* causes loading an address of the first MI into RAMI. At the same time, this pulse sets up the TF causing $Fetch = 1$. It allows fetching data from the control memory.

Let an address A^t be located into RAMI at time t. The corresponding MI is fetched from CM. Its operational part is transformed into microoperations $y_n \in Y$. After executing the microoperations, there are correct values of logical conditions $x_e \in X$ on the input of SFA. The second input of SFA receives information from the address part of MI. The rule (1.20) is used for obtaining the address of transition. This process is repeated till the last MI containing a special variable y_E into its operational part is fetched. It causes reset (zeroing) the TF leading to Fetch=0. It causes the termination of MCU operation.

Let us discuss an example of design for MCU (Fig. 1.9) using the GSA Γ_2 (Fig. 1.10).

Let $V(\Gamma_j)$ be a set of vertices from a GSA Γ_j, $V_0(\Gamma_j)$ be a set of operator vertices, and $V_c(\Gamma_j)$ be a set of conditional vertices. There are the following

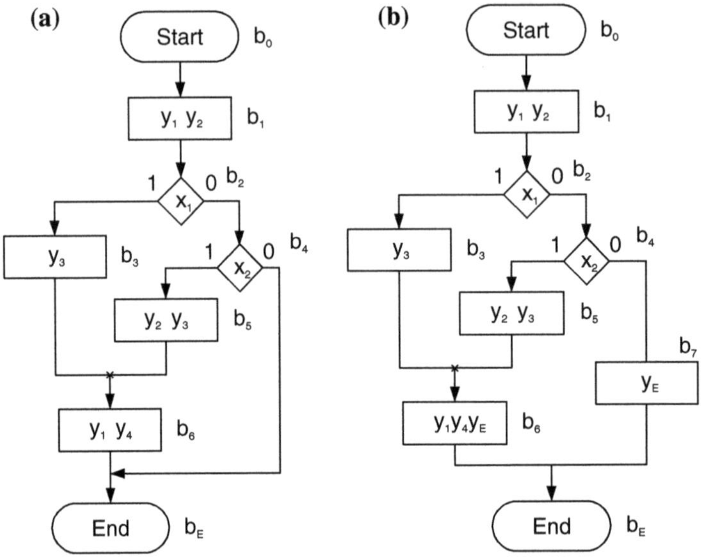

Fig. 1.10 Initial GSA Γ_2 before (**a**) and after (**b**) transformation

sets in the case of GSA Γ_2: $V_o(\Gamma_2) = \{b_1, b_3, b_5, b_6\}$, $V_c(\Gamma_2) = \{b_2, b_4\}$, and $V(\Gamma_2) = \{b_0, b_1, \ldots, b_6, b_E\}$. Obviously, the symbol b_0 denotes the *Start* vertex and b_E the *End* vertex. Also, there are the following sets and their parameters: $X = \{x_1, x_2\}$, $L = 2$, $Y = \{y_2, \ldots, y_4\}$ and $N = 4$.

The transformation of GSA Γ is reduced to providing the mode of termination. To do it, the variable y_E is inserted into vertices $b_q \in V_0(\Gamma_j)$ connected with b_E. If there are connections among vertices $b_q \in V_c(\Gamma_j)$ and b_E, then an additional vertex is inserted into a GSA. It contains y_E and it is connected with b_E. In the discussed case, the variable y_E is inserted into b_6. Also, the vertex b_7 is added (Fig. 1.10b).

Analysis of MCU operation shows that there are two stages in the execution of an MI. First, the microoperations are generated. Second, the logical conditions are checked. So, each MI corresponds to a pair $\langle b_q, b_t \rangle$ where $b_q \in V_0(\Gamma_j)$ and $b_t \in V_c(\Gamma_j)$. So, to find microinstructions, these pairs should be analysed. There are four possible configurations for vertices b_q, b_t such that the output of b_q is connected with the input of b_t (Fig. 1.11).

If both vertices are operator (Fig. 1.11a), then the vertex b_q corresponds to MI with empty fields FX and FA_1. The analysis should be continued for the vertex b_t. If $b_q \in V_0(\Gamma_j)$ and $b_t \in V_c(\Gamma_j)$, then they form an MI without empty fields. It corresponds to the configuration shown in Fig. 1.11b. If both vertices are conditional (Fig. 1.11c), then the vertex b_t corresponds to an MI with empty field FY. The analysis is continued for b_q. If $b_q \in V_c(\Gamma_j)$ and $b_t \in V_0(\Gamma_j)$, then the analysis should be continued for both vertices.

Using this approach, one can find the following MIs denoted as O_i: $O_1 = \langle b_1, b_2 \rangle$, $O_2 = \langle b_3, \emptyset \rangle$, $O_3 = \langle \emptyset, b_4 \rangle$, $O_4 = \langle b_5, \emptyset \rangle$, $O_5 = \langle b_6, \emptyset \rangle$ and $O_6 = \langle b_7, \emptyset \rangle$. There

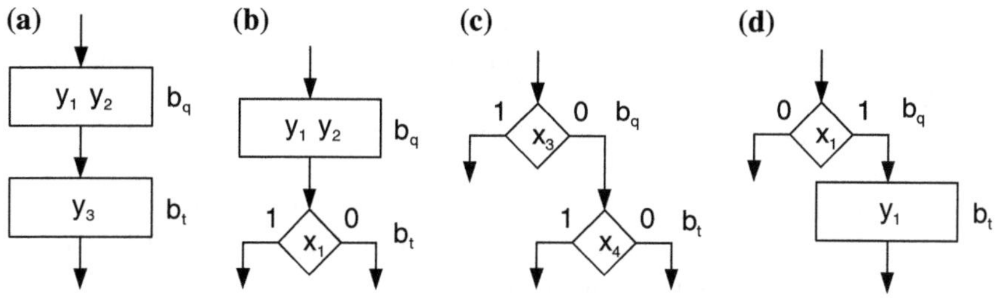

Fig. 1.11 Possible configurations of pairs of vertices

are $M_2 = 6$ microinstructions having the format (Fig. 1.8) in the case of GSA Γ_2. It is enough $R_2 = 3$ bits for addressing $M_2 = 6$ microinstructions. It gives the set $T = \{T_1, T_2, T_3\}$.

If *Start* $= 1$, then an address of the first MI is loaded into RAMI. It means that $A_1 = 000$, where A_m is an address of the MI $O_m (m = \overline{1, M_2})$. There is no influence of the outcome of addressing on the hardware amount of MCU. So, let us address them in the trivial way: $A_2 = 001, \ldots, A_6 = 101$.

Obviously, the CM can keep only binary codes. Let us encode the field *FY* using one-hot codes [16]. In the discussed case, there are $n_Y = 5$ bits in the field FY. Let us use the minimal possible amount of bits for encoding of logical conditions:

$$n_L = \lceil log_2(L + 1) \rceil. \tag{1.21}$$

In (1.21), one is added to take into amount unconditional transitions. In the discussed case, there is $n_L = 2$. Let us use the following codes: $K(x_1) = 01$, $K(x_2) = 10$ and $K(\emptyset) = 00$.

Now, there are codes and addresses for all fields of microinstructions. It allows constructing the content of CM (Table 1.4).

Table 1.4 Content of control memory for MCU with compulsory addressing of microinstructions

Address	*FY*	*FX*	FA_0	FA_1	*FT*
$T_1T_2T_3$	$y_1y_2y_3y_4y_E$	d_1d_2	$d_3d_4d_5$	$d_6d_7d_8$	
000	11000	01	010	001	$O_1 \rightarrow \bar{x}_1 O_3 \vee x_1 O_2$
001	00100	00	100	000	$O_2 \rightarrow O_5$
010	00000	10	101	011	$O_3 \rightarrow \bar{x}_2 O_6 \vee x_2 O_4$
011	01100	00	100	000	$O_4 \rightarrow O_5$
100	10011	00	000	000	$O_5 \rightarrow End$
101	00001	00	000	000	$O_6 \rightarrow End$

We use symbols $d_r(r = \overline{1, n_L + 2R_2})$ for the bits from the fields FX, FA_0 and FA_1. The last column contains the formula of transition (FP) for last vertex of the pair O_m.

Synthesis of the logic circuit of MCU is reduced to the synthesis of SFA with standard multiplexers and synthesis of CM with standard read-only memory (ROM) blocks. Let us use two variables, z_1 and z_2, to organize the loading RAMI from CM. If $z_1 = 1$, then $RAMI := CM[FA_1]$. If $z_1 = 1$, then $RAMI := CM[FA_0]$. So, the following function should be implemented:

$$A^{t+1} = z_1[FA_1]^t \vee z_2[FA_0]^t. \tag{1.22}$$

In the discussed case, the following expression can be used for z_1:

$$z_1 = \bar{d_1}d_2x_1 \vee d_1\bar{d_2}x_2. \tag{1.23}$$

That equation can be implemented using a multiplexer whose control inputs are connected with d_1 and d_2, and informational inputs with x_1 and x_2. For unconditional transitions, it should be $z_1 = 0$. Therefore, it is necessary to connect zero with informational input corresponding to $d_1 = d_2 = 0$.

Let us use D flip-flops to implement RAMI. It allows to transform (1.22) into R_2 of the following equations:

$$\begin{aligned} D_1 &= z_1d_6 \vee z_2d_3; \\ D_2 &= z_1d_7 \vee z_2d_4; \\ D_3 &= z_1d_8 \vee z_2d_5. \end{aligned} \tag{1.24}$$

It is enough R_2 of multiplexers to implement the system (1.24).

Obviously, that z_1 is complement to z_2. It allows to implement the following circuit of SFA (Fig. 1.12).

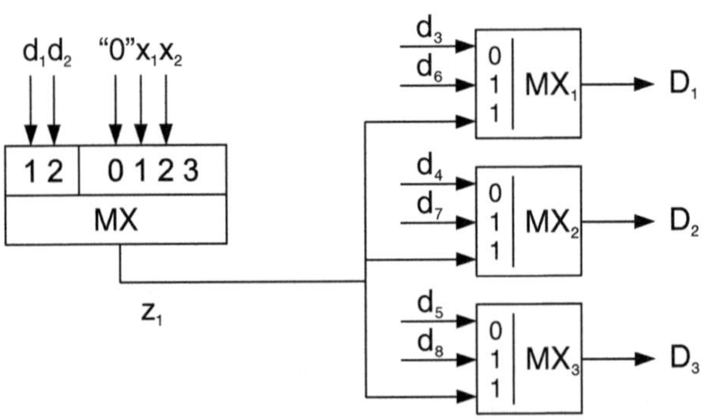

Fig. 1.12 Logic circuit of SFA

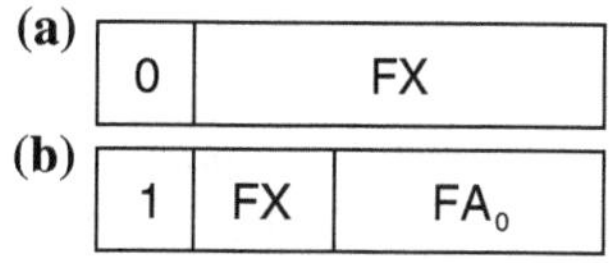

Fig. 1.13 Format of microinstructions with natural addressing

There are two microinstruction formats in the case of natural addressing [16]. An operational microinstruction (OMI) corresponds to an operator vertex (Fig. 1.13a). A conditional microinstruction (CMI) corresponds to a conditional vertex (Fig. 1.13b).

They determine the type of MI using the first bit of a format. Let us denote this bit as FA. The following rules are used to find the address A^t [4]:

$$A^{t+1} = \begin{cases} A^t + 1, & \textit{if} \;\; [FA]^t = 0; \\ A^t + 1, & \textit{if} \;\; (x_e^t = 1) \wedge ([FA]^t = 1); \\ [FA_0]^t, & \textit{if} \;\; (x_e^t = 0) \wedge ([FA]^t = 1); \\ [FA_0]^t, & \textit{if} \;\; ([FX]^t = \emptyset) \wedge ([FA]^t = 1). \end{cases} \tag{1.25}$$

If a current MI is an OMI, then the current address is incremented to get the address of transition. The same is true for conditional transitions with $x_e^t = 1$. The third line of (1.25) corresponds to the operation "go to".

Analysis of (1.25) shows that is should be a counter (CT) in the structure of MCU with natural addressing. We do not discuss the design methods based on (1.25). They can be found in [12, 16], for example. Counters are used in compositional microprogram control units [12]. Let us discuss these models of FSMs.

1.4 Logic Synthesis for Compositional MCUs

These models were proposed in 1983 [8]. They can be viewed as MPAs with a counter (CT) instead of the register. Compositional MCUs (CMCU) are based on the representing GSA Γ by a set of operational linear chains (OLC). Let us introduce some definitions.

Definition 1.1 An OLC α_g of a GSA Γ is a finite vector of operator vertices $\alpha_g = \langle bg_1, \ldots, bg_{F_g} \rangle$ such that each pair of adjacent vertices $\langle bg_i, bg_{i+1} \rangle$ corresponds to an edge of the GSA Γ. Here i is the component number of the vector $\alpha_g (i = \overline{1, F_g - 1})$.

Let V_0^g be a set of operator vertices, which are the components of α_g.

Definition 1.2 An operator vertex $b_g \in V_0^g$ is an input of OLC α_g if there is an edge $\langle b_t, b_g \rangle \in E$ such that $b_t \notin V_0^g$.

Here E is a set of edges of a GSA Γ. Each edge connects two vertices of GSA.

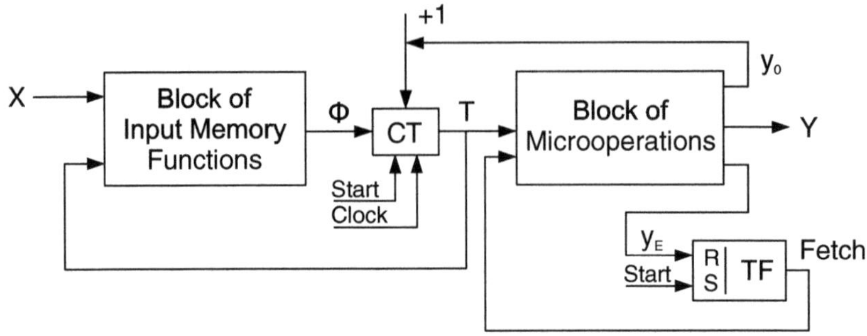

Fig. 1.14 Structural diagram of CMCU with common memory

Definition 1.3 An input $b_g \in V_0^g$ is a main input of OLC α_g if there is no edge $\langle b_t, b_g \rangle \in E$ such that $b_t \subset V_c(\Gamma)$.

Definition 1.4 An operator vertex $b_g \in V_0^g$ is an output of OLC α_g if there is an edge $\langle b_g, b_t \rangle \in E$ such that $B_t \notin V_0^g$.

Let us denote a set of OLC inputs as $I(\Gamma)$, a set of main inputs as $MI(\Gamma)$ and a set of outputs as $O(\Gamma)$. Let us find a set of OLC $C = \{\alpha_1, \ldots, \alpha_G\}$ satisfying the following conditions:

$$\begin{aligned} &\bigcup_{g=1}^{G} V_0^g = V_0; \\ &V_0^j \cap V_0^j = \emptyset \;\; (i \neq j;\; i, j \in \{1, \ldots, G\}); \\ &G \rightarrow min. \end{aligned} \tag{1.26}$$

Let the symbol $A(b_q)$ stand for the address of microinstruction corresponding to the vertex $b_q \in V_0(\Gamma)$. Let us execute the natural addressing of microinstructions for each OLC $\alpha_g \in C$. It corresponds to the following expression:

$$A(b_{gi+1}) = A(b_{gi}) + 1(i = \overline{1, F_g - 1};\, g = \overline{1, G}). \tag{1.27}$$

Now, a GSA Γ can be interpreted by a CMCU with common memory (Fig. 1.14).

This CMCU operates as the following. If $Start = 1$, then the address of the first MI is loaded into CT. At the same time, the flip-flop TF is set up giving $Fetch = 1$. It allows reading microinstructions out the memory represented by BMO. Let $CT = A(b_q)$ in the instant of time $t(t = 0, 1, \ldots)$. If b_q is not the output of any OLC, then BMO generates microoperations $y_n \in Y(b_q)$, where $Y(b_q) \subseteq Y$ is a collection of MOs from the vertex $b_q \in V_0(\Gamma)$. At the same time, the variable y_0 is generated causing execution the operation $CT := CT + 1$. The pulse *Clock* causes loading a next address of MI in accordance with (1.27). If $b_q \in O(\Gamma)$, then $y_0 = 0$. In this case, the content of CT is determined by functions (1.4) generated by BIMF. If there is an edge $\langle b_q, b_E \rangle \in E$, then $y_E = 1$. It causes $Fetch = 0$ and the operation is terminated.

Let us compare the CMCU (Fig. 1.14) with PY Moore FSM (Fig. 1.6). There are two different features: (a) the RG is replaced by the CT; (b) the flip-flop TF is added to organize the mode of microinstruction fetching. Let us compare the CMCU (Fig. 1.14) with MCU (Fig. 1.9). The SFA is replaced by the BIMF. It allows checking more than a single logical condition $x_e \in X$ during one cycle of operation. So, there are the features of MPA and MCU in CMCU. Obviously, states of MPA correspond to microinstructions; addresses of MIs correspond to state codes. In the case of CMCU, state codes obey the rule (1.27).

Let us denote the CMCU from Fig. 1.14 as U_1. Let the symbol $U_1(\Gamma_j)$ means that the CMCU U_1 is designed using a GSA Γ_j. Let us discuss an example of design for $U_2(\Gamma_3)$ where the GSA Γ_3 is shown in Fig. 1.15.

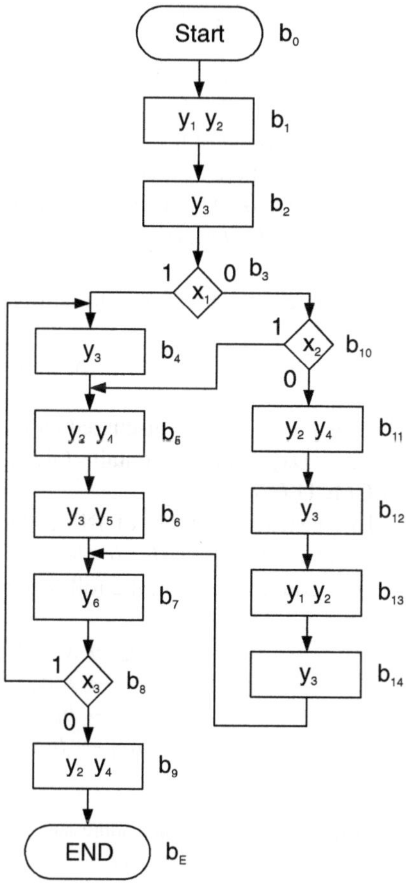

Fig. 1.15 Initial $GSA\,\Gamma_3$

There are the following steps in logic synthesis of CMCU U_1 [12]:

1. Constructing the set C of OLCs for a GSA Γ.
2. Executing the natural addressing (1.27).
3. Constructing the table of BIMF.
4. Constructing the table of BMO.
5. Implementing the CMCU logic circuit with particular logic elements.

There are two steps in executing the step 1. Firstly, it should be found the set of main inputs $MI(\Gamma_j)$. Secondly, a set C is constructed using elements of $MI(\Gamma_j)$ as main inputs of OLC $\alpha_g \in C$. To find the set $MI(\Gamma_j)$, they should use the Definition 1.3.

In the discussed case, there are the following sets: the set of vertices $V(\Gamma_3) = \{b_0, b_1, \ldots, b_{14}, b_E\}$; the set of operator vertices $V_0(\Gamma_3) = \{b_1, b_2, b_4, \ldots, b_7, b_9, b_{11}, \ldots, b_{14}\}$ having $M_1(\Gamma_3) = 11$ elements; the set of conditional vertices $V_c(\Gamma_3) = \{b_3, b_8, b_{10}\}$. Using the Definition 1.3, it can be found the set $MI(\Gamma_3) = \{b_1, b_4, b_{11}, b_9\}$.

Let us form the chain α_g starting from b_1. In the first step, there is $\alpha_1 = \langle b_1 \rangle$. There is an edge $\langle b_1, b_2 \rangle \in E$ where $b_2 \in V_0(\Gamma_3)$. The vertex b_2 is not included into already constructed OLCs. So, now we have $\alpha_1 = \langle b_1, b_2 \rangle \in E$. There is an edge $\langle b_2, b_3 \rangle \in E$ where $b_3 \in V_c(\Gamma_1)$. Using the Definition 1.4 we can find that b_3 is the output of OLC α_1. Using the same approach, the set $C = \{\alpha_1, \ldots, \alpha_4\}$ is constructed where $\alpha_1 = \langle b_1, b_2 \rangle$, $\alpha_2 = \langle b_4, \ldots, b_7 \rangle$, $\alpha_3 = \langle b_9 \rangle$, and $\alpha_4 = \langle b_{11}, \ldots, b_{14} \rangle$. So, there is $G = 4$.

It is enough R_2 bits for addressing microinstructions of $U_1(\Gamma_j)$:

$$R_2 = \lceil log_2 M_1(\Gamma_j) \rceil. \tag{1.28}$$

There is $R_2 = 4$ in the discussed case. It gives the sets $T = \{T_1, \ldots, T_4\}$ and $\Phi = \{D_1, \ldots, D_4\}$. Table 1.5 contains OLCs, vertices and addresses of corresponding MIs for CMCU $U_1(\Gamma_3)$.

To construct the table of BIMF, let us find the formulae of transitions for the vertices $b_g \in O(\Gamma_3) = \{b_2, b_7, b_9, b_{14}\}$. The rules from [6, 7] are used for executing this procedure. There are the following FTs in the discussed case:

$$\begin{aligned} &b_2 \rightarrow x_1 b_4 \vee \bar{x_1} x_2 b_5 \vee \bar{x_1} \bar{x_2} b_{11};\\ &b_7 \rightarrow x_3 b_4 \vee \bar{x_3} b_9;\\ &b_9 \rightarrow b_E;\\ &b_{14} \rightarrow b_7. \end{aligned} \tag{1.29}$$

Table 1.5 Outcome of natural addressing for $U_1(\Gamma_3)$

OLCs	α_1		α_2				α_3	α_4			
Vertices	b_1	b_2	b_4	b_5	b_6	b_7	b_9	b_{11}	b_{12}	b_{13}	b_{14}
Addresses	0000	0001	0010	0011	0100	0101	0110	0111	1000	1001	1010

Table 1.6 Table of BIMF for CMCU $U_1(\Gamma_3)$

b_q	$A(b_q)$	b_t	$A(b_t)$	X_h	Φ_h	h
b_2	0001	b_4	0010	x_1	D_3	1
		b_5	0011	$\bar{x_1}x_2$	D_3D_4	2
		b_{11}	0111	$\bar{x_1}\bar{x_2}$	$D_2D_3D_4$	3
b_7	0101	b_4	0010	x_3	D_3	4
		b_9	0110	$\bar{x_3}$	D_2D_3	5
b_{14}	1010	b_7	0101	1	D_2D_4	6

Let us point out that the transition $\langle b_9, b_E\rangle$ is executed using $y_E = 1$. So, this transition is not placed into the table of BIMF for $U_1(\Gamma_3)$. There are the following columns in the table of BIMF: b_q is the output of an OLC $\alpha_g \in C; A(b_q)$ is the address of an MI corresponding to the vertex b_q; b_t is the input of an OLC $\alpha_i \in C; A(b_t)$ is the address of MI corresponding to b_t; X_h is a conjunction of logical conditions $x_e \in X$ taken from the corresponding term of FT; Φ_h is a set of input memory functions equal to 1 to load the address $A(b_t)$ into CT; h is the number of transition.

There are 6 terms in the system (1.29) for FTs b_2, b_7 and b_{14}. So, there are 6 lines in the table of BIMF for the discussed case (Table 1.6). There is the obvious connection between the system (1.29) and Table 1.6. The addresses of MIs are taken from Table 1.5.

The BIMF implements the functions (1.4). They depend on terms F_h corresponding the rows of table of BIMF. It is possible to simplify the functions $D_r \in \Phi$ using the rules of Boolean algebra [31] and "don't care" combinations of variables $T_r \in T$. In the discussed case, there are the following "don't care" combinations: 1011, 1100, 1101, 1110, 1111. It follows from Table 1.5.

The following equations can be extracted from Table 1.6:

$$\begin{aligned}
D_1 &= T_1T_3;\\
D_2 &= \bar{T}_1\bar{T}_2\bar{T}_3T_4\bar{x_1}\bar{x_2} \vee T_2\bar{T}_3T_4\bar{x_3};\\
D_3 &= \bar{T}_1\bar{T}_3T_4;\\
D_4 &= \bar{T}_1\bar{T}_2\bar{T}_3T_4\bar{x_1} \vee T_1T_3.
\end{aligned} \tag{1.30}$$

There are the following columns in the table of BMO: b_q is a vertex of GSA Γ; $A(b_q)$ is an address of MI corresponding to the vertex $b_q \in V_0(\Gamma_j)$; y_0; $Y(b_q)$; y_E. There is $y_0 = 1$ for all MIs which do not correspond to outputs of OLC $\alpha_g \in C$. There is $y_E = 1$ for all MIs corresponding to vertices $b_q \in O(\Gamma_j)$ such that there is an edge $\langle b_q, b_E\rangle \in E$. Using these rules, we construct Table 1.7.

The circuit of BMO is implemented using memory blocks [12]. We do not discuss now this problem. The circuit of BIMF is implemented using expressions similar to (1.30).

Let us discuss how to control the counter CT. It has two inputs of synchronization. Let C_1 be an input such that pulse *Clock* causes the adding 1 to the content of CT.

Table 1.7 Table of BMO for CMCU $U_1(\Gamma_3)$

b_q	$A(b_q)$				y_0	$Y(b_q)$						y_E
	T_1	T_2	T_3	T_4		y_1	y_2	y_3	y_4	y_5	y_6	
b_1	0	0	0	0	1	1	1	0	0	0	0	0
b_2	0	0	0	1	0	0	0	1	0	0	0	0
b_4	0	0	1	0	0	0	0	1	0	0	0	0
b_5	0	0	1	1	1	0	1	0	1	0	0	0
b_6	0	1	0	0	1	0	0	1	0	1	0	0
b_7	0	1	0	1	0	0	0	0	0	0	1	0
b_9	0	1	1	0	0	0	1	0	1	0	0	1
b_{11}	0	1	1	1	1	0	1	0	1	0	0	0
b_{12}	1	0	0	0	1	0	0	1	0	0	0	0
b_{13}	1	0	0	1	1	1	1	0	0	0	0	0
b_{14}	1	0	1	0	0	0	0	1	0	0	0	0

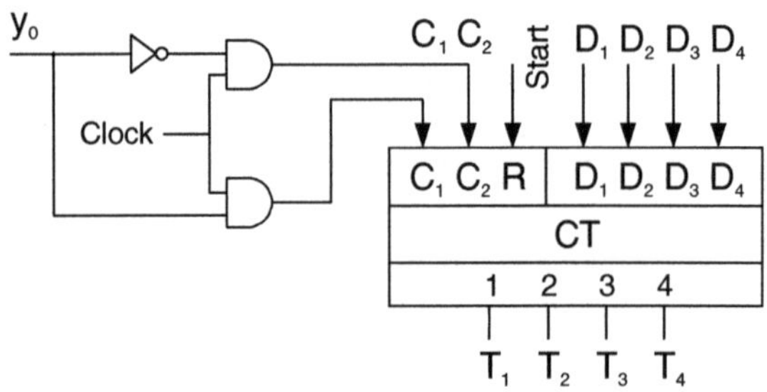

Fig. 1.16 Organization of counter for CMCU $U_1(\Gamma_3)$

Let C_2 be an input such that pulse *Clock* causes the loading an address $A(b_t)$ into CT. The following equations can be obtained for C_1 and C_2:

$$\begin{aligned} C_1 &= y_0 \cdot Clock; \\ C_2 &= \bar{y_0} \cdot Clock. \end{aligned} \quad (1.31)$$

Using these equations, the following circuit (Fig. 1.16) can be obtained for the CT.

There is a feed-back between CT and BIMF (Fig. 1.14). It means that variables y_0 and y_E can be generated by BIMF. For example, the following Karnaugh map (Fig. 1.17) corresponds to the variable y_0 of CMCU $U_1(\Gamma_3)$.

It is possible to find out the following minimized SOP of y_0: $y_0 = \bar{T}_3\bar{T}_4 \vee T_1\bar{T}_3 \vee \bar{T}_1 T_2 \bar{T}_3$. Using the same approach, it is possible to find that $y_E = T_2\bar{T}_3 T_4$.

Let us point out that CMCU is used only for interpreting so named linear GSAs [12]. There is more than 75 % of operator vertices in linear GSAs. There are many models of CMCU corresponding to different conditions [4]. Now, they do not use MCU in logic design [37]. But some methods are still used such as: (1) replacement

Fig. 1.17 Karnaugh map for variable y_0

T_3T_4 \ T_1T_2	00	01	11	10
00	1	1	*	1
01	0	0	*	1
11	1	1	*	*
10	1	0	*	0

of logical conditions(it is executed by SFA); (2) maximal encoding of CMOs $Y_t \subseteq Y$; (3) encoding of the fields of compatible microoperations [2, 3]. We will show a bit later how those methods are used in the modern logic design.

Our book is devoted to design of MPAs. We will call them FSMs. We will discuss some methods of hardware reduction targeting the technology of field-programmable gate arrays (FPGA). Some of these methods are taken from design of FSM targeting different devices of programmable logic. Let us discuss these methods.

1.5 Hardware Reduction for FPLD-Based FSMs

There is a connection among types of field-programmable logic devices (FPLD) and methods of hardware reduction [8]. In this section, we discuss the following FPLD: programmable read-only memory (PROM), programmable logic arrays (PLA), programmable array logic (PAL) and complex programmable logic devices (CPLD). An FPLD is a general purpose chip whose hardware can be configured by an end user.

PROMs are the first representatives of FPLDs. The first PROMs were produced by Harris Semiconductor in 1970 [12, 16]. They include a fixed array of AND-gates (AND-array) and a programmable array of OR-gates (OR-array). The AND-array is an address decoder (DC) of a chip. PROMs are used for implementing regular systems of Boolean functions [10] represented by truth tables. A PROM has a fixed amount of address inputs (S_A) and outputs (t_F). PROMs can be used in PY, PD and PH models of FSMs (Fig. 1.18) [12, 16].

In the case of P Moore FSM (Fig. 1.18a), PROM implements the circuit of BMO. It is denoted as PROM Y. The PY Mealy FSM (Fig. 1.18b) is based on the maximal encoding of CMOs taken from MCU design methods [12, 16].

Let it be T_0 different CMOs in the vertices $b_q \in V_0(\Gamma_j)$. Let us encode each CMO $y_t \subseteq Y$ by a binary code $C(Y_t)$ having R_Z bits:

$$R_Z = \lceil log_2 T_0 \rceil. \tag{1.32}$$

Let us use variables $z_r \in Z$ for the encoding, where $|Z| = R_Z$. In this case, the variables $z_r \in Z$ are connected with address inputs of PROMY, whereas the functions $y_n \in Y$ are the outputs of PROMY.

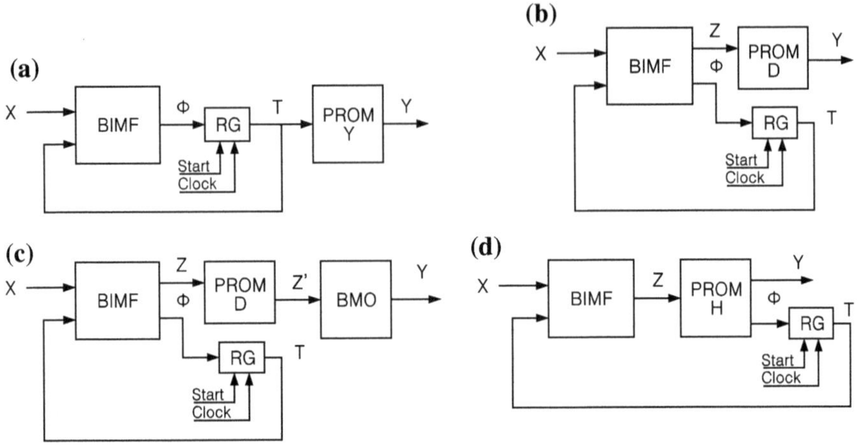

Fig. 1.18 Models of FSMs using PROM

They used the method of encoding of fields of compatible microoperations in microprogramming [20]. Microoperations $y_n, y_m \in Y$ are compatible if they do not belong to the same CMOs:

$$(y_n \in Y_t \rightarrow y_m \notin Y_t) = 1(t = \overline{1, T_0}). \tag{1.33}$$

The set Y is divided by subsets $Y^1, \ldots, Y^K$ of compatible MOs such that:

$$\begin{aligned} &\bigcup_{k=1}^{K} Y^k = Y; \\ &Y^i \cap Y^j = \emptyset \; (i \neq j; i, j \in \{1, \ldots, K\}); \\ &\sum_{k=1}^{K} log_2(|Y^t| + 1) \rightarrow min. \end{aligned} \tag{1.34}$$

Each set Y^k corresponds to a field in the control memory having R_k bits:

$$R_k = \lceil log_2(|Y^k| + 1) \rceil. \tag{1.35}$$

The third line of (1.34) means that the partition of Y should minimize the number of bits in the field FY.

Microoperations $y_n \in Y^k$ are encoded by binary codes $C(y_n)$ having R_k bits and using variables from set Z^k, where $|Z^k| = R_k (k = \overline{1, K})$. Next, the content of each field FY^k is decoded by a decoder DC_k. These decoders form the block D of an MCU.

Using this approach, the PD Mealy FSM can be organized (Fig. 1.18c). The CMOs $Y_t \subseteq Y$ are encoded by variables $z_r \in Z$, where $|Z| = R_Z$. The microoperations $y_n \in Y$ are encoded by variables $z_r \in Z$, where $|Z| = R_K$:

$$R_K = \sum_{k=1}^{K} R_k. \tag{1.36}$$

The PROM D transforms codes $C(Y_t)$ into codes $K(Y_t)$ where

$$K(Y_t) = C(y_n)^1 * C(y_n)^2 * \cdots * C(y_n)^K. \tag{1.37}$$

In (1.37), the symbol $C(y_n)^t$ stands for a code of MO from the field $Y^k(k = \overline{1, K})$.

The BMO executes functions of decoding for microoperations $y_n \in Y$. There are different approaches to implement this block. For example, they can use standard decoders [9].

In the case of PH Mealy FSM (Fig. 1.18d), they encode the functions F_h (1.8) corresponding to rows of ST. Let us encode each term F_h by a binary code $C(F_h)$ having R_H bits:

$$R_H = \lceil log_2 H_0 \rceil. \tag{1.38}$$

Let us use the variables $z_r \in Z$ for such encoding, where $|Z| = R_H$. These variable enter the PROMH generating functions

$$\Phi = \Phi(Z); \tag{1.39}$$

$$Y = Y(Z). \tag{1.40}$$

The PH Mealy FSM corresponds to the "one-hot" encoding of MOs [20]. In this case, exactly one bit of PROM corresponds to a single MO $y_n \in Y$. If CMOs are encoded, then two models of Mealy FSM are possible. A PHY FSM corresponds to the maximal encoding of MOs; a PHD FSM corresponds to encoding of the fields of compatible MOs.

The PH-models are not used in the case of Moore FSMs. It is connected with the following condition:

$$R_1 \leq \lceil log_2 H_1 \rceil. \tag{1.41}$$

Programmable logic arrays were introduced in the mid 1970s by Signetics. They were oriented towards the implementing systems of Boolean functions (SBF) represented in minimized SOP forms [6]. The peculiarity of PLA is a programmability of both its arrays (Fig. 1.19a).

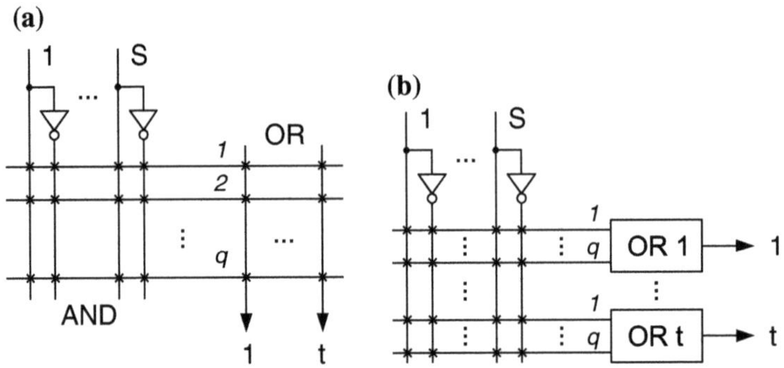

Fig. 1.19 Architectures of PLA (**a**) and PAL (**b**)

A PLA has S inputs and t outputs. It is possible to implement q conjunctions of inputs named "product terms". The sign "×" in Fig.1.19 stands for a connection between two wires. Let terms F_j depend on input variables x_i. Each term F_j can be represented as:

$$F_j = \bigwedge_{i=1}^{S} x_i^{e_i} \ \ (j = \overline{1,q};\, i = \overline{1,S}). \tag{1.42}$$

In (1.42), $l_i \in \{0, 1, *\}$ is a value of an input x_i, where $x_i^0 = \bar{x}_i$, $x_i^1 = x_i$, $x_i^* = 1$.

Let us denote output functions of PLA as $y_k(k = \overline{1,t})$. Each function y_k is represented as

$$y_k = \bigvee_{j=1}^{q} C_{jq} F_j \ \ (k = \overline{1,t}). \tag{1.43}$$

In (1.43), C_{jq} is a Boolean variable equal to 1 iff the term F_j enters the SOP of y_k.

So, PLAs are very flexible devices. As a rule, they have $S \leq 16$, $t \leq 8$, $q \leq 48$ [9]. They were used for implementing combinational parts of FSMs [6].

Let the following conditions take places:

$$\begin{aligned} &S \geq L + R_0;\\ &t \geq N + R_0;\\ &q \geq H_0. \end{aligned} \tag{1.44}$$

In this case, a P Mealy FSM is implemented in a trivial way using only a single PLA chip (Fig. 1.20a).

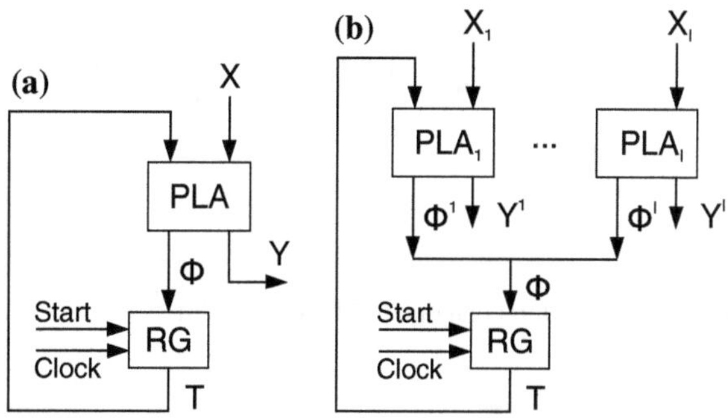

Fig. 1.20 Trivial (**a**) and single-level (**b**) circuits of P Mealy FSM

If conditions (1.44) are violated, then a single-level circuit (Fig. 1.20b) can be implemented [6]. It corresponds to a partition Π_F of set of SOP having minimal number of blocks $E_1, \ldots, E_I$ where each block satisfies to the following condition [6]:

$$\begin{aligned} &|X^i| + R_0 \leq S; \\ &|Y^i| + |\Phi^i| \leq t; \\ &|E^i| \leq q. \end{aligned} \tag{1.45}$$

In (1.45), $X^i \subseteq X$ is a set of logical conditions which are significant for functions implemented by $PLA_i(i = \overline{1,I})$. Also $\Phi^i \subseteq \Phi$, $Y^i \subseteq Y$ are subsets of input memory functions and microoperations depended on terms E^i and generated by $PLA_i(i = \overline{1,I})$.

The single-level P FSMs are the fastest among all possible solutions [6]. But they are the most expensive because different PLAs can have the same input and output variables.

To decrease the number of chips, multi-level circuits of FSMs were proposed. They used the following methods [6, 13, 17]:

1. Replacement of logical conditions.
2. Maximal encoding of the collections of MOs.
3. Maximal encoding of rows of ST.

Let us discuss the first approach. Let the transitions from a state $a_m \in A$ depend on logical conditions forming a set $X(a_m) \subseteq X$. Let it be $L_m = |X(a_m)|(m = \overline{1,M})$ and $G = max(L_1, \ldots, L_M)$. Let us form a new set P of additional variables $p_g \in P$, where $|P| = G$. Let us replace the variables $x_e \in X$ by the variables $p_g \in P$ using the following system of Boolean functions:

$$p_g = \bigvee_{m=1}^{M} \bigvee_{L=1}^{L} C_{me} A_m x_e (g = \overline{1,G}). \tag{1.46}$$

In (1.46), the Boolean variable $C_{me} = 1$ iff a variable $p_g \in P$ replaces the variable $x_e \in X$ in the set $X(a_m)$.

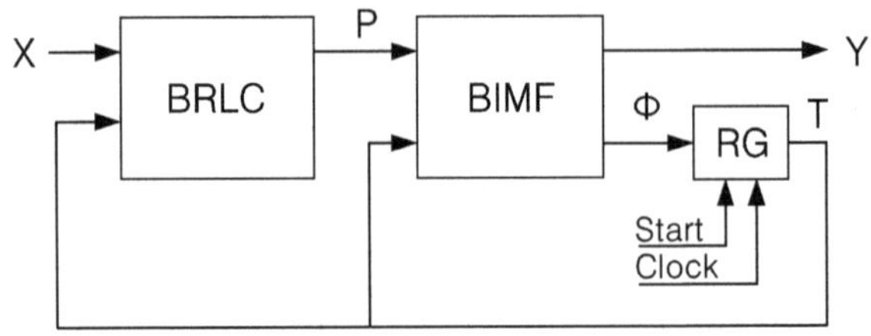

Fig. 1.21 Structural diagram of MP Mealy FSM

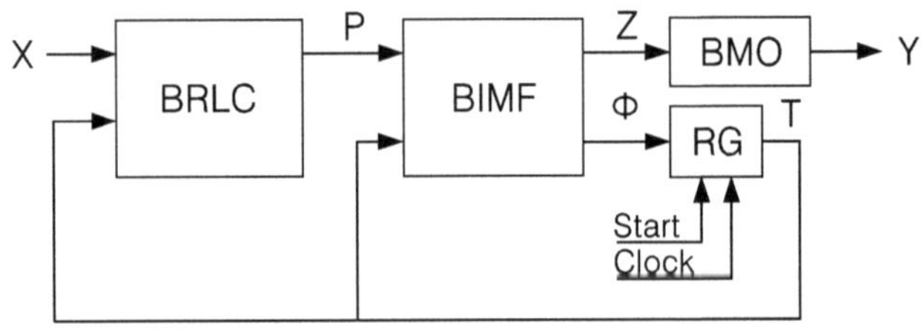

Fig. 1.22 Structural diagram of MPY Mealy FSM

Now it is possible to implement an MP Mealy FSM (Fig. 1.21) where the block of replacement of logical conditions (BRLC) implements the system (1.46).

They use PLAs for implementing circuits of BRLC and BIMF. The BRLC is represented by a system:

$$P = P(T, X). \tag{1.47}$$

The BIMF is represented by systems:

$$Y = Y(T, P); \tag{1.48}$$

$$\Phi = \Phi(T, P). \tag{1.49}$$

Using the maximal encoding of CMOs leads to PY Mealy FSMs. Using the maximal encoding of rows of ST leads to PH Mealy FSMs. Three-level FSM circuits are possible. For example, the organization of MPY Mealy FSM is shown in Fig. 1.22.

In this case, the BRLC implements the system (1.47). The BIMF implements the system (1.48) and the system

$$Z = Z(T, P). \tag{1.50}$$

The BMO implements the system (1.40).

Let us point out that such a representation of an FSM is named the structural decomposition [11, 37]. If the circuits of all blocks are implemented with PLAs, it is a homogeneous circuit [37]. But it is possible to diminish the amount of FPLDs in the FSM circuit using so called heterogeneous FSMs.

For example, there are the following chips in the heterogeneous MPY Mealy FSM: standard multiplexers (MX) implement the circuit of BRLC; PLAs implement the circuit of BIMF; the PROMs implement the circuit of BMO.

As it is pointed before, the programmability of AND-array makes PLAs very flexible. But, on the other hand, it leads to increasing a chip area because of a lot of fuses [12].

Functions (1.3) and (1.4) belong to the class of functions defined only for a small fraction of possible input assignments. Moreover, each of functions $y_n \in Y$ and $D_r \in \Phi$ appears very rarely in the structure table of FSM. It means that PLAs are very redundant for implementing, for example, input memory functions. But such functions can be very efficiently implemented using PAL chips [18].

PALs were introduced by Monolithic Memories in 1978 [16]. They include the programmable AND-array and t fixed OR-gates (Fig. 1.19b). One of the new conceptions connected with PAL is the conception of a macrocell. The macrocell is a part of the PAL chip connected with a single output of the chip. Each macrocell has S inputs and one output; a PAL chip has S inputs and t outputs (macrocells). Each macrocell has q terms, as a rule, there is $q \leq 8$.

The same FSM structures can be used for both PAL and PLA-based solutions. But it should be executed the joint minimization of SBF representing PLA-based circuits. In the case of PAL, each function is minimized separately, to diminish the number of its product terms up to q.

It is necessary to have the external state register to keep the state codes. It requires additional connections on a printed board with an FSM circuit. But this register can be eliminated if some part of a chip area is occupied by flip-flops. It leads to programmable logic sequencers (PLS) in the case of PLAs and to generic array logic (GAL) in the case of PALs [16]. There is a GAL-oriented P Mealy FSM shown in Fig. 1.23.

Using GALs and PLSs, it is possible to represent an FSM as a network of interconnected partial FSMs. It is a decomposition of FSM circuit [6, 7]. For example, some FSM S is decomposed by three FSMs ($S_1 - S_3$). Each FSM S_i is characterized by sets $X^i \subseteq X$ and $Y^i \subseteq Y$. To evoke the FSM S_i, they use the control signal $E_i (i = \overline{1, 3})$ (Fig. 1.24).

Further evolution of electronic technology allows tremendous increasing for the numbers of transistors per chip [35, 38]. It results in appearance of such FPLDs

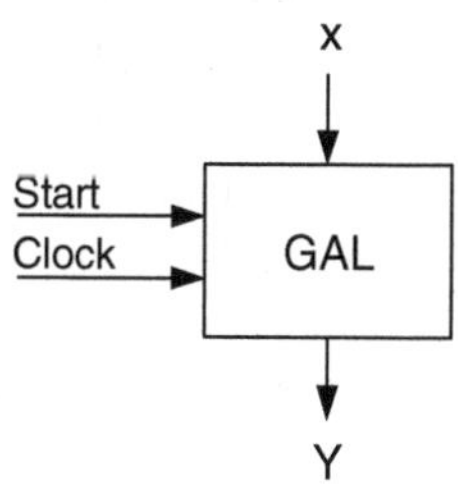

Fig. 1.23 Structural diagram of GAL-oriented P Mealy FSM

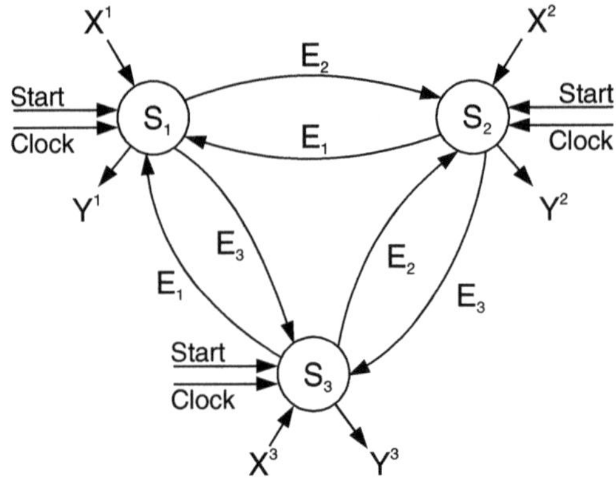

Fig. 1.24 Mealy FSM as a network of FSMs

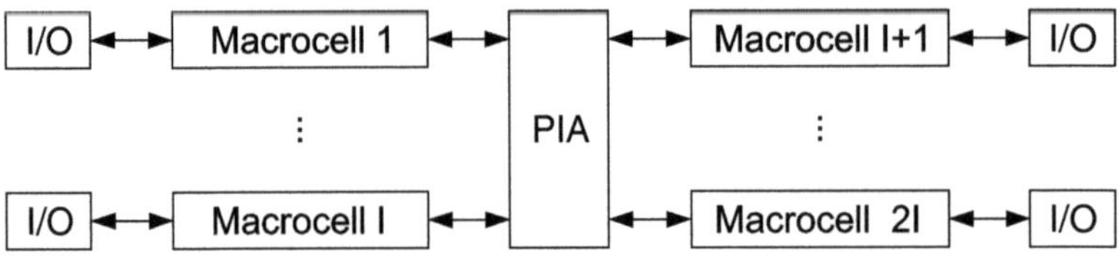

Fig. 1.25 Structural diagram of CPLD

as FPGAs and CPLDs. Now, they are the most popular families of FPLDs used in logic design [21, 29, 30, 33, 34]. The next chapter, as the rest of the book, is devoted to FPGA-based design methods. Here we just briefly describe CPLDs. The vast majority of CPLDs are organized as a simple PAL-cell matrix, a programmable interconnection array (PIA) and blocks of input-outputs connected with macrocells (Fig. 1.25).

The typical macrocell includes three main blocks: programmable AND-array, logic allocator and programmable flip-flop (Fig. 1.26). The logic allocator connects product terms with OR-gate. It allows having different amount of terms in macrocells. This property makes macrocells more flexible in comparison with macrocells with fixed amount of product terms. There are two possibilities for organization of a macrocells. It can have the combinational output (if a flip-flop is bypassed) or registered output. The output can be connected with I/O blocks or with PIA. The connection with PIA allows creating feed-backs in the final project. As a rule, there is a tri-stable buffer between the output of a macrocell and I/O block. It is controlled by a combinational circuit belonging to a macrocell.

In Fig. 1.26, the MX1 executes the choice between the combinational (f_c) and registered (f_R) outputs of a macrocell. The MX2 makes choice between the output f_m, the output of tri-stable buffer and the data from I/O block.

A lot of technology-dependent methods of FSM design can be found in [14, 19, 22–28, 39–41]. For example, the tri-stable buffer can be used for organization of

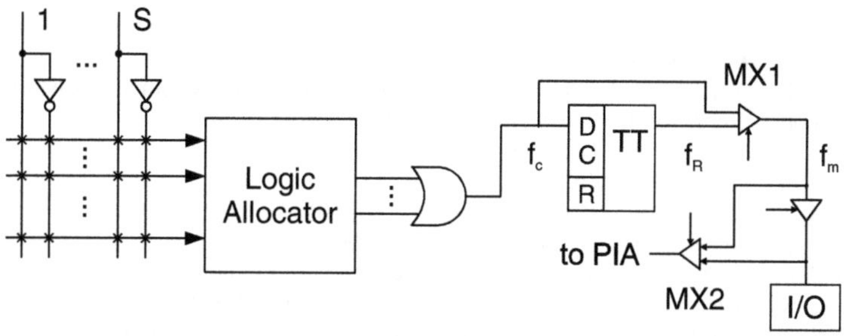

Fig. 1.26 Typical structure of PAL-based macrocell

term expansion without creating multi-level FSM circuits [19]. The wide fan-in of PAL-macrocells allows using more than one source of state codes [15]. We discuss this approach in other parts of this book.

So, to design an FSM circuit with desirable characteristics, it is necessary to take into account the following specifics:

1. Specific features of logic elements used for implementing logic circuits. For example, it is possible to use two or three sources of state codes due to the wide fan-in of PAL macrocells.
2. Specific features of an FSM model. For example, it is possible to use memory blocks for implementing the system of output functions for Moore FSMs.
3. Specific features of a GSA Γ representing the FSM behaviour. For example, it is possible to use a counter instead of the state register for linear GSAs.

In further chapters, we show how to use these specifics in the case of FPGA-based FSMs.

References

1. M. Adamski, A. Barkalov, *Architectural and Sequential Synthesis of Digital Devices* (University of Zielona Góra Press, Zielona Góra, 2006)
2. T. Agerwala, Microprogram optimization: a survey. IEEE Trans. Comput. **25**(10), 962–973 (1976)
3. A. Agrawala, T. Rauscher, *Foundations of Microprogramming* (Academic Press, New York, 1976)
4. F. Anceau, *The Architecture of Microprocessors* (Addison-Wesley, Workingham, 1986)
5. P. Asahar, S. Devidas, A. Newton, *Sequential Logic Synthesis* (Kluwer Academic Publishers, Boston, 1992)
6. S. Baranov, *Logic Synthesis of Control Automata* (Kluwer Academic Publishers, Dordrecht, 1994)
7. S. Baranov, *Logic and System Design of Digital Systems* (TUT Press, Tallinn, 2008)
8. A. Barkalov, Microprogram control unit as composition of automate with programmable and hardwired logic. Autom. Comput. Sci. **17**(4), 36–41 (1983)
9. A. Barkalov, Multilevel PLA schemes for microprogram automata. Cybern. Syst. Anal. **31**(4), 489–495 (1995)

10. A. Barkalov, Principles of logic optimization for Moore microprogram automaton. Cybern. Syst. Anal. **34**(1), 54–60 (1998)
11. A. Barkalov, A. Barkalov Jr, Design of Mealy finite-state machines with the transformation of object codes. Int. J. Appl. Math. Comput. Sci. **15**(1), 151–158 (2005)
12. A. Barkalov, L. Titarenko, *Logic Synthesis for Compositional Microprogram Control Units*, vol. 22 (Springer, Berlin, 2008)
13. A. Barkalov, L. Titarenko, *Logic Synthesis for FSM-Based Control Units*. Lecture Notes in Electrical Engineering, vol. 53 (Springer, Berlin, 2009)
14. A. Barkalov, L. Titarenko, S. Chmielewski, Reduction in the number of PAL macrocells int the circuit of a Moore FSM. Int. J. Appl. Math. Comput. Sci. **17**(4), 565–675 (2007)
15. A. Barkalov, L. Titarenko, S. Chmielewski, Hardware reduction in CPLD-based Moore FSM. J. Circuits, Syst., Comput. **23**(6), 1450086–1–1450086–21 (2014)
16. A. Barkalov, M. Węgrzyn, *Design of Control Units With Programmable Logic* (University of Zielona Góra Press, Zielona Góra, 2006)
17. A. Barkalov, I. Zelenjova, Optimization of replacement of logical conditions for an automaton with bidirectional transitions. Autom. Control Comput. Sci. **34**(5), 48–53. Allerton Press Inc
18. S. Chmielewski, Using structural pecularities of Moore FSM for reduction of number of PALS. PhD thesis, University of Zielona Góra (2014)
19. R. Czerwinski, D. Kania, *Finite State Machine Logic Synthesis for Complex Programmable Logic Devices*. Lecture Notes in Electrical Engineering, vol. 23 (Springer, Berlin, 2013)
20. S. Husson, *Microprogramming: Principles and Practices* (Prentice Hall, Englewood Cliffs, 1970)
21. J. Jenkins, *Design with FPGAs and CPLDs* (Prentice Hall, New York, 1995)
22. D. Kania, Two-level logic synthesis on PAL-based CPLD and FPGA using decomposition, in *Proceedings of 25th Euromicro Conference* (1999), pp. 278–281
23. D. Kania, Two-level logic synthesis on PALs. Electron. Lett. **17**, 879–880 (1999)
24. D. Kania, Coding capacity of PAL-based logic blocks included in CPLDs and FPGAs, in *Proceedings of IFAC Workshop on Programmable Devices and Sysytems (PDS'2000)* (Elsevier Science, 2000), pp. 164–169
25. D. Kania, Decomposition-based synthesis and its application in PAL-oriented technology mapping, in *Proceedings of 26th Euromicro Conference* (IEEE Compuetr Society Press, Maastricht, 2000), pp. 138–145
26. D. Kania, An efficient algorithm for output coding in PAL-based CPLDs. Int. J. Eng. **15**(4), 325–328 (2002)
27. D. Kania, Logic synthesis of multi–output functions for PAL-based CPLDs, in *Proceedings of IEEE International Conference on Field-Programmable Technology* (2002), pp. 429–432
28. D. Kania, An efficient approach to synthesis of multi-output Boolean functions on PAL-based devices. IEEE Proc.—Comput. Digital Tech. **150**, 143–149 (2003)
29. C. Maxfield, *The Design Warrior's Guide to FPGAs* (Academic Press Inc., Orlando, 2004)
30. C. Maxfield, *FPGAs: Instant Access* (Newnes, 2008)
31. E. McCluskey, *Logic Design Principles* (Prentice Hall, Englewood Cliffs, 1986)
32. G. De Micheli, *Synthesis and Optimization of Digital Circuits* (McGraw–Hill, New York, 1994)
33. P. Minns, I. Elliot, *FSM-Based Digital Design Using Verilog HDL* (Wiley, Chichester, 2008)
34. Z. Navabi, *Embedded Core Design with FPGAs* (McGraw–Hill, New York, 2007)
35. K. Sakamura, Future SoC possibilities. IEEE Micro. (5), 7 (2002)
36. A. Salisbury, *Microprogrammable Computer Architectures* (Elsevier Science, New York, 1976)
37. V. Sklyarov, I. Skliarova, A. Barkalov, L. Titarenko, *Synthesis and Optimization of FPGA-Based Systems*. Lecture Notes in Electrical Engineering, vol. 294 (Springer, Berlin, 2014)
38. M. Smith, *Application-Specific Integrated Circuits* (Addison-Wesley, Boston, 1997)
39. V. Solovjev, M. Czyzy, Refined CPLD macrocells architecture for effective FSM implementation, in *Proceedings of the 25th EUROMICRO Conference*, vol. 1 (Milan, Italy, 1999), pp. 102–109
40. V. Solovjev, M. Czyzy, The universal algorithm for fitting targeted unit to complex programmable logic devices, in *Proceedings of the 25th EUROMICRO Conference*, vol. 1 (Milan, Italy, 1999), pp. 286–289

41. V. Solovjev, M. Czyzy, Synthesis of sequential circuits on programmable logic devices based on new models of finite state machines, in *Proceedings of the EUROMICRO Conference, Milan* (2001), pp. 170–173
42. S. Tucker, Microprogram control for system/360. IBM Syst. J. **6**(4), 222–241 (1967)
43. T. Villa, T. Kam, R. Brayton, A. Sangiovanni-Vincentelli, *A Synthesis of Finite State Machines: Logic Optimization* (Kluwer Academic Publishers, Boston, 1998)
44. M. Wilkes, The best way to design an automatic calculating machine, in *Proceedings of Manchester University Computer Inaugural Conference* (1951)
45. M. Wilkes, J. Stringer, Microprogramming and the design of the control circuits in an electronic digital computer. Proc. Camb. Philos. Soc. **49**, 230–238 (1953)
46. S. Yanushkevich, V. Shmerko, *Introduction to Logic Design* (CRC Press, Boca Raton, 2008)

Chapter 2
Field Programmable Gate Arrays in FSM Design

2.1 General Characteristic of FPGAs

The field programmable gate arrays (FPGA) were introduced in 1984 [53] by Xilinx. It is very hard to overestimate the FPGA influence on different areas of electrical engineering [77]. These chips are very flexible due to their property of reconfigurability. Contemporary FPGA chip can replace billions of 2NAND gates [54], called system gates. In the beginning, FPGAs were used for implementing such parts of electrical circuits as glue logic. Now, these chips have up to 7 billion transistors [75], possess clock frequency exceeding the Gigahertz. The most advanced FPGAs have the technology less than 20 nm [2, 52, 58, 64, 89].

First FPGAs, XC2064, were produced by Xilinx in 1985. They included up to 85 000 transistors, 128 logic cells, 64 configurable logic blocks; they have the clock frequency up to 50 MHz. But they made huge progress and now they are very powerful devices.

To show this progress, let us start from the family Spartan-3 of Xilinx [89]. These chips were introduced in 2002, were powered by 1, 2 V and used the 90 nm technology. They included look-up table (LUT) elements having 4 inputs. These chips included up to 104 embedded memory blocks named blocks of RAMs (BRAM). So, it was up to 1,87 Mbits of BRAMs in these FPLDs. They operated with the frequency from 25 MHz till 325 MHz. Each chip was equivalent from 50 000 to 5×10^6 system gates (SG). Some characteristics of Spartan-3 family are shown in Table 2.1.

The last column of Table 2.1 includes the capacity of memory distributed among LUTs of the chip. It is named distributed RAM (DRAM).

In this chapter we discuss only the basic features of FPGAs relevant to implementing logic circuits of control units. Let us analyse peculiarities of LUT-based FPGAs. As a rule, typical FPGAs include four main elements: configurable logic blocks (CLB) based on LUTs; matrix of programmable interconnections (MPI); input-output blocks (IOB) and embedded memory blocks. The organization of an FPGA chip is shown in Fig. 2.1.

A. Barkalov et al., *Logic Synthesis for FPGA-Based Finite State Machines*,
Studies in Systems, Decision and Control 38, DOI 10.1007/978-3-319-24202-6_2

Table 2.1 Characteristics of Spartan-3 family

Device	Number of CLBs	Number of SGs (K)	Capacity of BRAMs, bits (K)	Capacity of DRAM, bits (K)
XC3S50	1 728	50	72	12
XC3S200	4 320	200	216	30
XC3S400	8 064	400	288	56
XC3S1000	17 280	1000	432	120
XC3S1500	29 952	1500	576	208
XC3S2000	46 080	2000	720	320
XC3S4000	62 208	4000	1728	432
XC3S5000	74 880	5000	1872	520

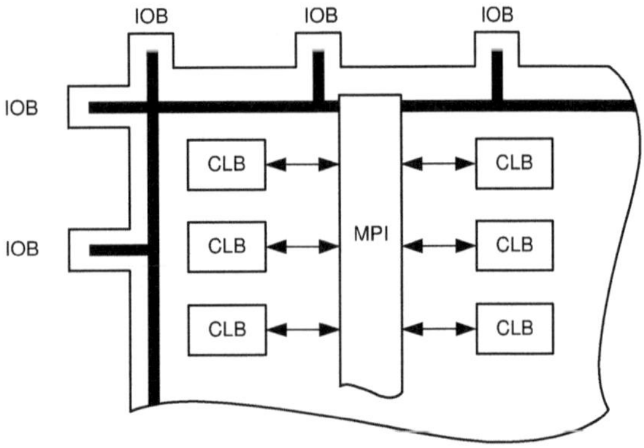

Fig. 2.1 Simplified organization of FPGA

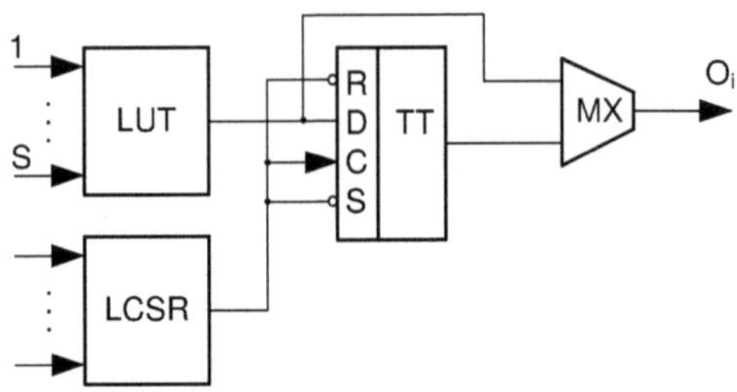

Fig. 2.2 Simplified organization of CLB

As a rule, LUTs are based on RAM having limited amount of inputs S ($S \leq 6$). A single LUT can implement an arbitrary Boolean function depended on L input variables ($L \leq S$) and represented by a truth table. A typical CLB included a single LUT, programmable flip-flop (FF), multiplexer (MX) and logic of clock and set-reset (LCSR). The simplified structure of CLB is shown in Fig. 2.2.

The output of LUT is connected with FF which could be programmed as D, JK or T flip-flop. The FF could be by-passed due to programmable MX. So, the output O_i of a CLB can be either combinational or registered. The existence of flip-flops allows organization of either registers or counters. Both these devices are used for FSM implementation.

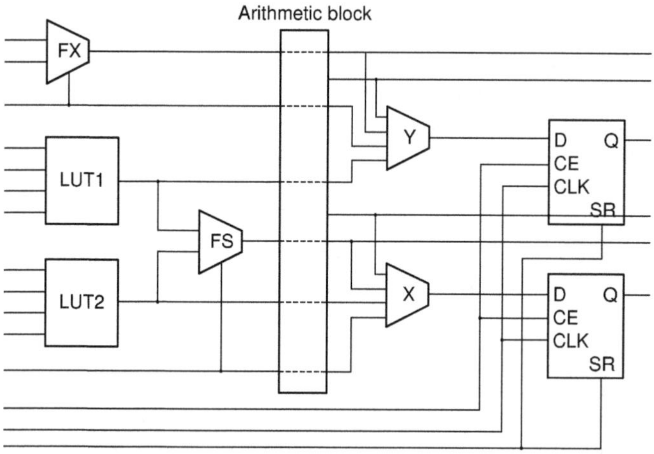

Fig. 2.3 Structure diagram of a slice of Virtex-4 family

With development of technology, more and more sophisticated CLBs were introduced. Let us discuss, for example the CLB of Virtex-7. It includes 4 slices having fast interconnections. A slice includes 2 LUTs, 4 multiplexers, arithmetic logic and 2 programmable flip-flops (Fig. 2.3).

This slice includes 2 LUTs; each of them has $S = 4$ inputs. Each LUT can implement an arbitrary logic function depended on 4 variables. Using the multiplexer FS, both LUTs are viewed as a single LUT having $S = 5$. The multiplexer FX combines together outputs of FS and FX from other slices. So, a slice can implement a Boolean functions depending on 5 variables; two slices on 6 variables; four slices (a CLB) on 7 variables. The arithmetic block allows organizing adders and multipliers. Multiplexers Y and X determine input data for programmable flip-flops. So, each CLB can include either RG or CT.

The number of inputs per a LUT is increased up to 5 for Virtex-5 family, whereas CLBs of Virtex-6 and Virtex-7 include LUTs having $S = 6$. There are different modifications of FPGAs for each family. We do not discuss them. Some characteristics of modern FPGA chips by Xilinx are shown in Table 2.2.

Analysis of Tables 2.1 and 2.2 proves our statement about the tremendous progress in FPGAs. Let us point out that modern chips include blocks of digital signal processors and central processing units. But these blocks are not used for FSM design. So, we do not discuss them.

At present, five companies dominate on the FPGA market: Xilinx, Altera, Lattice Semiconductor, Microsemi and Quicklogic. It can be found all necessary data about products of these companies on their homepages [2, 52, 58, 64, 89].

As follows from Table 2.2, there are powerful EMBs in the modern FPGAs. They have a property of configurability. It means that such parameters as number of memory cells (V) and their outputs (t_F) can be changeable. Of course, there is the constant size (V_o) of an EMB. Now, there are the following typical configurations of EMBs:

Table 2.2 Characteristics of FPGAs by Xilinx

Family	Modification	Number of Slices	Capacity of BRAMs, k bits	Capacity of DRAM, k bits	Technology, nm
Virtex-4	LX	10752–89088	1296–6048	168–1392	90
	SX	10240–24576	2304–5760	160–384	
	FX	5472–63168	648–9936	86–987	
Virtex-5	LX	4800–51840	1152–10368	320–3420	65
	LXT	3120–51840	936–11664	210–3420	
	SXT	5440–37440	3024–18576	520–4200	
	TXT	17280–24320	8208–11664	1500–2400	
	FXT	5120–30720	2448–16416	380–2280	
Virtex-6	LXT	11640–118560	5616–25920	1045–8280	40
	SXT	49200–74400	25344–38304	5090–7640	
	HXT	39360–88560	18144–32832	3040–6370	
	CHT	11640–37680	5616–14976	1045–3650	
Virtex-7	T	44700–305400	14760–46512	3475–21550	28
	XT	64400–135000	31680–64800	6525–13275	
	HT	45000–135000	21600–64800	4425–13275	

Fig. 2.4 Karnaugh map for system Y

T_1 \ T_2T_3	00	01	11	10
0	0	y_1y_2	y_2y_3	y_3
1	y_1y_2	y_1y_4	*	*

36 K ×1, 18 K ×2, 8 K ×4, 4 K ×8, 2 K ×16, 1 K ×32 and 512 ×64 (bits) [2, 52, 58, 64, 89]. Let us discuss how these blocks can be used in logic design.

Let an EMB contain V cells having t_F outputs. Let V_o be a number of cells if there is $t_F = 1$. So, the number V can be determined as

$$V = \lceil V_o/t_F \rceil. \tag{2.1}$$

Let us point out that decreasing t_F by 1 leads to doubling for the number of cells (and vice versa).

Embedded memory blocks could be used for implementing regular functions [8]. Let us discuss the Karnaugh map (Fig. 2.4) corresponding to Table 1.3.

The following functions can be found from this map:

$$\begin{aligned} y_1 &= T_1 \vee \bar{T}_2 T_3; \\ y_2 &= \bar{T}_1 T_3 \vee T_1 \bar{T}_3; \\ y_3 &= T_2;\ y_4 = T_1 T_3. \end{aligned} \tag{2.2}$$

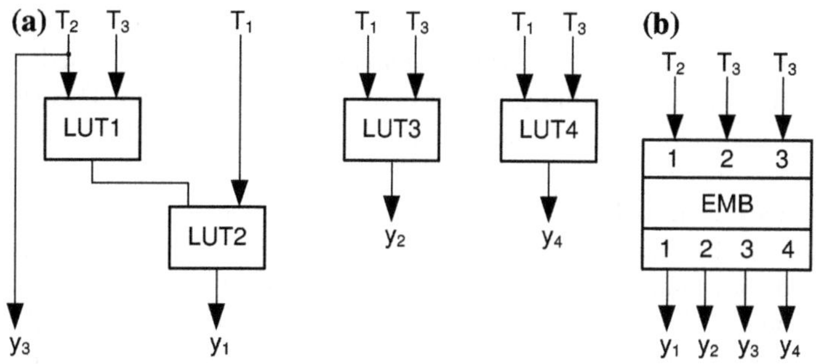

Fig. 2.5 Logic circuit for system (2.2)

Let us use system gates (LUTs with s = 2) for implementing the system (2.2).

From Fig. 2.2 the following negative features could be seen: (1) different propagation time for different output functions; (2) input variables should be connected with more than one logic element. The second feature leads for necessity of bigger value of fan-out for inputs T_1–T_3 than in the case when each input is connected only with a single logic element. Also, it leads to complications in the routing process.

If an EMB is used for implementing the system (2.2), all these problems are absent (Fig. 2.5b). Each input T_r $(r = \overline{1,3})$ is connected only with a single input of EMB. All functions y_n $(n = \overline{1,4})$ have the same propagation time. This example is very simple. But a simple EMB having the configuration 512×64 could replace at least 64 LUTs. It is possible if a system Y depends on 9 inputs and includes up to 64 different functions. Of course, the circuit includes 64 LUTs having $S = 9$. But there is no such LUTs in modern FPGAs. If minimization allows dependence of each function $y_n \in Y$ on 8 variables, then 256 LUTs with $S = 6$ are necessary for creating a logic circuit.

In a typical FPGA 60 % of power is consumed by the programmable interconnects, 16 % is consumed by programmable logic and 14 % by the clock distribution [14]. Replacement of LUTs by EMBs allows decreasing of the number of interconnections. So, it is very important to use EMBs in implementing FSM circuits.

The exceptional complexity of FPGA requires using computer-aided design (CAD) tools for designing logic circuits [41]. It assumes development of formal methods for synthesis and verification of control units [36, 37, 43, 59]. For example, a design process for FPGAs from Xilinx includes the following steps:

1. **Specification of a project**. A design entry can be executed by the schematic editor (if a design is represented by a circuit), or the state editor (if a design is represented by a STG) or a program written with some hardware description language (HDL). The most popular HDLs are VHDL and Verilog [6, 7]. This initial specification is verified and corrected if necessary.
2. **Logic synthesis**. During this step, the package FPGA Express executes synthesis and optimization of an FSM logic circuit. As an outcome of this step, an FPGA

Netlist file is generated. This file is represented in either EDIF or XNF format. During this step, library cells from system and user libraries are used.

3. **Simulation**. The functional correctness of an FSM is checked. This step is executed without taking into account real propagation times in a chip. If the outcome of simulation is negative, then the previous steps should be repeated.
4. **Implementation of logic circuit**. Now the Netlist is translated into an internal format of CAD system. Such physical objects as CLBs and chip pins are assigned for initial physical elements. This step is named the packing. The step of mapping is the first stage of the packing. The mapping refers to the process of associating entities such as the gate-level functions in the gate-level netlist with the LUT-level functions available on the FPGA [53]. It is not a one-to-one mapping because each LUT can be used to represent a number of logic gates [42]. The mapping step gives results for executing the packing. During this step the LUTs and flip-flops are packed into the CLBs. Both mapping and packing steps are very difficult because there are too many variants of their solutions. Following packing the step of place-and-route is executed. Now we know the connection between CLBs and parts of a logic function to be implemented. But there are many ways how these CLBs could be placed in the FPGA. The placement problem is very difficult because hundreds of thousands or even millions CLBs should be placed. During the routing it is necessary to decide how to connect all CLBs for a particular project. This step should be executed in a way giving the maximum possible performance. Obviously, the outcome of placement affects tremendously the outcome of routing.
   When routing is finished, the real performance could be found. Also, the BitStream is formed which will be used for chip programming.
5. **Project verification**. The final simulation is performed where the actual values of delays among the physical elements of a chip are used. If outcome of this step is negative (the actual performance of an FSM is less than it is necessary), then the previous steps of the design process should be repeated.
6. **Chip programming**. This step is connected with the writing of the final bit stream into the chip.

One of the most important roles in the design process plays the step of logic synthesis. Let us analyse the steps and possible solutions for FPGA-based FSMs.

2.2 Trivial Implementing FPGA-Based FSMs

The process of FSM synthesis can be viewed as a transformation of FSM initial specification into structural specifications where elements of lower abstraction levels are used [1, 17, 21]. This process is repeated till each element to be designed is represented by some library element. In this book, we start from GSAs and finish by LUTs and EMBs.

Fig. 2.6 Trivial structural diagrams of LUT-based Mealy (**a**) and Moore (**b**) FSMs

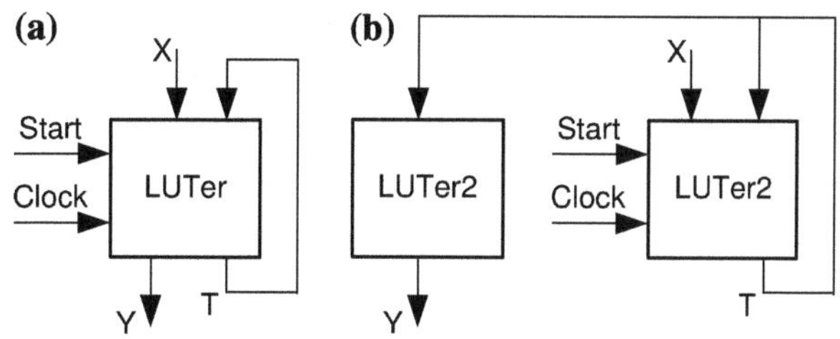

Let us start from LUT-based P Mealy FSM (Fig. 2.6a). It includes a LUTer which is a circuit implemented with LUTs. There is no register in the explicitly as it is in Fig. 1.4. Now, the register is formed by flip-flops distributed among LUT-elements.

Let us discuss an example of design for Mealy FSM $P(\Gamma_1)$. There is the marked GSA Γ_1 shown in Fig. 1.5a. To design the circuit of LUTer, it is necessary to find the functions (1.3) and (1.4). In the case of $P(\Gamma_1)$, these functions are represented as (1.13). Let the symbol $L(D_r)$ stand for the number of literals in the SOP of function $D_r \in \Phi$. Let the symbol $L(y_n)$ stand for the number of literals in the SOP of function $y_n \in Y$. Let the following condition take place:

$$S \geq max(L(D_1), \ldots, L(D_R), L(y_1), \ldots, L(y_N)). \tag{2.3}$$

In (2.3), the symbol S stand for the number of inputs of a LUT. In this case, each function of FSM is implemented using only a single LUT.

There are the following values of $L(D_r)$ and $L(y_n)$ in the case of $P(\Gamma_1)$: $L(D_1) = L(y_4) = 1, L(D_2) = 2, L(D_1) = 4, L(y_2) = L(y_3) = 3$. Let us implement the circuit of $P(\Gamma_1)$ using LUTs having $S = 4$ inputs. It leads to the circuit shown in Fig. 2.7. Let us point out that we use Eq. (1.13).

There are three LUTs in the part implementing functions $y_n \in Y$. There is only a wire T_2 in the circuit for $y_4 = T_2$. There are two LUTs in the part of the circuit implementing input memory functions $D_r \in \Phi$. They have additional inputs for pulses Start and Clock. Of course, they are not address inputs as it follows, for example, from Fig. 2.2. There is $D_1 = T_2$, but it is still necessary a LUT to implement this function.

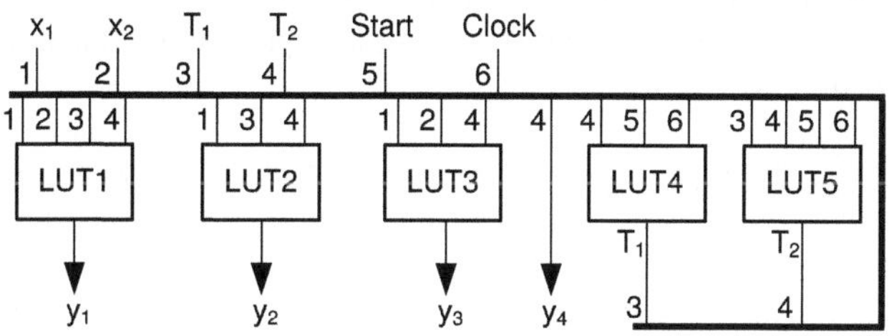

Fig. 2.7 Logic circuit of Mealy FSM $P(\Gamma_1)$

Table 2.3 Table for programming LUT1

T_1	T_2	x_1	x_2	y_1	T_1	T_2	x_1	x_2	y_1	T_1	T_2	x_1	x_2	y_1	T_1	T_2	x_1	x_2	y_1
0	0	0	0	1	0	1	0	0	1	1	0	0	0	1	1	1	0	0	1
0	0	0	1	1	0	1	0	1	0	1	0	0	1	1	1	1	0	1	1
0	0	1	0	1	0	1	1	0	0	1	0	1	0	1	1	1	1	0	1
0	0	1	1	1	0	1	1	1	0	1	0	1	1	1	1	1	1	1	1

To program LUT1–LUT5, they should construct truth tables corresponding to functions y_1, y_2, y_3, D_1 and D_2. Let us discuss this step for the function y_1 from (1.13). It is represented by the following SOP:

$$y_1 = \bar{T}_1\bar{T}_2 \vee T_2\bar{x}_1\bar{x}_2 \vee T_1. \tag{2.4}$$

To form a truth table, this SOP should be transformed into the perfect SOP [55] where each term has exactly $L(y_1)$ literals. To do it, we should find the following functions: $\bar{T}_1\bar{T}_2(x_1 \vee \bar{x}_1)(x_2 \vee \bar{x}_2)$; $T_2\bar{x}_1\bar{x}_2(T_1 \vee \bar{T}_1)$; $T_1(T_2 \vee \bar{T}_2)(x_1 \vee \bar{x}_1)(x_2 \vee \bar{x}_2)$. After transformation (2.4), the table of LUT1 can be constructed (Table 2.3).

The term $\bar{T}_1\bar{T}_2$ is represented by input assignments from 0000 till 0011; the term $T_2\bar{x}_1\bar{x}_2$ by 0100 and 1100; the term T_1 by input assignments from 1000 till 1111.

Using this approach, it can be found tables of programming for each function of FSM. Next, the corresponding bit-streams should be created and loaded into LUTs [77].

The circuit from Fig. 2.7 is a single-level circuit, because there is only a single level of logic among inputs $x_e \in X$, $T_r \in T$ and outputs of FSM. It is possible only if the condition (2.3) takes place. But it is violated very often in practical cases of FSM design [77].

A small amount of inputs per LUT creates a big problem for logic design. Let us consider some input memory function D_1 depending on $L(D_1) = 7$ Boolean variables:

$$D_1 = T_1\bar{T}_2T_3x_1\bar{x}_2 \vee T_1T_2T_3x_3x_4 \vee \bar{T}_1\bar{T}_2\bar{T}_3\bar{x}_1x_3. \tag{2.5}$$

Let LUTs in use have $S = 7$ inputs. In this case, the logic circuit for D_1 includes only a single LUT Fig. 2.8a. Now, let LUT have $S = 6$. In this case the function (2.5) should be transformed. It should be represented by some functions $f_1, f_2, \ldots$ having $L(f_1) \leq 6$, $L(f_2) \leq 6$, and so on. Let us represent the function (2.5) in the following form:

$$D_1 = T_1(\bar{T}_2T_3x_1\bar{x}_2 \vee T_2T_3x_3x_4) \vee \bar{T}_1\bar{T}_2T_3\bar{x}_1x_3 = T_1A \vee B. \tag{2.6}$$

The function D_1 represented as (2.6) requires two LUTs to be implemented. Moreover, this circuit Fig. 2.8b includes two levels of LUTs. It means that the solution corresponding to (2.6) is twice slower.

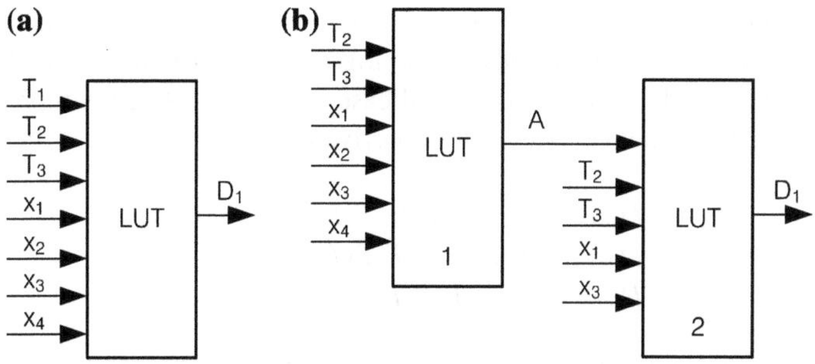

Fig. 2.8 Implementing function D_1 with LUTs having $S = 7$ (**a**) and $S = 6$ (**b**)

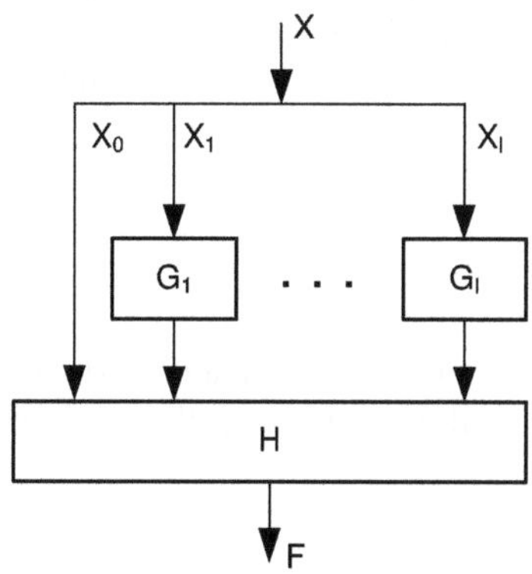

Fig. 2.9 Illustration of the principle of functional decomposition

This approach is named functional decomposition. The principle of functional decomposition is the basic one for FPGA-based design [65, 71]. There are basically approaches targeting only LUTs. But methods exists using EMBs as tools for implementing some subfunctions [18, 66].

In general, the method of functional decomposition is based on representation of a Boolean function $F(X)$ in the following form:

$$F(X) = H(X_0, G_1(X_2), \ldots, G_I(X_I)). \tag{2.7}$$

The Eq. (2.7) corresponds to the implementation of the circuit shown in Fig. 2.9.

The negative influence of functional decomposition is increasing of the propagation time in comparison with a single-level circuit. It follows from comparison of the circuits from Fig. 2.8. As it is mentioned in [65], the methods of functional decomposition are far from ideal. Let us point out that it is very important to decrease the numbers of arguments and product terms in Boolean functions to be implemented. We discuss these methods a bit further.

Now, let us discuss the trivial LUT-based implementing P Moore FSM. Its structural diagram is shown in Fig. 2.6b. The LUTer1 implements the functions (1.5), the LUTer2 the functions (1.4). There is a distributed register hidden among LUTs of LUTer2.

Let us discuss an example of design for Moore FSM $P(\Gamma_1)$. There is the marked GSA Γ_1 shown in Fig. 1.5b. To design the logic circuit, it is necessary to find the functions (1.4) and (1.5). In the discussed case, the system (1.5) is represented by (1.17), whereas the system (1.4) should be derived from Table 1.2. It is the following system:

$$\begin{aligned} D_1 &= \bar{T}_1\bar{T}_2T_3\bar{x}_1\bar{x}_2 \vee \bar{T}_1T_2 \vee T_1\bar{T}_2\bar{T}_3; \\ D_2 &= \bar{T}_1\bar{T}_2\bar{T}_3x_1 \vee \bar{T}_1\bar{T}_2\bar{T}_3x_2; \\ D_3 &= \bar{T}_2\bar{T}_3 \vee \bar{T}_1\bar{T}_2T_3\bar{x}_1x_2 \vee \bar{T}_1T_2. \end{aligned} \quad (2.8)$$

If the condition (2.3) takes place, then circuits of LUTer1 and LUTer2 have only a single level of logic.

In the discussed case, it should be $S = 5$. Let us point out that functions (1.5) can be optimized using a proper state assignment.

The following system can be derived from Fig. 2.5b:

$$\begin{aligned} y_1 &= A_2 \vee A_5 \vee A_6; \\ y_2 &= A_2 \vee A_4 \vee A_5; \\ y_3 &= A_3 \vee A_4;\ y_4 = A_6. \end{aligned} \quad (2.9)$$

Let us encode states of Moore FSM $P(\Gamma_1)$ as it is shown in Fig. 2.10.

After minimizing, the following system of equations can be found instead of (2.9):

$$\begin{aligned} y_1 &= T_1; \\ y_2 &= T_2; \\ y_3 &= \bar{T}_1T_3; \\ y_4 &= T_1\bar{T}_2. \end{aligned} \quad (2.10)$$

It is necessary 2 of LUTs having $S = 2$ to implement the system (2.10). The functions $y_1, y_2 \in Y$ are represented just by the outputs of LUTs from the LUTer2. Let us point out that it is necessary 3 of LUTs having $S = 3$ to implement the system (1.17). It shows the big importance of the state assignment step.

FPGA-based devices can be found in many areas where the consumed energy is a critical factor [53]. It concerns portable computing devices, wireless telecommunication equipment, space-based applications [53]. The rising of FPGA complexity leads to increasing the power consumed by FPGA-based devices. It is well known that FSMs consume a significant amount of power in any FPGA-based project [22]. Therefore, minimizing power consumed by the FSMs can significantly reduce the total power consumed by a device.

Fig. 2.10 State codes of Moore FSM $P(\Gamma_1)$

T_1 \ T_2T_3	00	01	11	10
0	a_1	$*$	a_2	a_6
1	a_3	a_4	a_5	$*$

The dynamic power dissipated in CMOS circuits can be expressed by the well-known formula [81]:

$$P = \sum_{n=1}^{N} C_n f_n V_{DD}^2. \tag{2.11}$$

In (2.11), N is the number of elements, C_n is the load capacitance at the output of the element number n, f_n is the frequency of its switching, and V_{DD} is the supply voltage. One of the ways for decreasing the power dissipation is decreasing of the switching activity of flip-flops [81].

There is a very interesting result of investigations conducted by the authors of the article [81]. They found that the smaller FSM circuit consumes less power than its bigger version. It is clear because a smaller circuit needs less interconnections than its bigger counterpart. One of the ways leading to smaller FSM circuits is application of EMBs for implementing some parts of FSM circuits [76]. It is shown that FSM implementation with EMBs provides some benefits compared to synthesis with LUTs [34, 72]. The maximum clock frequency of an FSM implemented in a ROM block is independent on its complexity of course, it is possible if the whole circuit is implementing using just a single EMB. The memory blocks of FPGAs provide control signals that allow for module deactivation when the FSM is inactive. It provides an efficient mechanism for power saving. It has been proved [83] that complex FSMs consume less power when implemented as memory blocks. Let us consider some EMB-based models of FSMs.

In the simplest case, P FSMs can be implemented using a single EMB (Fig. 2.11). There is an external register shown in Fig. 2.11. If an EMB is synchronized, then pulses Clock and Start enter the EMB. But it does not change the design method. The structural diagram (Fig. 2.11) is general for both Mealy and Moore FSMs. We use symbol EMBer to define the circuit implemented with EMBs.

In the case of P Mealy FSM, the EMBer implements systems (1.3) and (1.4). To design it, it is necessary to construct the table of EMBer having columns T, X, Y, Φ, q. Here sets T and X determine addresses of cells, the sets Y and Φ content of cells, q is the number of a cell.

In the case of Mealy FSM $P(\Gamma_1)$, there are the following sets $T = \{T_1, T_2\}$, $X = \{x_1, x_2\}$, $Y = \{y_1, \ldots, y_4\}$ and $\Phi = \{D_1, D_2\}$. So, there are $R_0 = 2$, $L = 2$, $N = 4$. It is necessary to have an EMB with 16 cells having up to 6 bits (Table 2.4).

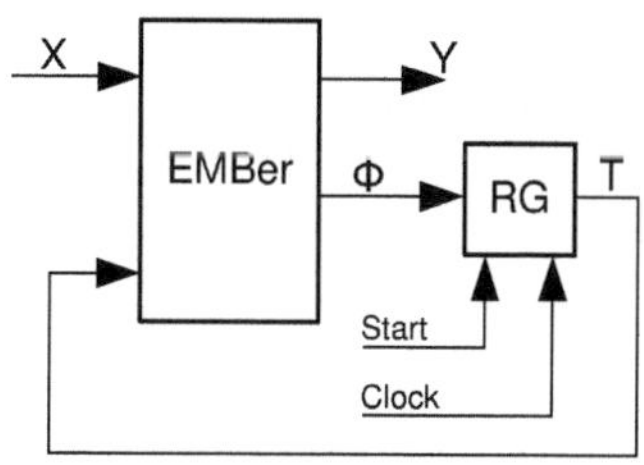

Fig. 2.11 Structural diagram of EMB-based P FSM

Table 2.4 Content of EMBer of Mealy FSM $P(\Gamma_1)$

T	X	Y	Φ	q	T	X	Y	Φ	q
T_1T_2	x_1x_2	$y_1y_2y_3y_4$	D_1D_2		T_1T_2	x_1x_2	$y_1y_2y_3y_4$	D_1D_2	
00	00	1100	01	1	10	00	1001	00	9
00	01	1100	01	2	10	01	1001	00	10
00	10	1100	01	3	10	10	1001	00	11
00	11	1100	01	4	10	11	1001	00	12
01	00	1100	10	5	11	00	0000	00	13
01	01	0110	10	6	11	01	0000	00	14
01	10	0010	10	7	11	10	0000	00	15
01	11	0010	10	8	11	11	0000	00	16

The macrocells 1–4 correspond to row 1 of Table 1.1. The macrocell 5 corresponds to row 4, the macrocell 6 to row 3, macrocells 7, 8 to row 2 of Table 1.1. The macrocells 9–12 correspond to row 5 of Table 1.1. There is no state a_m having the code 11. So, the macrocells 13–16 have no counterparts in Table 1.1. Their content is ignored and can be any. They are filled by zeros in Table 2.4.

EMBs of EMBer (Fig. 2.11) should satisfy to the following condition:

$$2^{R+L} \leq V_0. \tag{2.12}$$

In the case of Mealy FSM, there is $R = R_0$. In the case of Moore FSM, there is $R = R_1$. The condition (2.12) shows that an EMB has enough cells to implement a table similar to Table 2.4.

The number of EMBs in EMBer can be found using the following expression:

$$n_{EMB} = \left\lceil \frac{N + R}{t_F} \right\rceil. \tag{2.13}$$

In (2.13) the symbol R has the same meaning as in (2.12). The value of t_F is determined using (2.1). Obviously, it is very important to minimize the number of EMBs in the circuit of EMBer.

There are homogenous circuits corresponding to the diagram shown in Fig. 2.11. In the case of Moore FSM, it is possible to use the heterogeneous approach (Fig. 2.12).

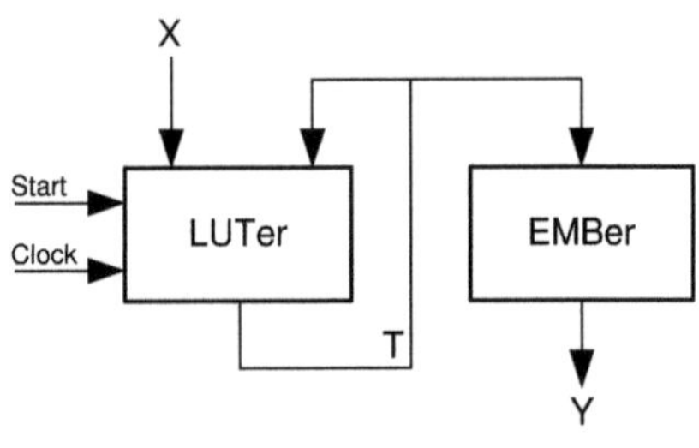

Fig. 2.12 Heterogeneous structure of P Moore FSM

In this circuit, the LUTer implements the system (1.4), whereas the EMBer the system (1.5). It allows implementing the final circuits using less amount of EMBs than in the case of completely EMB-based approach (Fig. 2.11).

Now, the following condition should be true:

$$2^{R_1} \cdot N \leq V_0. \tag{2.14}$$

In this case, it is enough only a single EMB in the EMBer.

For a heterogeneous P Moore FSM, there are the following columns in the table of EMBer: T (address of the cell) and Y (content of the cell). There is $R_1 = 3$ and $N = 4$ for Moore FSM $P(\Gamma_1)$. So it is enough to use an EMB having $V = 8 \times 4 = 32$ bits to implement the circuit of EMBer.

It is very important to diminish the number of LUTs in circuits of LUTers. One of the possible ways is a proper state assignment. Let us discuss this step in more details.

2.3 Methods of State Assignment

If an FSM is specified by a GSA, then such sets as X, Y and A are known. But there are no state codes. To obtain them, the step of state assignment is executed [57]. This step is very important because its outcome has a tremendous influence on the hardware amount (the number of LUTs) in the FSM logic circuit [1]. A strategy of state assignment could target optimization for area, performance, power consumption, or testability.

One of the most popular state assignment algorithms is JEDI which is distributed with the system SIS [73, 74]. JEDI targets a multi-level logic implementation. It is based on the weight assignment for states $a_m \in A$.

The input dominant algorithm assigns higher weights to pairs of present states which assert similar inputs and produce sets of next states. It allows maximizing the size of common cubes in the implemented logic function. The output dominated algorithm assigns higher weights to pairs of next states which are generated by similar input combinations and similar sets of present states. It maximizes the number of common cubes in the logic function. This method can be used for the optimal state assignment based on PES [8]. We discuss this approach a bit later.

In modern industrial packages different state assignment strategies are used. For example, two optimization criteria are used in the design tool XST of Xilinx: maximum performance and minimum hardware [90]. Seven different approaches are used for state assignment. The automatic state assignment is based on some special algorithm proposed by Xilinx. It has been never published. The method of one-hot encoding is based on the following expression:

$$R = M. \tag{2.15}$$

This method is very popular because: (1) it is very simple and (2) each LUT is connected with a flip-flop. So, this conception is implemented very easy in FPGAs. In this case, there is a lot of input memory functions but each of them is relatively small. The compact state assignment is based on the formula (1.6) for a Mealy FSM and the formula (1.14) for a Moore FSM. In this case the number of input memory functions is minimum possible, but they are rather complex. In this book we mostly use this approach and name it a binary state assignment. Two other methods are based on codes either Gray or Johnson. At last, there is so named speed encoding. When the performance is maximized and the sequential encoding based on using of the counter instead of state register.

The master thesis [82] is devoted to investigation of influence of the state assignment method on characteristics of Mealy FSM. The benchmarks from [91] are used in the investigation. The results obtained for Mealy FSM are represented in Table 2.5.

The investigations are executed for the FPGA XC5VLX30 of Xilinx. The first column of Table 2.6 shows the name of a benchmark. The columns "LUT" show numbers of LUTs in the final circuit. The columns "MHz" represent the maximal frequency of operation for final Mealy FSMs.

The best results are produced when the automatic state assignment is used. It gives the best outcomes for area (58.54 % of all benchmarks) and performance (39.02 %). The binary state assignment possesses the second place in this competition. As follows from Table 2.5, the automatic state assignment produces the best results when both area and performance are optimized (29.27 %). The same results are produced for the compact (binary) state assignment. It is interesting that the one-hot state assignment can optimize only one parameter of FSM circuit.

Of course, these results are true only for the chip XC5VLX30. But similar conclusions are made, for example, in [51]. It allows to suggest that these conclusions have a rather common nature.

As it was pointed before, it is very important to diminish the power dissipation in the circuit of FSM [77].

One of the approaches leading to decreasing the power dissipation in FSMs is the energy-saving state assignment [67]. Main works in low-power FSMs compute first the switching activity and transition probabilities [84]. The key idea of these methods is the reduction of the average activity by minimizing the bit changes during state transitions [16, 60]. The state assignment should minimize the Hamming distance between states with high transition probability. Different variations of this approach can be found in many works [3, 23, 32, 61]. There are hundreds of articles devoted to this approach.

Now, let us discuss an approach based on using classes of pseudoequivalent states in Moore FSMs. Remind, that states $a_m, a_s \in A$ belong to the same class of PES if corresponding vertices of a GSA are connected with an input of the same vertex of this GSA [5]. It allows constructing the partition $\Pi_A = \{B_1, \ldots, B_I\}$, where $B_i \in \Pi_A$ is a class of PES. After constructing the partition Π_A, an initial GSA Γ can be transformed into a block GSA $B(\Gamma)$.

Table 2.5 Outcomes of investigations

FSM	Auto		One-hot		Compact		Sequential		Gray		Johnson		Speed1	
	LUT	MHz	LUT	MHz	LUT	MHz	LUT	MHz	LUT	MHz	LUT	MHz	LUT	MHz
bbara	11	639	9	966	13	635	13	639	19	589	24	545	13	962
bbsse	29	559	29	559	29	582	29	538	31	538	36	408	38	556
bbtas	5	962	8	966	5	966	5	962	5	955	5	962	9	966
beecount	7	952	19	639	7	952	7	952	7	948	21	625	30	583
ese	49	480	52	477	46	463	50	487	46	454	71	434	72	453
dk14	8	945	29	522	8	945	8	945	8	945	19	623	40	512
dk15	7	1062	19	737	7	1062	7	1062	7	1062	7	1062	19	659
dk16	46	556	46	556	15	625	19	506	27	554	86	355	70	399
dk17	6	952	14	669	6	952	6	952	6	952	7	895	27	571
dk27	5	900	8	906	5	897	5	959	5	955	6	899	10	903
dk512	17	730	17	730	7	899	7	895	7	899	21	437	19	790
ex1	64	586	64	586	74	447	67	478	66	406	106	340	72	605
ex4	15	962	15	962	16	626	15	598	14	748	33	546	15	962
ex6	29	553	30	580	20	621	23	615	22	616	36	426	31	598
keyb	56	384	56	384	65	358	71	382	66	447	62	435	85	374
kirkman	51	874	84	1058	53	569	48	880	51	874	112	451	84	1058
lion	3	1084	5	962	3	1080	3	1080	3	1084	3	1084	5	962
mark1	27	726	27	726	19	708	22	622	18	623	27	574	29	959
me	5	1071	8	1071	5	1071	6	1071	5	1071	5	1071	8	1071
opus	22	596	22	754	21	628	26	585	22	596	26	576	26	671
planet	100	888	100	888	138	389	145	417	149	375	192	346	106	637
planet1	100	888	100	888	138	389	145	417	149	375	192	346	106	637

(continued)

Table 2.5 (continued)

FSM	Auto		One-hot		Compact		Sequential		Gray		Johnson		Speed1	
	LUT	MHz	LUT	MHz	LUT	MHz	LUT	MHz	LUT	MHz	LUT	MHz	LUT	MHz
pma	73	554	73	554	115	438	108	367	112	375	121	405	88	559
s1	77	550	77	550	75	447	89	328	105	368	114	361	81	552
s1488	140	425	140	425	141	432	130	394	147	433	192	334	162	458
s1494	124	412	124	412	143	442	135	383	145	383	192	333	152	462
s208	28	559	28	559	13	669	12	716	15	639	29	483	50	386
s27	4	962	21	636	4	962	7	679	4	962	12	664	21	631
s298	362	406	362	406	330	313	264	311	274	314	716	244	399	397
s386	26	577	31	586	28	581	28	558	29	429	43	22	36	441
s420	28	559	28	559	14	629	12	716	15	639	29	483	36	510
s510	42	900	42	900	39	448	53	440	50	427	123	388	42	900
s820	63	429	63	429	85	395	92	441	93	438	98	366	93	399
s832	63	429	63	429	73	412	77	431	87	394	97	335	108	444
sand	99	569	99	569	121	426	125	421	125	438	189	306	103	490
scf	179	676	179	676	202	338	205	349	197	389	337	327	180	561
shiftreg	0	1584	9	1080	0	1584	4	959	4	959	5	902	4	903
sse	29	559	29	559	28	543	37	548	32	540	44	394	36	612
styr	118	430	118	430	127	369	138	363	138	353	181	323	161	454
tav	6	1556	6	1556	6	911	6	911	5	914	5	914	6	1556
tbk	55	406	179	360	71	465	129	342	137	290	295	276	444	342

Fig. 2.13 State codes for Moore FSM $P_0(\Gamma_1)$

T_1 \ T_2T_3	00	01	11	10
0	a_1	a_3	a_4	a_2
1	a_6	*	a_5	*

Table 2.6 Reduced structure table of Moore FSM $P_0(\Gamma_1)$

B_i	$K(B_i)$	a_s	$K(a_s)$	X_h	Φ_h	h
B_1	*00	a_2	010	1	D_2	1
		a_3	001	x_1	D_3	2
B_2	*10	a_4	011	$\bar{x}_1x_2$	D_2D_3	3
		a_5	111	$\bar{x}_1\bar{x}_2$	$D_1D_2D_3$	4
B_3	**1	a_6	100	1	D_1	5

Let us encode states $a_m \in A$ in such a way that each class $B_i \in \Pi_A$ be represented by the minimal possible amount of generalized intervals of R_1-dimensional Boolean space. Let us name such an approach "optimal state assignment". For example, there is the partition $\Pi_A = \{B_1, \ldots, B_4\}$ in the case of Moore FSM represented by GSA Γ_1 Fig. 1.5b. It corresponds to the BGSA $B(\Gamma_1)$ shown in Fig. 1.7.

Let us point out that the JEDI algorithm [73] can be used for the optimal state assignment. Let us use the symbol $P_0(\Gamma_j)$ to show that a Moore FSM is synthesized using a GSA Γ_j and the principle of optimal state assignment. One of the possible outcomes is shown in Fig. 2.13 for the Moore FSM $P_0(\Gamma_1)$.

Analysis of $B(\Gamma_1)$ shows that there is only an unconditional transition from state $a_6 \in B_4$ into state a_1. There are no functions $D_r = 1$ $(r = \overline{1, R_1})$ during this transition. It means that $K(a_6)$ can be treated as a "don't care" state code.

Taking it into account, the following generalized intervals $K(B_i)$ can be found from the Karnaugh map (Fig. 2.13): $K(B_1) = *00$, $K(B_2) = *10$, $K(B_3) = **1$. These cubes can be treated as codes of classes $B_i \in \Pi_A$. Using these codes and BGSA $B(\Gamma_1)$, the reduced structure table of Moore FSM $P_0(\Gamma_1)$ can be constructed (Table 2.6).

We hope that there is obvious connection among BGSA $B(\Gamma_1)$, the state codes (Fig. 2.13) and Table 2.6. There are the following terms F_h in the case of this table: $F_1 = \bar{T}_2\bar{T}_3$; $F_2 = T_2\bar{T}_3x_1$; $F_3 = T_2\bar{T}_3\bar{x}_1x_2$; $F_4 = T_2\bar{T}_3\bar{x}_1\bar{x}_2$; $F_5 = T_3$.

It allows finding the following system of input memory functions (after minimization):

$$\begin{aligned} D_1 &= T_2\bar{T}_3\bar{x}_1\bar{x}_2 \vee T_3; \\ D_2 &= \bar{T}_2\bar{T}_3 \vee T_2\bar{T}_3\bar{x}_1; \\ D_3 &= T_2\bar{T}_3. \end{aligned} \quad (2.16)$$

Each function of the system (2.16) is implemented using only a single LUT having $S \leq 4$.

Fig. 2.14 State codes for Moore FSM S_1

T_1 \ T_2T_3	00	01	11	10
0	a_1	a_2	a_3	a_4
1	a_8	a_5	a_6	a_7

In the common case, the terms F_h are represented as:

$$F_h = \bigwedge_{r=1}^{R_1} T_r^{e_{ir}} \cdot X_h \quad (h = \overline{1, H_0}). \tag{2.17}$$

In (2.17), $e_{ir} \in \{0, 1, *\}$ is the value of the r-th bit of $K(B_i)$ from the h-th row of a reduced structure table, $T_r^0 = \bar{T}_r$, $T_r^1 = T_r$, $T_r^* = 1$.

It follows from comparison of Tables 1.1 and 2.6 that there is the same amount of rows in both of them. It is true if each class $B_i \in \Pi_A$ is represented by a single cube. Because of it, the symbol H_0 is used in (2.17).

Let us point out that sometimes it is necessary more than one cube to represent some class $B_i \in \Pi_A$. For example, let it be the partition $\Pi_A = \{B_1, \ldots, B_4\}$ for some FSM S_1. Let it be $B_1 = \{a_1\}$, $B_2 = \{a_2, a_3, a_4\}$, $B_3 = \{a_5, a_6, a_7\}$, $B_4 = \{a_8\}$. There is one of the possible outcomes of the optimal state assignment for S_1 shown in Fig. 2.14.

Let us treat as "don't care" the code of the state $a_8 \in A$. Even in this case, the class B_2 is represented by two cubes: 00$*$ and 01$*$. If there are conditional transitions from the state $a_8 \in B_4$, its code cannot be treated as "don't care". In this case, two classes are represented using two cubes each. The class B_3 is represented by the cubes 11 and 11$*$. In this case, there are more than H_0 rows in the reduced structure table. We deal with this case a bit later.

So, the existed methods of state assignment can be divided using the logic elements they target. For example, there are PLA-oriented methods [4, 19, 20, 27, 56, 68–70, 86, 87], as well as PAL-oriented methods [9–14, 24–26, 28, 44–50, 78–80]. A lot of methods deals with FPGA-based FSMs [29–31, 33, 38, 39, 63]. There are a lot of state assignment methods targeting saving the energy consumed by the FSM circuit [3, 40, 62, 85, 88], just to name a few.

Our book is devoted to FPGA-based FSMs. The main goal of all discussed methods is to save the chip area occupied by an FSM circuit. So, we use JEDI-based algorithms for executing the state assignment.

2.4 Hardware Reduction for FPGA-Based FSMs

As it is mentioned in many works, there are the positive back effects of hardware reduction. The solution of this problem leads to increasing of performance and decreasing of power consumption of FSM logic circuits [35, 92]. To decrease

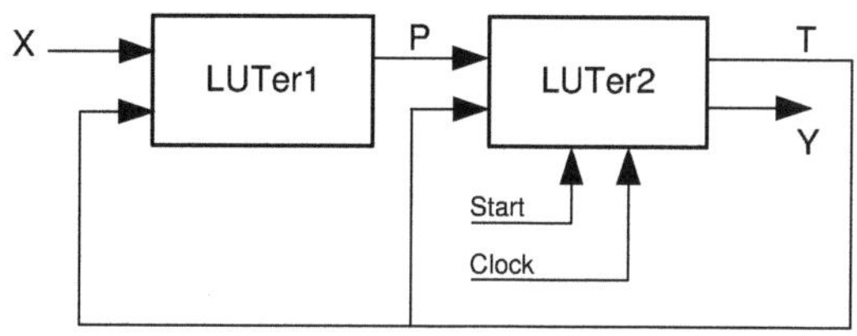

Fig. 2.15 Structural diagram of LUT-based MP Mealy FSM

the amount of hardware, we use three groups of methods: (1) structural decomposition; (2) heterogeneous implementation and (3) optimal encoding of states and other objects of FSMs.

We discussed the method of replacement of logical conditions (RLC) in Sect. 1.5. Let us start with LUT-based MP Mealy FSM (Fig. 2.15).

In this circuit, the LUTer1 is equivalent to BRLC from Fig. 1.21. It implements the system (1.47). The LUTer2 executes functions of BIMF from Fig. 1.21. It generates microoperations (1.48) and input memory functions (1.49). The register RG is distributed among LUTs of LUTer2, so, there are state variables $T_r \in T$ shown as the outputs of LUTer2 (Fig. 2.15).

There are the following steps in design method for MP Mealy FSM:

1. Constructing the set of states for a GSA Γ.
2. Executing the state assignment.
3. Constructing the tables of LUTer1.
4. Constructing the structure table of MP Mealy FSM.
5. Constructing the tables of LUTer2.
6. Implementing the FSM logic circuit.

Let us discuss an example of design for Mealy FSM $MP(\Gamma_4)$, where a GSA Γ_4 is shown in Fig. 2.16. There are marks of states on GSA Γ_4. So, it can be found the set $A = \{a_1, a_2, a_3\}$ with $M_0 = 3$. It is enough $R_0 = 2$ of state variables to encode the states $a_m \in A$. Let us encode them in the following way: $K(a_1) = 00$, $K(a_2) = 01$ and $K(a_3) = 10$.

To construct the table of LUTer1 it is necessary to execute the RLC. There are the following sets $X(a_m)$ in the discussed case: $X(a_1) = \{x_1, x_2\}$, $X(a_2) = \{x_3, x_4\}$ and $X(a_3) = \emptyset$. So, there is $G = 2$ and $P = \{p_1, p_2\}$.

Table 2.7 represents the table of replacement of logical conditions for $MP(\Gamma_4)$. It includes the columns with states, their codes and logical conditions replaced by corresponding variables $p_g \in P$.

There are G circuits in LUTer1. Each of them corresponds to a single function $p_g \in P$. Let $X(p_g) \subseteq X$ be a set of logical conditions replaced by the variable $p_g \in P$. There are G tables representing functions $p_g \in P$. Each table includes the columns $K(a_m)$, $X(p_g)$, p_g. First two columns create the address of the cell, the last column shows its content. There are the following functions derived from Table 2.7.

$$\begin{aligned} p_1 &= \bar{T}_1\bar{T}_2 x_1 \vee \bar{T}_1 T_2 x_3; \\ p_2 &= \bar{T}_1\bar{T}_2 x_2 \vee \bar{T}_1 T_2 x_4. \end{aligned} \quad (2.18)$$

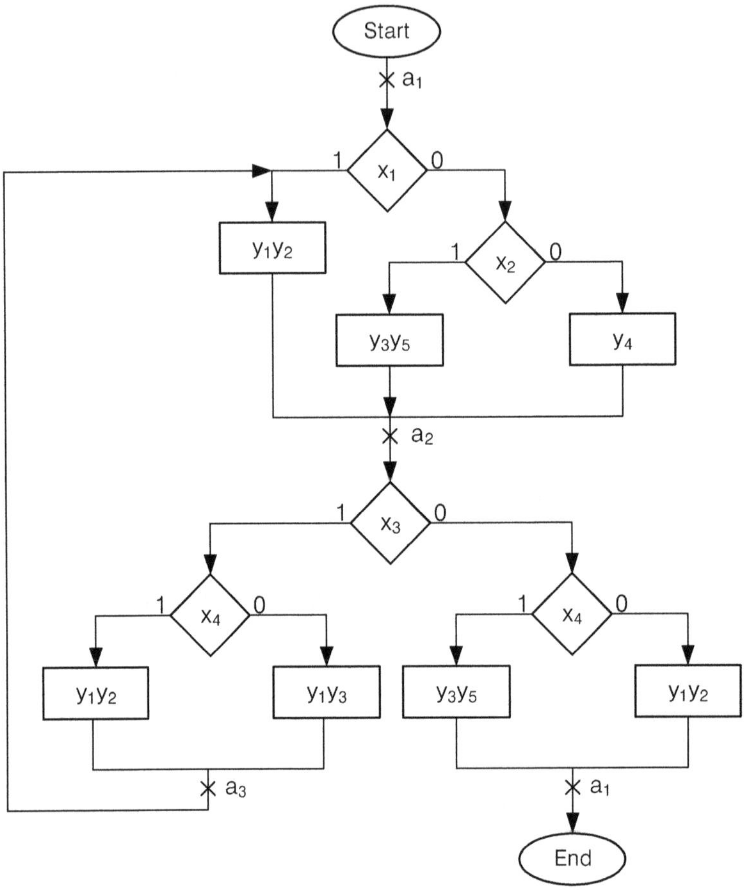

Fig. 2.16 Initial GSA Γ_4

Table 2.7 Table of RLC for Mealy FSM $MP(\Gamma_4)$

a_m	a_1	a_2	a_3
$K(a_m)$	00	01	10
p_1	x_1	x_3	–
p_2	x_2	x_4	–

Table 2.8 represents the function p_1, whereas Table 2.9 the function p_2 from the system (2.18). There are the following sets $X(p_g)$ used in these tables: $X(p_1) = \{x_1, x_3\}$ and $X(p_2) = \{x_2, x_4\}$. There is the obvious connection between system (2.18) and Tables 2.8 and 2.9.

The structure table of MP Mealy FSM includes the same columns as in the case of P Mealy FSM. But the column X_h is replaced by a column P_h. For example, the conjunction x_1x_2 (it determines the transition $\langle a_1, a_2\rangle$) is replaced by p_1p_2, the

Table 2.8 Table of LUTer1 for function p_1

T	$X(p_1)$	p_1	T	$X(p_1)$	p_1	T	$X(p_1)$	p_1	T	$X(p_1)$	p_1
T_1T_2	x_1x_3		T_1T_2	x_1x_3		T_1T_2	x_1x_3		T_1T_2	x_1x_3	
00	00	0	01	00	0	10	00	0	11	00	0
00	01	0	01	01	1	10	01	0	11	01	0
00	10	1	01	10	0	10	10	0	11	10	0
00	11	1	01	11	1	10	11	0	11	11	0

Table 2.9 Table of LUTer1 for function p_2

T	$X(p_2)$	p_2	T	$X(p_2)$	p_2	T	$X(p_2)$	p_2	T	$X(p_2)$	p_2
T_1T_2	x_2x_4		T_1T_2	x_2x_4		T_1T_2	x_2x_4		T_1T_2	x_2x_4	
00	00	0	01	00	0	10	00	0	11	00	0
00	01	0	01	01	1	10	01	0	11	01	0
00	10	1	01	10	0	10	10	0	11	10	0
00	11	1	01	11	1	10	11	0	11	11	0

Table 2.10 Structure table of Mealy FSM $MP(\Gamma_4)$

a_m	$K(a_m)$	a_s	$K(a_s)$	P_h	Y_h	Φ_h	h
a_1	00	a_2	01	p_1	y_1y_2	D_2	1
		a_2	01	$\bar{p}_1p_2$	y_3y_5	D_2	2
		a_2	01	$\bar{p}_1p_2$	y_4	D_2	3
a_2	10	a_3	10	p_1p_2	y_1y_2	D_1	4
		a_3	10	$p_1\bar{p}_2$	y_1y_3	D_1	5
		a_1	00	$\bar{p}_1p_2$	y_3y_5	–	6
		a_1	00	$\bar{p}_1\bar{p}_2$	y_1y_2	–	7
a_3	10	a_2	01	1	y_1y_2	D_2	8

conjunction $x_1\bar{x}_2$ by $p_1\bar{p}_2$ and so on. Using these rules and table of RLC, it is possible to construct the table of MP Mealy FSM. It is Table 2.10 in the discussed case.

Similar tables are used for constructing tables of LUTer2. There are R_0+N tables representing LUTer2. Each of them includes columns $K(a_m)$, P (address of the cell) and either $y_n \in Y$ or $D_r \in \Phi$. All these tables are represented by Table 2.11.

The address 0000 corresponds to the row 3 of Table 2.10. So, the corresponding cell contains two ones (y_4 and D_2). The addresses 0010 and 0011 correspond to the row 1 of Table 2.10. Using the same approach, we fill all rows of Table 2.11.

It is possible to implement systems (1.48) and (1.49) by EMBer. It leads to heterogeneous MP Mealy FSM (Fig. 2.17).

In this case, the LUTer implements the system (1.47), whereas the EMBer the systems (1.48) and (1.49). There are the same steps in design of both MP FSMs, but in the heterogeneous case the step 5 is "Constructing the table of EMBer". In the discussed case, table of EMBer is the same as Table 2.11.

Table 2.11 Table of LUTer2 for Mealy FSM $MP(\Gamma_4)$

T	P	y_1	y_2	y_3	y_4	y_5	D_1	D_2	T	P	y_1	y_2	y_3	y_4	y_5	D_1	D_2
T_1T_2	p_1p_2								T_1T_2	p_1p_2							
00	00	0	0	0	1	0	0	1	10	00	1	1	0	0	0	0	1
00	01	0	0	1	0	1	0	1	10	01	1	1	0	0	0	0	1
00	10	1	1	0	0	0	0	1	10	10	1	1	0	0	0	0	1
00	11	1	1	0	0	0	0	1	10	11	1	1	0	0	0	0	1
01	00	1	1	0	0	0	0	0	11	00	0	0	0	0	0	0	0
01	01	0	0	1	0	1	0	0	11	01	0	0	0	0	0	0	0
01	10	1	0	1	0	0	1	0	11	10	0	0	0	0	0	0	0
01	11	1	1	0	0	0	0	1	11	11	0	0	0	0	0	0	0

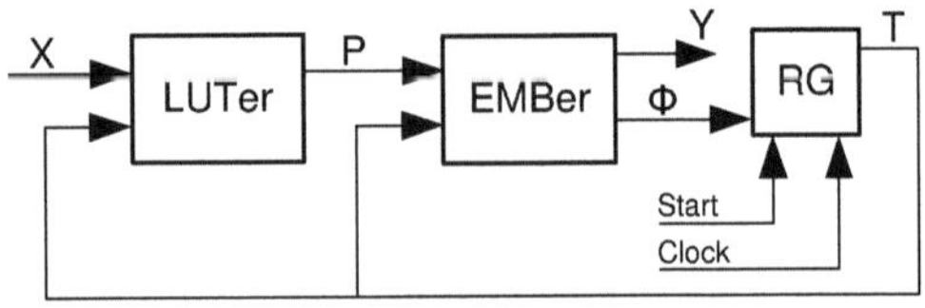

Fig. 2.17 Heterogeneous structure of MP Mealy FSM

Let the following condition take place:

$$R_0 + |X(p_g)| \leq S \quad (g = \overline{1, G}). \tag{2.19}$$

In this case, each function $p_g \in P$ is implemented using a single LUT. If this condition is violated, then it is possible to encode logical conditions [15].

Let us explain this approach on the base of Table 2.7. There are two pairs of conditions in the columns of this table: $\langle x_1, x_2 \rangle$ and $\langle x_3, x_4 \rangle$. Let us denote this number as n_P. Let us encode each pair using R_P variables, where:

$$R_P = \lceil \log_2 n_P \rceil. \tag{2.20}$$

Let us use the variables $z_r \in Z$ for such encoding where $|Z| = R_P$.

In the discussed case, there are $n_P = 2$, $R_P = 1$ and $Z = \{z_1\}$. Let us encode the pairs in the following way: $K(x_1, x_2) = 0$, $K(x_3, x_4) = 1$. Now, the system (1.47) can be replaced by the following system:

$$P = P(Z, X). \tag{2.21}$$

It leads to $M_P P$ Mealy FSM (Fig. 2.18).

In this case, the EMBer implements systems (1.48) and (1.49) and the system:

$$Z = Z(T, P). \tag{2.22}$$

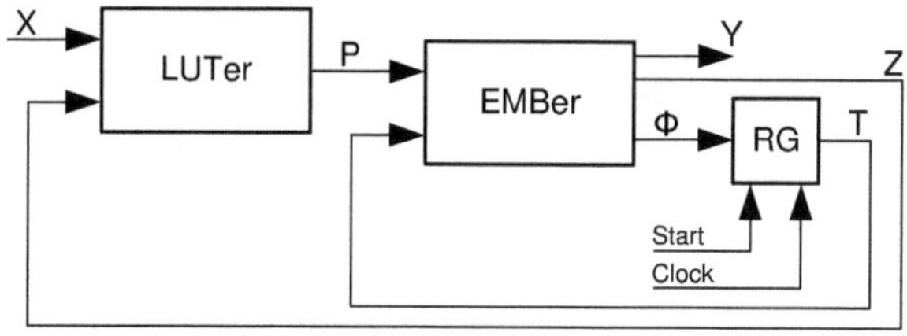

Fig. 2.18 Structural diagram of $M_P P$ Mealy FSM

Table 2.12 Table of RLC for Mealy FSM $M_P P(\Gamma_4)$

$\langle x_e, x_m \rangle$	$\langle x_1, x_2 \rangle$	$\langle x_3, x_4 \rangle$
$K(x_e, x_m)$	0	1
p_1	x_1	x_3
p_2	x_2	x_4

There is the following table of RLC in the case of $M_P P(\Gamma_1)$ (Table 2.12). Now, there are the following functions implemented by the LUTer:

$$
\begin{aligned}
p_1 &= \bar{z_1} x_1 \vee z_1 x_3;\\
p_2 &= \bar{z_1} x_2 \vee z_1 x_4.
\end{aligned}
\tag{2.23}
$$

Each function (2.21) is implemented by a single LUT if the following condition takes place:

$$R_P + |X(p_g)| \leq S. \tag{2.24}$$

Of course, this approach has sense if there is

$$R_P < R_0. \tag{2.25}$$

Let us discuss the MP Moore FSM (Fig. 2.19) implemented as a heterogeneous circuit. Here the LUTer1 implements the system (1.47), the LUTer2 the system (1.49) and the EMBer the system (1.5).

To minimize the circuit of LUTer1, it is possible to apply the encoding of logical conditions. It leads to $M_P P$ Moore FSM (Fig. 2.20).

This approach is a general one. It can be used in any model of FSM to diminish the hardware amount. It uses two main issues: (1) the structural decomposition and (2) the heterogeneous implementation.

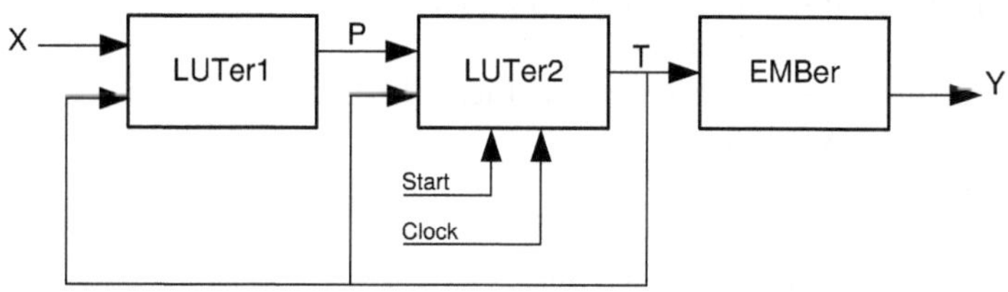

Fig. 2.19 Structural diagram of MP Moore FSM

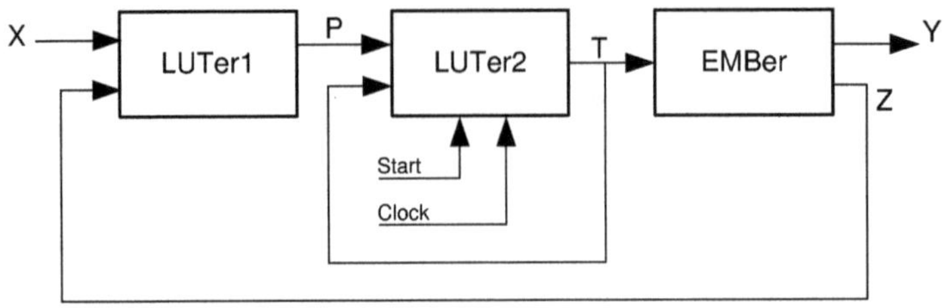

Fig. 2.20 Structural diagram of $M_P P$ Moore FSM

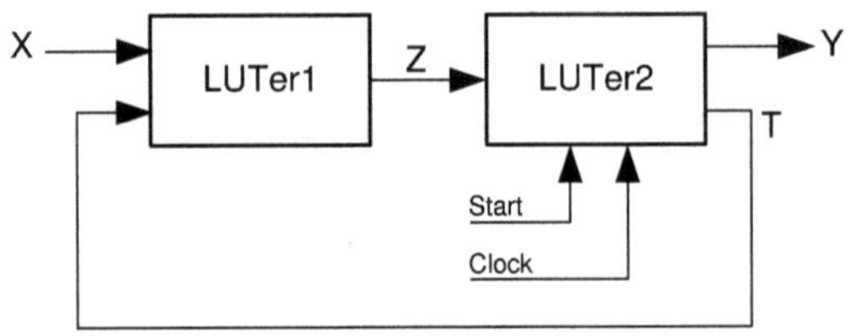

Fig. 2.21 Homogeneous structural diagram of PH Mealy FSM

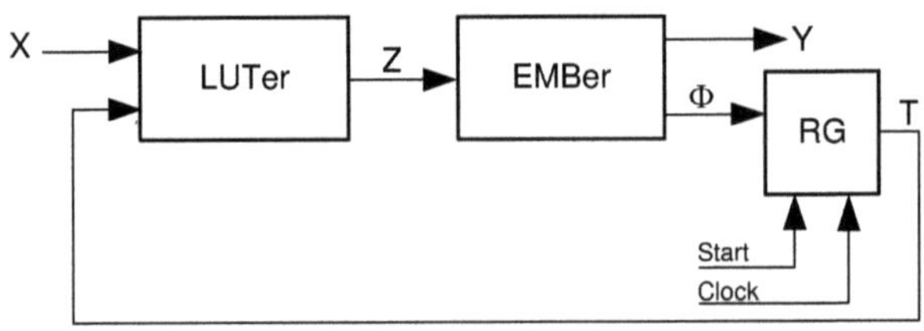

Fig. 2.22 Heterogeneous structural diagram of PH Mealy FSM

The second approach is based on the encoding of the rows of ST. It can be used only for Mealy FSMs. It is reduced to encoding of each function F_h by a binary code $C(F_h)$ having R_H bits ($h = \overline{1, H_0}$). The value of R_H is determined by (1.38). It is possible either homogeneous (Fig. 2.21) or heterogeneous (Fig. 2.22) structures of PH Mealy FSMs.

In both cases, the first block corresponds to BIMF (Fig. 1.18d), whereas the second to BMO (Fig. 1.18d). For example, the design method for the homogeneous variant includes the following steps:

1. Constructing the set of states for a GSA Γ.
2. Executing the state assignment.
3. Constructing the structure table of P Mealy FSM.
4. Encoding the rows of ST.
5. Constructing the table of LUTer1.
6. Constructing the table of LUTer2.
7. Implementing the FSM logic circuit.

Table 2.13 Structure table of Mealy FSM $P(\Gamma_4)$

a_m	$K(a_m)$	a_s	$K(a_s)$	X_h	Y_h	Φ_h	h
a_1	00	a_2	01	x_1	y_1y_2	D_2	1
		a_2	01	$\bar{x}_1x_2$	y_3y_5	D_2	2
		a_2	01	$\bar{x}_1\bar{x}_2$	y_4	D_2	3
a_2	01	a_3	10	x_3x_4	y_1y_2	D_1	4
		a_3	10	$x_3\bar{x}_4$	y_1y_3	D_1	5
		a_1	00	$\bar{x}_3x_4$	y_3y_5	–	6
		a_1	00	$\bar{x}_3\bar{x}_4$	y_1y_2	–	7
a_3	10	a_2	01	1	y_1y_2	D_2	8

Let us discuss an example of design for Mealy FSM $PH(\Gamma_4)$. There is the GSA Γ_4 shown in Fig. 2.16. As in the previous case, the following sets can be found for $P(\Gamma_4)$: $X = \{x_1, \ldots, x_4\}$, $Y = \{y_1, \ldots, y_5\}$, $A = \{a_1, a_2, a_3\}$, $\Phi = \{D_1, D_2\}$, $T = \{T_1, T_2\}$. Let us encode the states in the following way: $K(a_1) = 00$, $K(a_2) = 01$, $K(a_3) = 10$. There are $H_0 = 8$ rows in the structure table of FSM $P(\Gamma_4)$ (Table 2.13).

Let the following condition take place:

$$R_H \leq S. \tag{2.26}$$

In this case, there is no influence of codes $C(F_h)$ on the hardware amount in LUTer2. It is implemented using $R + N$ of LUTs. If this condition is violated, then the outcome of row encoding can influence the hardware amount in LUTer2.

In the discussed case, there are $H_0 = 8$ and $R_H = 3$. So, there is the set $Z = \{z_1, z_2, z_3\}$. Let us use LUTs having $S = 2$. In this case, the condition (2.26) is violated. There are the following system of Boolean functions derived from Table 2.13:

$$\begin{aligned}
D_1 &= F_4 \vee F_5;\\
D_2 &= F_1 \vee F_2 \vee F_3 \vee F_8;\\
y_1 &= F_1 \vee F_4 \vee F_5 \vee F_7 \vee F_8;\\
y_2 &= F_1 \vee F_4 \vee F_7 \vee F_8;\\
y_3 &= F_2 \vee F_5 \vee F_6;\\
y_4 &= F_3;\\
y_5 &= F_2 \vee F_6.
\end{aligned} \tag{2.27}$$

In the worst case, each equation of (2.27) requires two LUTs to be implemented. Let us encode the rows as it is shown in Fig. 2.23.

Fig. 2.23 Codes of rows for FSM $PH(\Gamma_4)$

z_1 \ z_2z_3	00	01	11	10
0	F_1	F_4	F_7	F_8
1	F_3	F_5	F_6	F_2

The following equations can be obtained from the Karnaugh map (Fig. 2.23):

$$\begin{aligned}
D_1 &= \bar{z}_2 z_3;\\
D_2 &= \bar{z}_3;\\
y_1 &= \bar{z}_1 \vee \bar{z}_2 z_3;\\
y_2 &= \bar{z}_1;\\
y_3 &= z_1 z_3 \vee z_1 z_2;\\
y_4 &= z_1 \bar{z}_2 \bar{z}_3;\\
y_5 &= z_1 z_2.
\end{aligned} \tag{2.28}$$

The system (2.28) corresponds to the following circuit of LUTer2 (Fig. 2.24).

In the worst case, it is necessary 14 of LUTs, whereas there are only 10 of LUTs in Fig. 2.24. This number can be decreased if there are the following codes of rows (Fig. 2.25).

Using this Karnaugh map, we can find the following system of functions:

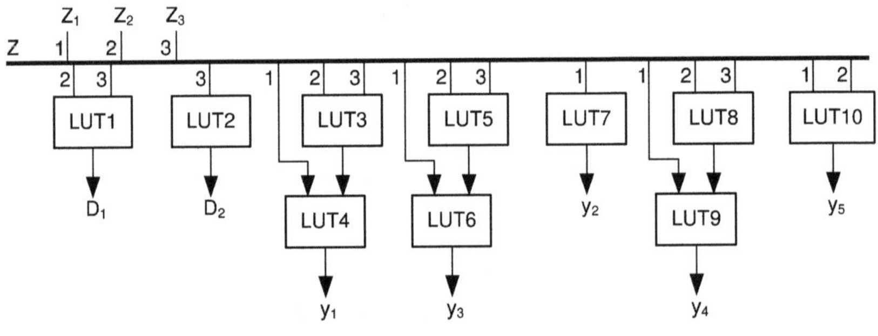

Fig. 2.24 Logic circuit of LUTer2 for codes from Fig. 2.23

Fig. 2.25 Outcome of row encoding for $PH(\Gamma_1)$

z_1 \ z_2z_3	00	01	11	10
0	F_5	F_3	F_2	F_6
1	F_4	F_1	F_8	F_7

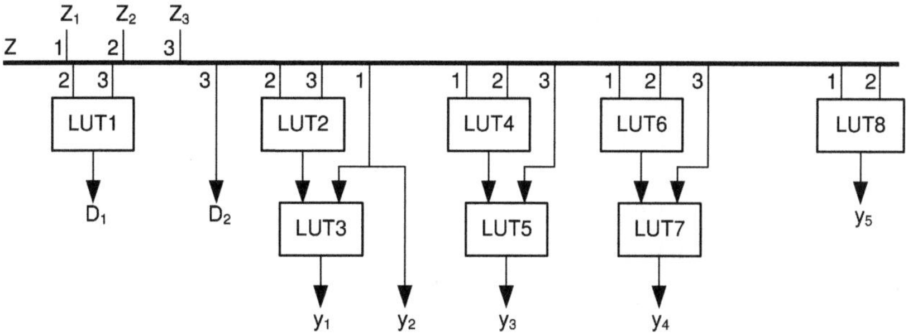

Fig. 2.26 Logic circuit of LUTer2 corresponding to codes from Fig. 2.25

$$\begin{aligned} D_1 &= \bar{z}_2\bar{z}_3; \\ D_2 &= z_3; \\ y_1 &= z_1 \vee \bar{z}_2\bar{z}_3; \\ y_2 &= z_1; \\ y_3 &= \bar{z}_1\bar{z}_3 \vee \bar{z}_1 z_2; \\ y_4 &= z_1\bar{z}_2 z_3; \\ y_5 &= \bar{z}_1 z_2. \end{aligned} \tag{2.29}$$

There is the circuit (Fig. 2.26) corresponding to the system (2.29). It requires only 8 of LUTs having $S = 2$.

The table of LUTer1 is constructed on the base of the structure table. The columns a_s, $K(a_s)$, Y_h and Φ_h are deleted from the ST. They are replaced by columns $C(F_h)$, Z_h. There are the variables $z_r \in Z$ in the row h of the table, if there is 1 in the position number r of $K(F_h)$. In the discussed case, it is the Table 2.14.

The table of LUTer1 is constructed using the codes from Fig. 2.25. It is used for deriving the system

$$Z = Z(T, X). \tag{2.30}$$

Table 2.14 Table of LUTer1 for Mealy FSM $PH(\Gamma_4)$

a_m	$K(a_m)$	X_h	$C(F_h)$	Z_h	h
a_1	00	x_1	101	z_1z_2	1
		$\bar{x}_1x_2$	011	z_2z_3	2
		$\bar{x}_1\bar{x}_2$	001	z_3	3
a_2	01	x_3x_4	100	z_1	4
		$x_3\bar{x}_4$	000	–	5
		$\bar{x}_3x_4$	010	z_2	6
		$\bar{x}_3\bar{x}_4$	110	z_1z_2	7
a_3	10	1	111	$z_1z_2z_3$	8

Table 2.15 Table of LUTer2 for Mealy FSM $PH(\Gamma_4)$

$C(F_h)$	Y_h	Φ_h	h
$z_1z_2z_3$	$y_1y_2y_3y_4y_5$	D_1D_2	
000	10100	10	5
001	00010	01	3
010	00101	00	6
011	00101	01	2
100	11000	10	4
101	11000	01	1
110	11000	00	7
111	11000	01	8

It is the following system in the discussed case:

$$\begin{aligned} z_1 &= \bar{T}_1\bar{T}_2x_1 \vee \bar{T}_1T_2x_3x_4 \vee \bar{T}_1T_2\bar{x}_3\bar{x}_4 \vee T_1\bar{T}_2; \\ z_2 &= \bar{T}_1\bar{T}_2x_1 \vee \bar{T}_1\bar{T}_2x_2 \vee \bar{T}_1T_2\bar{x}_3 \vee T_1\bar{T}_2; \\ z_3 &= \bar{T}_1\bar{T}_2\bar{x}_1 \vee T_1\bar{T}_2. \end{aligned} \quad (2.31)$$

The table of LUTer2 is constructed on the base of the ST. The columns a_m, $K(a_m)$, X_h are replaced by columns $C(F_h)$. In the discussed case, it is Table 2.15.

This table is used to find the systems (1.39) and (1.40). They are represented by the system (2.28) in the discussed case.

To design this circuit, we used optimal codes for variables F_h. This example shows the importance of this step and its influence on the hardware amount in LUT-based design.

There are three groups of methods discussed in the next chapters of the book:

1. EMB-based FSMs with RLC and other methods of structural decomposition. There are two main goals in our approach. First, decreasing for the number of EMBs in the FSM logic circuit. Second, decreasing the number of LUTs in the circuit replacing logical conditions.

2. EMB-based FSMs with transformation of object codes. These models are based on transformation of state codes into codes of collections of microoperations and vice versa.

3. EMB-based FSMs with more than one source of codes of classes of pseudoequivalent states. These methods concern only Moore FSMs.

References

1. M. Adamski, A. Barkalov, *Architectural and Sequential Synthesis of Digital Devices* (University of Zielona Góra Press, Zielona Góra, 2006)
2. Altera: http://www.altera.com. Accessed Jan 2015

3. P. Bacchetta, L. Daldos, D. Sciuto, C. Silvano, Low-power state assignment techniques for finite state machines, in *Proceedings of the 2000 IEEE International Symposium on Circuits and Systems (ISCAS'2000)*, vol. 2, Geneva. IEEE (2000), pp. 641–644
4. S. Baranov, *Logic Synthesis of Control Automata* (Kluwer Academic Publishers, Dordrecht, 1994)
5. A. Barkalov, Principles of logic optimization for Moore microprogram automaton. Cybern. Syst. Anal. **34**(1), 54–60 (1998)
6. A. Barkalov, A. Barkalov Jr., Synthesis of finite-state machines with transformation of the object's codes, in *Proceedings of the International Conference TCSET'2004*, Lviv, Ukraina (2004), pp. 61–64. Lviv Polytechnic National University, Lviv, Publishing House of Lviv Polytechnic
7. A. Barkalov, A. Barkalov Jr, Design of mealy finite-state machines with the transformation of object codes. Int. J. Appl. Math. Comput. Sci. **15**(1), 151–158 (2005)
8. A. Barkalov, L. Titarenko, *Logic Synthesis for FSM-Based Control Units*. Lecture Notes in Electrical Engineering, vol. 53 (Springer, Berlin, 2009)
9. A. Barkalov, L. Titarenko, S. Chmielewski, Optimization of logic circuit of Moore FSM on CPLD. Pomiary Automatyka Kontrola **53**(5), 18–20 (2007)
10. A. Barkalov, L. Titarenko, S. Chmielewski, Optimization of Moore FSM on CPLD, in *Proceedings of the Sixth International Conference CAD DD'07*, vol. 2, Minsk (2007), pp. 39–45
11. A. Barkalov, L. Titarenko, S. Chmielewski, Optimization of Moore FSM on system-on chip, in *Proceedings of IEEE East-West Design and Test Symposium—EWDTS'07*, Yerevan, Armenia, Kharkov (2007), pp. 105–109
12. A. Barkalov, L. Titarenko, S. Chmielewski, Reduction in the number of PAL macrocells int the circuit of a Moore FSM. Int. J. Appl. Math. Comput. Sci. **17**(4), 565–675 (2007)
13. A. Barkalov, L. Titarenko, S. Chmielewski, Decrease of hardware amount in logic circuit of Moore FSM. Przegląd Telekomunikacyjny i Wiadomości Telokomunikacyjne (6), 750–752 (2008)
14. A. Barkalov, L. Titarenko, S. Chmielewski, Optimization of Moore FSM on System-on-chip using PAL technology, in *Proceedings of the International Conference TCSET 2008*, Lviv-Slavsko, Ukraina (2008), pp. 314–317. Ministry of Education and Science of Ukraine, Lviv Polytechnic National University, Lviv, Publishing House of Lviv Polytechnic
15. A. Barkalov, I. Zelenjova, Optimization of replacement of logical conditions for an automaton with bidirectional transitions. Autom. Control Comput. Sci. **34**(5), 48–53. Allerton Press Inc
16. L. Benini, G. De Micheli, State assignment for low power dissipation. IEEE J. Solid-State Circuits **30**(3), 258–268 (1995)
17. M. Bolton, *Digital System Design with Programmable Logic* (Addison-Wesley, Boston, 1990)
18. G. Borowik, Synthesis of sequential devices into FPGA with embedded memory blocks. Ph.D. thesis, Warszawa: WUT (2007). Ph.D. Thesis
19. R. Brayton, G. Hatchel, C. McMullen, A. Sangiovanni-Vincentelli, *Logic Minimization Algorithms for VLSI Synthesis* (Kluwer Academic Publishers, Boston, 1984)
20. R. Brayton, R. Rudell, A. Sangiovanni-Vincentelli, A. Wang, MIS: a multi-level logic optimization system. IEEE Trans. Comput. -Aided Des. **6**(11), 1062–1081 (1987)
21. S. Brown, Z. Vernesic, *Fundamentals of Digital Logic with VHDL Design* (McGraw–Hill, New York, 2000)
22. S. Chattopadhyay, P. Chaudhuri, Genetic algorithm based approach for integrated state assignment and flipflop selection in finite state machines synthesis, in *Proceedings of the IEEE International Conference on VLSI Design*, Los Alamitos. IEEE Computer Society (1998), pp. 522–527
23. C. Chen, J. Zhao, M. Ahmadi, A semi-gray encoding algorithm for low-power state assignment, in *Proceedings of the 2003 International Symposium on Circuits and Systems, ISCAS'03*, vol. 5. IEEE (2003), pp. 389–392
24. R. Czerwiński, D. Kania, State assignment method for high speed FSM, in *Proceedings of Programmable Devices and Systems* (2004), pp. 216–221

25. R. Czerwiński, D. Kania, State assignment for PAL-based CPLDs, in *Proceedings of 8th Euromicro Symposium on Digital System Design* (2005), pp. 127–134
26. R. Czerwinski, D. Kania, *Finite State Machine Logic Synthesis for Complex Programmable Logic Devices*. Lecture Notes in Electrical Engineering, vol. 23 (Springer, Berlin, 2013)
27. D. Debnath, T. Sasao, Doutput phase optimization for AND-OR-EXOR PLAs with decoders and its application to design of adders. IFICE Trans. Inf. Syst. **E88-D**(7), 1492–1500 (2005)
28. S. Deniziak, K. Sapiecha, An efficient algorithm of perfect state encoding for CPLD based systems, in *Proceedings of IEEE Workshop on Design and Diagnostic of Electronic Circuits and Systems (DDECS'98)* (1998), pp. 47–53
29. S. Devadas, H. Ma, A. Newton, A. Sangiovanni-Vincentelli, MUSTANG: state assignment of finite state machines targeting multilevel logic implementation. IEEE Trans. Comput.-Aided Des. **7**(12), 1290–1300 (1988)
30. S. Devadas, A. Newton, Exact algorithms for output encoding, state assignment, and four-level boolean minimization. IEEE Trans. Comput.-Aided Des. **10**(1), 143–154 (1991)
31. X. Du, G. Hachtel, B. Lin, A. Newton, MUSE: a multilevel symbolic encoding algorithm for state assignment. IEEE Trans. Comput.-Aided Des. Integr. Circuits Syst. **10**(1), 28–38 (1991)
32. A. El-Maleh, S. Sait, F. Khan, Finite state machine state assignment for area and power minimization, in *2006 IEEE International Symposium on Circuits and Systems, ISCAS 2006. Proceedings*. IEEE (2006), pp. 5303–5306
33. B. Escherman, State assignment for hardwired VLSI control units. ACM Comput. Surv. **25**(4), 415–436 (1993)
34. I. Garcia-Vargas, R. Senhadji-Navarro, G. Jiménez-Moreno, A. Civit-Balcells, P. Guerra-Gutierrez, ROM-based finite state machine implementation in low cost FPGAs, in *IEEE International Symposium on Industrial Electronics ISIE 2007*. IEEE (2007), pp. 2342–2347
35. I. Grout, *Digital Systems Design with FPGAs and CPLDs* (Elsevier Science, Oxford, 2008)
36. S. Hassoun, T. Sasao, *Logic Synthesis and Verification* (Kluwer Academic Publishers, Boston, 2002)
37. G. Hatchel, F. Somenzi, *Logic synthesis and verification algorithms* (Kluwer Academic Publishers, Boston, 2000)
38. H. Hu, H. Xue, J. Bian, A heuristic state assignment algorithm targeting area, in *Proceedings of 5th International Conference on ASIC*, vol. 1 (2003), pp. 93–96
39. J. Huang, J. Jou, W. Shen, ALTO: an iterative area/performance algorithms for LUT-based FPGA technology mapping. IEEE Trans. VLSI Syst. **18**(4), 392–400 (2000)
40. A. Iranli, P. Rezvani, M. Pedram, Low power synthesis of finite state machines with mixed D and T flip-flops, in *Proceedings of the Asia and South Pacific—DAC* (2003), pp. 803–808
41. J. Jenkins, *Design with FPGAs and CPLDs* (Prentice Hall, New York, 1995)
42. A. Kahng, *VLSI Physical Design: From Graph Partitioning to Timing Closure* (Springer, Berlin, 2011)
43. T. Kam, T. Villa, R. Brayton, A. Sangiovanni-Vincentelli, *A Synthesis of Finie State Machines: Functional Optimization* (Kluwer Academic Publishers, Boston, 1998)
44. D. Kania, Two-level logic synthesis on PAL-based CPLD and FPGA using decomposition, in *Proceedings of 25th Euromicro Conference* (1999), pp. 278–281
45. D. Kania, Two-level logic synthesis on PALs. Electron. Lett. (17), 879–880 (1999)
46. D. Kania, Coding capacity of PAL-based logic blocks included in CPLDs and FPGAs, in *Proceedings of IFAC Workshop on Programmable Devices and Sysytems (PDS'2000)*. Elsevier Science (2000), pp. 164–169
47. D. Kania, Decomposition-based synthesis and its application in PAL-oriented technology mapping, in *Proceedings of 26th Euromicro Conference*. Maastricht: IEEE Compuetr Society Press (2000), pp. 138–145
48. D. Kania, An efficient algorithm for output coding in PAL-based CPLDs. Int. J. Eng. **15**(4), 325–328 (2002)
49. D. Kania, Logic synthesis of multi–output functions for PAL-based CPLDs, in *Proceedings of IEEE International Conference on Field-Programmable Technology* (2002), pp. 429–432

50. D. Kania, An efficient approach to synthesis of multi-output boolean functions on PAL-based devices. IEEE Proc.—Comput. Digital Tech. **150**, 143–149 (2003)
51. H. Kubatova, *Design of Embedded Control Systems*, chapter Finie State Machine Implementation in FPGAs (Springer, New York, 2005), pp. 177–187
52. Lattice: http://www.lattice.com. Accessed Feb 2015
53. C. Maxfield, *The Design Warrior's Guide to FPGAs* (Academic Press Inc., Orlando, 2004)
54. C. Maxfield, *FPGAs: Instant Access* (Newnes, 2008)
55. E. McCluskey, *Logic Design Principles* (Prentice Hall, Englewood Cliffs, 1986)
56. G. De Micheli, Symbolic design of combinational and sequential logic implemented by two-level macros. IEEE Trans. Comput.-Aided Des. **5**(9), 597–616 (1986)
57. G. De Micheli, *Synthesis and Optimization of Digital Circuits* (McGraw–Hill, New York, 1994)
58. Microsemi: http://www.microsemi.com. Accessed Jan 2015
59. Z. Navabi, *Embedded Core Design with FPGAs* (McGraw–Hill, New York, 2007)
60. W. Nöth, R. Kolla, Spanning tree based state encoding for low power dissipation, in *Proceedings of the Conference on Design, Automation and Test in Europe*. Association for Computing Machinery (1999), p. 37
61. S. Park, S. Cho, S. Yang, M. Ciesielski, A new state assignment technique for testing and low power, in *Proceedings of the 41st annual Design Automation Conference*. Association for Computing Machinery (2004), pp. 510–513
62. C. Pedram, A. Despain, Low-power state assignment targeting two- and multilevel logic implementations. IEEE Trans. Comput.-Aided Des. Integr. Circuits Syst. **17**(12), 1281–1291 (1998)
63. I. Pomerancz, K. Cheng, STOIC: state assignment based on output/input functions. IEEE Trans. Comput.-Aided Des. Integr. Circuits Syst. **12**(8), 1123–1131 (1993)
64. QuickLogic: http://www.quicklogic.com. Accessed Jan 2015
65. M. Rawski, T. Łuba, Z. Jachna, P. Tomaszewicz, *Design of Embedded Control Systems*, chapter The influence of functional decomposition on modern digital design process (Springer, Boston, 2005), pp. 193–203
66. M. Rawski, H. Selvaraj, T. Łuba, An application of functional decomposition in ROM-based FSM implementation in FPGA devices. J. Syst. Architect. **51**(6–7), 423–434 (2005)
67. K. Roy, S. Prasad, SYCLOP: synthesis of CMOS logic for low power applications, in *Proceedings, IEEE 1992 International Conference on Computer Design: VLSI in Computers and Processors, ICCD'92*. IEEE (1992), pp. 464–467
68. R. Rudell, A. Sangiovanni-Vincentelli, Multiple-valued minimization for PLA optimization. IEEE Trans. Comput.-Aided Des. **6**(5), 727–750 (1987)
69. G. Saucier, M. Depaulet, P. Sicard, ASYL: a rule-based system for controller synthesis. IEEE Trans. Comput.-Aided Des. **6**(11), 1088–1098 (1987)
70. G. Saucier, P. Sicard, L. Bouchet, Multi-level synthesis on programmable devices in the ASYL system, in *Proceedings of Euro ASIC* (1990), pp. 136–141
71. C. Scholl, *Functional Decomposition with Application to FPGA Synthesis* (Kluwer Academic Publishers, Boston, 2001)
72. R. Senhadji-Navarro, I. Garcia-Vargas, G. Jiménez-Moreno, A. Civit-Balcells, P. Guerra-Gutierrez, ROM-based FSM implementation using input multiplexing in FPGA devices. Electron. Lett. **40**(20), 1249–1251 (2004)
73. E. Sentowich, K. Singh, L. Lavango, C. Moon, R. Murgai, A. Saldanha, H. Savoj, P. Stephan, R. Bryton, A. Sangiovanni-Vincentelli, SIS: a system for sequential circuit synthesis. Technical Report, University of California, Berkely (1992)
74. E. Sentowich, K. Singh, L. Lavango, C. Moon, R. Murgai, A. Saldanha, H. Savoj, P. Stephan, R. Bryton, A. Sangiovanni-Vincentelli, SIS: a system for sequential circuit synthesis, in *Proceedings of the International Conference of Computer Design (ICCD'92)* (1992), pp. 328–333
75. I. Skliarova, V. Sklyarov, A. Sudnitson, *Design of FPGA-Based Circuits Using Hierarchical Finite State Machines* (TUT Press, Tallinn, 2012)
76. V. Sklyarov, Synthesis and implementation of RAM-based finite state machines in FPGAs, in *Proceedings of Field-Programmable Logic and Applications: The Roadmap to Reconfigurable Computing*, Villach. Springer (2000), pp. 718–728

77. V. Sklyarov, I. Skliarova, A. Barkalov, L. Titarenko, *Synthesis and Optimization of FPGA-Based Systems*. Lecture Notes in Electrical Engineering, vol. 294 (Springer, Berlin, 2014)
78. V. Solovjev, M. Czyzy, Refined CPLD macrocells architecture for effective FSM implementation, in *Proceedings of the 25th EUROMICRO Conference*, vol. 1, Milan, Italy (1999), pp. 102–109
79. V. Solovjev, M. Czyzy, The universal algorithm for fitting targeted unit to complex programmable logic devices, in *Proceedings of the 25th EUROMICRO Conference*, vol. 1, Milan, Italy (1999), pp. 286–289
80. V. Solovjev, M. Czyzy, Synthesis of sequential circuits on programmable logic devices based on new models of finite state machines, in *Proceedings of the EUROMICRO Conference, Milan* (2001), pp. 170–173
81. G. Sutter, E. Todorovich, S. López-Buedo, E. Boemo, Low-power FSMs in FPGA: encoding alternatives, in *Integrated Circuit Design. Power and Timing Modeling, Optimization and Simulation*. Springer (2002), pp. 363–370
82. E. Tatalov, Synthesis of compositional microprogram control units for programmable devices. Master's thesis, Donetsk National Technical University, Donetsk (2011)
83. A. Tiwari, K. Tomko, Saving power by mapping finite-state machines into embedded memory blocks in FPGAs, in *Proceedings of the Conference on Design, Automation and Test in Europe—Volume 2*. IEEE Computer Society (2004), pp. 916–921
84. C. Tsui, M. Pedram, A. Despain, Exact and approximate methods for calculating signal and transition probabilities in FSMs, in *31st Conference on Design Automation*. IEEE (1994), pp. 18–23
85. G. Venkatamaran, S. Reddy, I. Pomerancz, GALLOP: genetic algorithm based low power FSM synthesis by simultaneous partitioning and state assignment, in *Proceedings of 16th International Conference on VLSI Design* (2003), pp. 533–538
86. T. Villa, T. Saldachna, R. Brayton, A. Sangiovanni-Vincentelli, Symbolic two-level minimization. IEEE Trans. Comput.-Aided Des. **16**(7), 692–708 (1997)
87. T. Villa, A. Sangiovanni-Vincentelli, NOVA: state assignment of finite state machines for optimal two-level logic implememntation. IEEE Trans. Comput.-Aided Des. **9**(9), 905–924 (1990)
88. Y. Xia, A. Almani, Genetic algorithm based state assignment for power and area optimization. IEEE Proc. Comput. Digital Tech. **149**, 128–133 (2002)
89. Xilinx: http://www.xilinx.com. Accessed Jan 2015
90. Xilinx: XST User Guide. V. 11.3, http://www.xilinx.com/support/documentation/sw_manuals/xilinx11/xst.pdf. Accessed Jan 2015
91. S. Yang, Logic synthesis and optimization benchmarks user guide. Technical report, Microelectronic Center of North Carolina (1991)
92. B. Zeidman, *Designing with FPGAs and CPLDs* (CMP Books, Lawrence, 2002)

Chapter 3
Object Codes Transformation for Mealy FSMs

3.1 Principle of OCT for Mealy FSMs

The principle of object codes transformation (OCT) was proposed in [1] and developed in [2–6]. In accordance with this principle, there are two kinds of objects in an FSM. They are internal states $a_m \in A$ and CMOs $Y_t \subseteq Y$. Let us point out that the states and collections of microoperations have different nature. So, they are heterogeneous objects, in contrast to, for example, state a_m and state a_s. In this Chapter, we discuss some hardware reduction methods based on transformation of codes for heterogeneous objects.

The discussed methods are based on finding the one-to-one correspondence between heterogeneous objects. When it is found, a code of one object is generated as a function depended on logical conditions $x_e \in X$ and state variables $T_r \in T$. It is a primary object. A code of the second object is generated using variables, representing the first object, and additional variables, encoding the identifiers. It is a second object.

Let us find a one-to-one match $A \rightarrow Y$ among the states and CMOs, where states are primary and CMOs are secondary objects. In the case of Mealy FSM, the BIMF generates functions (1.4) and some additional variables $v_r \in V$ used for encoding of the identifiers. Next, these variables are transformed into codes of CMOs (it could be unitary codes where each MO $y_n \in Y$ corresponds to a single bit of $C(Y_t)$).

Let us start from P_A Mealy FSM (Fig. 3.1). Here BIMF generates the system (1.4) and the system

$$V = V(T, X). \tag{3.1}$$

The variables $v_r \in V$ are used for encoding of identifiers. We discuss how to form this set a bit later .

A. Barkalov et al., *Logic Synthesis for FPGA-Based Finite State Machines*,
Studies in Systems, Decision and Control 38, DOI 10.1007/978-3-319-24202-6_3

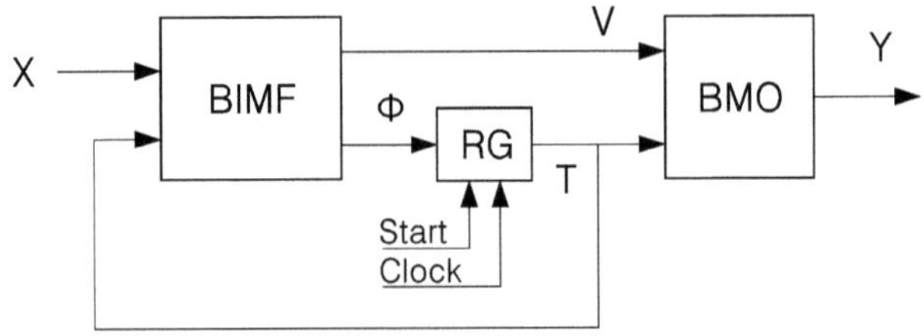

Fig. 3.1 Structural diagram of P_A Mealy FSM

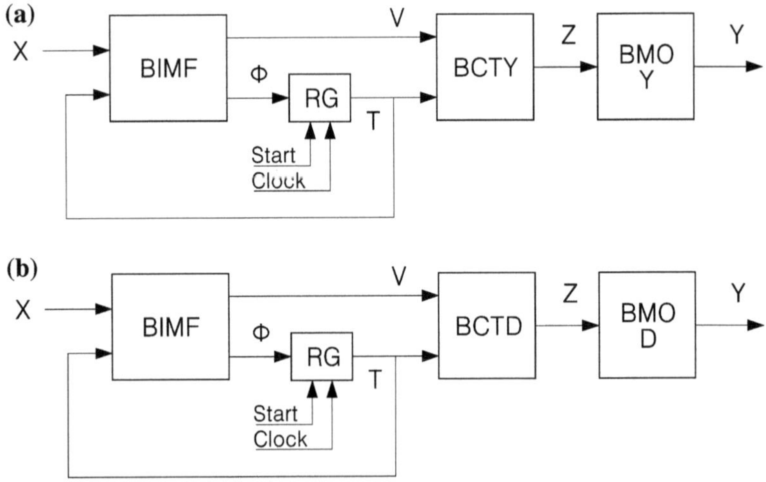

Fig. 3.2 Structural diagram of P_AY (**a**) and P_AD (**b**) Mealy FSM

There is the following system of functions generated by BMO:

$$Y = Y(T, V). \tag{3.2}$$

If we compare expressions (1.3) and (3.2), we can make a conclusion that the variables $v_r \in V$ replace the logical conditions $x_e \in X$. Let it be $R_V = |V|$. If $R_V \ll L$, then functions (3.2) require less amount of literals in their terms in comparison with functions (1.3). This allows to hope that it is necessary fewer hardware in BMO in comparison with the corresponding part of the circuit of P Mealy FSM.

It is possible to encode the collections of MOs. Using the maximal encoding of CMOs leads to P_AY Mealy FSM (Fig. 3.2a), whereas the encoding of the fields of compatible microoperations leads to P_AD Mealy FSM (Fig. 3.2b). A block of code transformer (BCT) implements the system

$$Z = Z(T, V). \tag{3.3}$$

Next, the BMO generates functions (1.40).

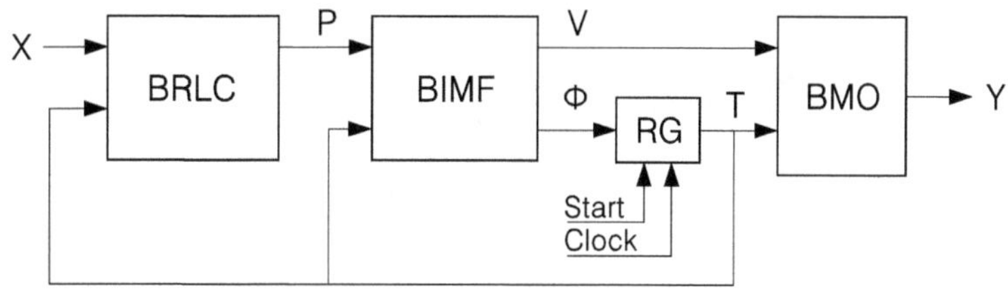

Fig. 3.3 Structural diagram of MP_A Mealy FSM

To show the difference between P_AY and P_AD FSMs, we use a letter Y (for the maximal encoding of CMOs) and D (for the encoding of the fields of compatible microoperations).

It is possible to use the RLC in the cases of P_A, P_AY and P_AD FSMs. It leads to MP_A, MP_AY and MP_AD FSMs. For example, there is a structural diagram of MP_A Mealy FSM shown in Fig. 3.3.

In MP_A FSM, the BRLC implements the system (1.47), the BMO the system (3.2). The BIMF implements the system (1.49) and the system

$$V = V(T, P). \tag{3.4}$$

Let us find a one-to-one match $Y \to A$ among the microoperations as primary objects and the states as secondary objects. It leads to P_Y (with unitary encoding of MOs) or P_YY (with maximal encoding of CMOs) or P_YD (with encoding of the fields of compatible microoperations) Mealy FSM (Fig. 3.4).

In P_Y Mealy FSM, the BMO implements systems (1.3) and (3.1). They are transformed by BIMF into the input memory functions

$$\Phi = \Phi(V, Y). \tag{3.5}$$

In P_YY Mealy FSM, the BMOY implements functions (3.1) and (2.30). They are transformed by BIMF in the functions

$$\Phi = \Phi(V, Z). \tag{3.6}$$

The functions (1.40) are implemented by BMO.

In P_YD Mealy FSM, the BMOD implements the functions (3.1) and (2.30). The BIMF implements the system (3.6), whereas the BMO the system (1.40).

Using RLC leads to MP_Y, MP_YY and MP_YD FSMs. For example, the structural diagram of MP_Y Mealy FSM is shown in Fig. 3.5.

Let us discuss design methods for FPGA-based Mealy FSMs based on OCT. As always, there are possible homogeneous or heterogeneous approaches for implementing FSM logic circuits.

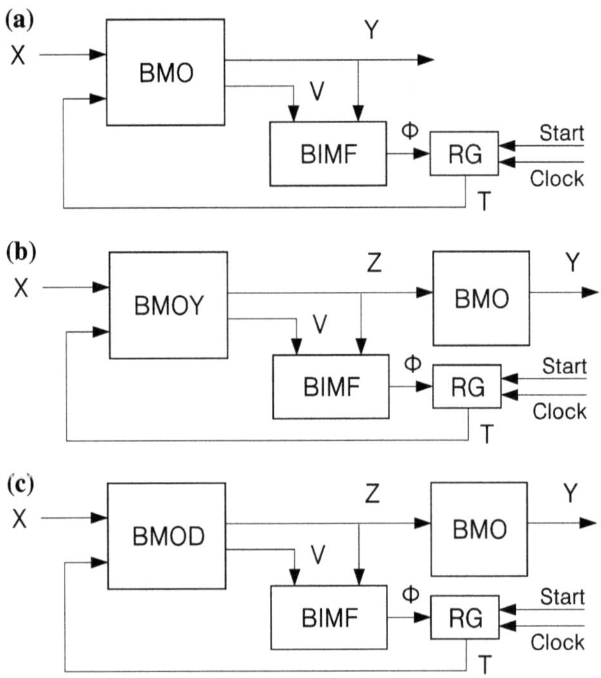

Fig. 3.4 Structural diagram of P_Y (**a**), $P_Y Y$ (**b**) and $P_Y D$ (**c**) Mealy FSM

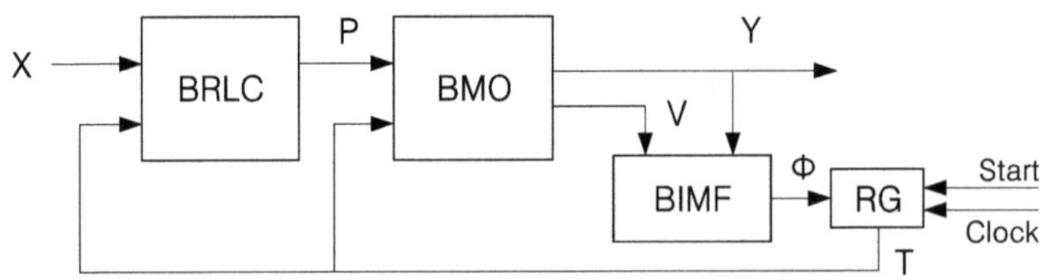

Fig. 3.5 Structural diagram of MP_Y Mealy FSM

Basing on results from [2], the following method can be proposed for synthesis of P_A Mealy FSM.

1. Marking the initial GSA Γ and creating a set A.
2. Constructing the FSM structure table.
3. Finding the collections of microoperations $Y_t \subseteq Y$.
4. Constructing the set of identifiers I.
5. Representing CMOs by pairs $\langle identifier, state \rangle$.
6. Constructing the table of BIMF.
7. Constructing the table of BMO.
8. Implementing FSM logic circuit.

Table 3.1 Structure table of Mealy FSM S_1

a_m	$K(a_m)$	a_s	$K(a_s)$	X_h	Y_h	Φ_h	h
a_1	000	a_2	010	x_1	y_1y_2	D_2	1
		a_3	011	$\bar{x_1}$	y_3	D_2D_3	2
a_2	010	a_2	010	x_2	y_1y_2	D_2	3
		a_3	011	$\bar{x_2}x_3$	y_3	D_2D_3	4
		a_4	100	$\bar{x_2}\bar{x_3}$	y_1y_2	D_1	5
a_3	011	a_4	100	x_1	y_2y_5	D_1	6
		a_5	101	$\bar{x_1}$	y_6	D_1D_3	7
a_4	100	a_5	101	1	y_3y_7	D_1D_3	8
a_5	101	a_2	010	x_2x_3	y_1y_2	D_2	9
		a_3	011	$x_2\bar{x_3}$	y_3	D_2D_3	10
		a_5	101	$\bar{x_2}x_4$	y_3y_7	D_1D_3	11
		a_1	000	$\bar{x_2}\bar{x_4}$	–	–	12

Let us discuss an example for design of P_A Mealy FSM S_1 represented by the structure table (Table 3.1).

There are $T_0 = 6$ different CMOs in Table 3.1. They are $Y_1 = \emptyset$, $Y_2 = \{y_1, y_2\}$, $Y_3 = \{y_3\}$, $Y_4 = \{y_2, y_5\}$, $Y_5 = \{y_6\}$, $Y_6 = \{y_3, y_7\}$.

Let us use the following approach to find the set I. Let $T(a_s)$ be a set of CMOs generated under transitions into the state $a_s \in A$. Let it be $n_s = |T(a_S)|$. Obviously it is enough n_s of identifiers to select one of those CMOs.

To identify any of CMOs $Y_t \in Y$, it is enough K identifiers, where $K = max(n_1, \ldots, n_M)$. They form a set of identifiers $I = \{I_1, \ldots, I_K\}$.

There are the following sets $T(a_s)$ in Table 3.1: $T(a_1) = \{y_1\}$, $T(a_2) = \{y_2\}$, $T(a_3) = \{y_3\}$, $T(a_4) = \{y_2, y_4\}$, $T(a_5) = \{y_5, y_6\}$. So, there is $K = 2$. It gives the set $I = \{I_1, I_2\}$.

Let us find the pairs $\langle I_k, a_s \rangle$ determining the CMOs $Y_t \in Y$. If there is only a single Y_t in a set $T(a_s)$, then $I_k = \emptyset$. If there are more than one pair determining a collection Y_t, then these pairs are connected by the sign of disjunction. There are the following correspondence among the CMOs and pairs in the discussed example: $Y_1 = \langle \emptyset, a_1 \rangle$; $Y_2 = \langle \emptyset, a_2 \rangle \vee \langle I_1, a_4 \rangle$; $Y_3 = \langle \emptyset, a_3 \rangle$; $Y_4 = \langle I_2, a_4 \rangle$; $Y_5 = \langle I_1, a_5 \rangle$; $Y_6 = \langle I_2, a_5 \rangle$.

Let us encode the identifiers $I_k \in I$. It is enough R_V bits for the encoding:

$$R_V = \lceil log_2 K \rceil. \tag{3.7}$$

In the discussed case, there is $R_V = 1$. It determines the set $V = \{v_1\}$. Let it be $K(I_1) = 0$ and $K(I_2) = 1$.

There are the following columns in the table of BIMF: a_m, $K(a_m)$, a_s, $K(a_s)$, X_h, I_h, $K(I_h)$, V_h, Φ_h, h. It is constructed on the base of a ST. To do it, the column Y_h is replaced by the columns I_h (identifier for the collection $Y_t \in Y$), $K(I_h)$ (the code of I_h) and V_h (the variables $v_r \in V$ in the code $K(I_h)$). There is the same amount

Table 3.2 Table of BIMF for P_A Mealy FSM S_1

a_m	$K(a_m)$	a_s	$K(a_s)$	X_h	I_h	$K(I_h)$	V_h	Φ_h	h
a_1	000	a_2	010	x_1	–	–	*	D_2	1
		a_3	011	$\bar{x_1}$	–	–	*	D_2D_3	2
a_2	010	a_2	010	x_2	–	–	*	D_2	3
		a_3	011	$\bar{x_2}x_3$	–	–	*	D_2D_3	4
		a_4	100	$\bar{x_2}\bar{x_3}$	I_1	0	–	D_1	5
a_3	011	a_4	100	x_1	I_2	1	v_1	D_1	6
		a_5	101	$\bar{x_1}$	I_1	0	–	D_1D_3	7
a_4	100	a_5	101	1	I_2	1	v_1	D_1D_3	8
a_5	101	a_2	010	x_2x_3	–	–	*	D_2	9
		a_3	011	$x_2\bar{x_3}$	–	–	*	D_2D_3	10
		a_5	101	$\bar{x_2}x_4$	I_2	1	v_1	D_1D_3	11
		a_1	000	$\bar{x_2}\bar{x_4}$	–	–	*	–	12

of rows in a ST and a corresponding table of BIMF. In the discussed case, there is $H_0 = 12$ rows in the table of BIMF for P_A Mealy FSM S_1 (Table 3.2). If some CMO is determined by a pair $\langle \emptyset, a_s \rangle$, then there is a sign "–" in the corresponding row of the table in the column $K(I_h)$.

The sign "*" means that the variables V_h are of no concern for functions (3.2). The table of BIMF is used to find functions (1.4) and (3.1). The functions (3.1) depend on the same terms as the functions (1.4). There is the following equation derived from Table 3.2:

$$v_1 = \bar{T_1}T_2T_3 \vee T_1\bar{T_2}. \tag{3.8}$$

To get the term $T_1\bar{T_2}$ from (3.8), we used the "don't care" assignment from Table 3.2.

There are the following columns in the table of BMO: a_s, $K(a_s)$, I_h, $K(I_h)$, Y_h, h. This table is constructed on the base of Tables 3.1 and 3.2 in the discussed case. It is Table 3.3 in the discussed case. This table has 7 rows, which is the same as the number of pairs representing CMOs $Y_t \subseteq Y$. In the common case, the number H_{BMO} is determined as:

$$H_{BMO} = \sum_{s=1}^{M_0} n_s. \tag{3.9}$$

In (3.5), the symbol H_{BMO} stands for the number of rows in the table of BMO.

This table is used for deriving the system (3.2). It is represented as a system having N following functions:

$$y_n = \bigvee_{h=1}^{H_{BMO}} C_{nh}A_mV_h \; (n = \overline{1, N}). \tag{3.10}$$

Table 3.3 Table of BMO for P_A Mealy FSM S_1

a_m	$K(a_m)$	I_h	$K(I_h)$	Y_h	h
a_1	000	–	–	–	1
a_2	010	–	–	y_1y_2	2
a_3	011	–	–	y_3	3
a_4	100	I_1	0	y_1y_2	4
a_4	100	I_2	1	y_2y_5	5
a_5	101	I_1	0	y_6	6
a_6	101	I_2	1	y_3y_7	7

In (3.10), the symbol C_{nh} stands for Boolean variable equal to 1 iff the MO $y_n \in Y$ is written in the row number $h(h = \overline{1, H_{BMO}})$. Let us point out that there is $V_h = 1$ if $K(Y_h) = ''-''$. It is possible to derive the following equations from Table 3.3:

$$
\begin{aligned}
y_1 &= \bar{T}_1 T_2 \bar{T}_3 \vee T_1 \bar{T}_2 \bar{T}_3 \bar{v}_1;\\
y_2 &= \bar{T}_1 T_2 \bar{T}_3 \vee T_1 \bar{T}_2 \bar{T}_3 \bar{v}_1 \vee T_1 \bar{T}_2 \bar{T}_3 v_1 = \bar{T}_1 T_2 \bar{T}_3 \vee T_1 \bar{T}_2 \bar{T}_3;\\
y_3 &= \bar{T}_1 T_2 T_3 \vee T_1 \bar{T}_2 T_3 v_1;\\
y_5 &= T_1 \bar{T}_2 \bar{T}_3 v_1;\\
y_6 &= T_1 \bar{T}_2 T_3 \bar{v}_1;\\
y_7 &= T_1 \bar{T}_2 T_3 v_1.
\end{aligned} \tag{3.11}
$$

The same approach can be used for logic synthesis of other models based on the one-to-one match of states and CMOs. They are the models P_AY, P_AD, MP_A, MP_AY and MP_AD. Of course, the corresponding methods should include such steps as encoding of CMOs, executing of RLC, constructing tables of BRLC, BCTY, BCTD, BMOY and BMOD. Now let us discuss the methods based on the correspondence $Y \rightarrow A$.

There are the same steps in design methods for P_A and P_Y Mealy FSMs [2]. But there is a difference in execution of some steps. Moreover, the step 5 is the following: "Representing the state $a_s \in A$ by the pairs $\langle identifier, CMO \rangle$".

Let us discuss an example of synthesis for P_Y Mealy FSM S_1 (Table 3.1). There are the set A and collections $Y_t \subseteq Y$ which were found before.

Let $A(Y_t)$ be a set of states such that the microoperations $y_n \in Y_t$ are generated under transitions into these states $a_s \in A$. It is possible to find the following sets $A(Y_t)$ for FSM S_1: $A(Y_1) = \{a_1\}$, $A(Y_2) = \{a_2, a_4\}$, $A(Y_3) = \{a_3\}$, $A(Y_4) = \{a_4\}$, $A(Y_5) = \{a_5\}$ and $A(Y_6) = \{a_5\}$. Let it be $n_t = |A(Y_t)|$. Now, the number of identifiers can be found as $K = max(n_1, \ldots, n_{T_0})$. They form a set $I = \{I_1, \ldots, I_K\}$. The formula (3.7) is used to find the number of elements in the set V.

In the discussed case, there is $K = 2$. It determines the set $V = \{v_1\}$. Let it be $K(I_1) = 0$ and $K(I_2) = 1$.

Table 3.4 Table of BIMF for P_Y Mealy FSM S_1

Y_h	V_h	a_m	$K(a_m)$	Φ_h	h
$\emptyset$	*	a_1	000	–	1
y_1y_2	0	a_2	010	–	2
y_3	*	a_3	011	–	3
y_2y_5	*	a_4	100	–	4
y_1y_2	1	a_4	100	–	5
y_6	*	a_5	101	–	6
y_3y_7	*	a_6	101	–	7

Let us find the pairs $\langle I_K, Y_t\rangle$ corresponding the states $a_m \in A$. There are the following formulae in the discussed case: $a_1 = \langle\emptyset, Y_1\rangle$, $a_2 = \langle I_1, Y_2\rangle$, $a_3 = \langle\emptyset, Y_3\rangle$, $a_4 = \langle\emptyset, Y_4\rangle \vee \langle I_2, Y_2\rangle$, $a_5 = \langle\emptyset, Y_5\rangle \vee \langle\emptyset, Y_6\rangle$.

There are the following columns in the table of BIMF: Y_h, V_h, a_m, $K(a_m)$, Φ_h, h. This table is constructed on the base of an ST and the system of pairs discussed above. The table includes H_{BIMF} rows:

$$H_{BIMF} = \sum_{t=1}^{T_0} n_t. \tag{3.12}$$

In the discussed case, the BIMF is represented by Table 3.4.

This table is used for constructing the system (3.4). For example, the following equations can be derived from Table 3.4:

$$\begin{aligned} D_1 &= y_2y_5 \vee y_1y_2v_1 \vee y_6 \vee y_3y_7; \\ D_2 &= y_1y_2\bar{v}_1 \vee y_3; \\ D_3 &= y_3 \vee y_6 \vee y_3y_7 = y_3 \vee y_6. \end{aligned} \tag{3.13}$$

There are the following columns in the table of BMO for P_Y Mealy FSM: a_m, $K(a_m)$, a_s, X_h, Y_h, V_h, h. It is constructed on the base of an ST and the system of pairs for states $a_m \in A$. In the discussed case, the BMO is represented by Table 3.5.

This table is used for deriving the systems (1.3) and (3.1). There is only a single equation in the system (3.1) for the discussed case:

$$v_1 = \bar{T}_1T_2\bar{T}_3\bar{x}_2\bar{x}_3. \tag{3.14}$$

Using the "don't care" state assignment 110, this equation can be minimized up to $v_1 = T_2\bar{T}_3\bar{x}_2\bar{x}_3$.

Let us point out that we use the law of Boolean algebra for optimizing the equation D_3 from (3.13). It is the following law: $a \vee ab = a$ [7].

Now let us discuss FPGA-based approaches for designing FSMs based on the OCT.

Table 3.5 Table of BMO for P_Y Mealy FSM S_1

a_m	$K(a_m)$	a_s	X_h	Y_h	V_h	h
a_1	000	a_2	x_1	y_1y_2	–	1
		a_3	$\bar{x}_1$	y_3	–	2
a_2	010	a_2	x_2	y_1y_2	–	3
		a_3	$\bar{x}_2x_3$	y_3	–	4
		a_4	$\bar{x}_2x_3$	y_1y_2	v_1	5
a_3	011	a_4	x_1	y_2y_5	–	6
		a_5	$\bar{x}_1$	y_6	–	7
a_4	100	a_5	1	y_3y_7	–	8
a_5	101	a_2	x_2x_3	y_1y_2	–	9
		a_3	$x_2\bar{x}_3$	y_3	–	10
		a_5	$\bar{x}_2x_4$	y_3y_7	–	11
		a_1	$\bar{x}_2\bar{x}_4$	–	–	12

3.2 Synthesis of FPGA-Based Mealy FSMs with Transformation of States

Let us start from P_A Mealy FSMs (Fig. 3.1). Let us start from the homogeneous approach shown in Fig. 3.6. Let us denote this FSM as P_{A1} Mealy FSM.

In P_{A1} Mealy FSM, the LUTer1 implements systems (1.4) and (3.1), the LUTer2 the system (3.2). There are the following steps in the design method of P_{A1} Mealy FSM:

1. Marking the initial GSA Γ and creating the set A.
2. Constructing the structure table of P Mealy FSM.
3. Finding the collections of microoperations $Y_t \subseteq Y$.
4. Constructing the set of identifiers $I = \{I_1, \ldots, I_K\}$.
5. Representing CMOs by pairs $\langle identifier, state \rangle$.
6. Constructing the table of LUTer1.
7. Constructing the table of LUTer2.
8. Implementing FSM logic circuit.

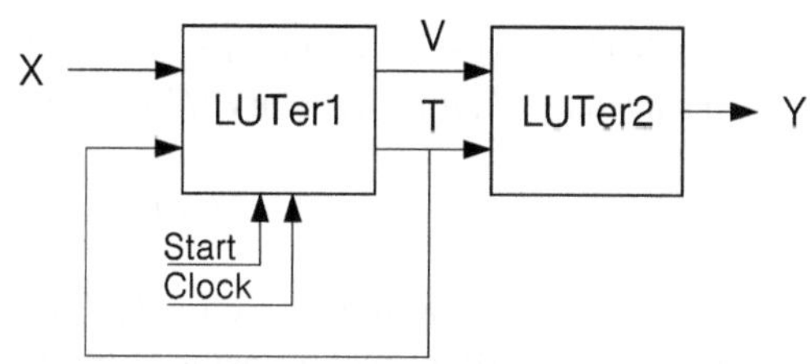

Fig. 3.6 Structural diagram of P_{A1} Mealy FSM

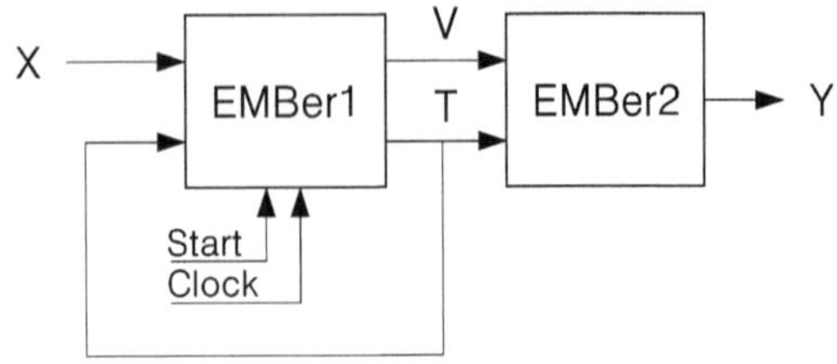

Fig. 3.7 Structural diagram of P_{A2} Mealy FSM

There are $R_0 + R_V$ different subcircuits in the circuit of LUTer1. Let $L(f)$ be the number of literals in SOP of some function f. Let the following condition take place:

$$max(L(D_1), \ldots, L(D_{R_0}), L(v_1), \ldots, L(v_{R_V})) \leq S. \quad (3.15)$$

In this case, there are exactly $R_0 + R_v$ of LUTs having S inputs in the circuit of LUTer1. It is a single-level circuit.

There are N of different subcircuits in the circuit of LUTer2. Let the following condition take place:

$$max(L(y_1), \ldots, L(Y_n)) \leq S. \quad (3.16)$$

In this case, there are exactly N of LUTs in the circuit of LUTer2.

Let us discuss the P_{A1} Mealy FSM S_1 starting from Table 3.1. It is clear that there are the same steps in the design methods for P_A and P_{A1} Mealy FSMs. So, the LUTer1 is represented by Table 3.2 and the LUTer2 by Table 3.3. We hope that everything is clear with this example.

It is possible to implement both blocks of P_A FSM with EMBs. It leads to P_{A2} Mealy FSM (Fig. 3.7). In this case, the EMBer1 corresponds to BIMF and the EMBer2 to BMO. There are the same steps in the design of P_A and P_{A2} FSMs. But the step 6 is reduced to constructing the table of EMBer1, whereas the step 7 to constructing the table of EMBer2.

There are the following columns in the table of EMBer1 of P_{A2} Mealy FSM: $K(a_m), X$ (they determine the address of a cell), V, Φ (they determine the content of a cell), q (the number of a cell). The table has $Q_1(P_{A2})$ rows, where

$$Q_1(P_{A2}) = 2^{R_0+L}. \quad (3.17)$$

There is $Q_1(P_{A2}) = 64$ in the case of P_{A2} Mealy FSM S_1. To describe transitions from a state $a_m \in A$, it is necessary $H(a_m)$ rows, where

$$H(a_m) = 2^L. \quad (3.18)$$

There is $H(a_m) = 8$ in the case of P_{A2} Mealy FSM S_1. Table 3.6 represents the part of table of EMBer1 with transitions from the state $a_3 \in A$. It is constructed on

Table 3.6 Part of table of EMBer1 for P_{A2} Mealy FSM S_1

$K(a_m)$	X	V	Φ		
$T_1T_2T_3$	$x_1x_2x_3$	v_1	$D_1D_2D_3$	q	h
011	000	0	110	24	7
011	001	0	110	25	7
011	010	0	110	26	7
011	011	0	110	27	7
011	100	1	100	28	6
011	101	1	100	29	6
011	110	1	100	30	6
011	111	1	100	31	6

the base of Table 3.2. We add the column h in this table to show the correspondence among rows of Tables 3.2 and 3.6.

There are cells having addresses from 011000 till 011111 in Table 3.6. It corresponds to values of q from 24 till 31. The row 6 of Table 3.2 corresponds to the rows 28–31 of Table 3.6. The row 7 of Table 3.2 corresponds to the rows 24–27 of Table 3.6. The content of corresponding cells is taken from Table 3.2.

Using the same approach, it is possible to find all subtables corresponding to table of BIMF. They are combined together to form the table of EMBer1.

There are the following columns in the table of EMBer2: $K(a_m)$, V (they determine the address of a cell), Y (it determines the content of a cell), q. The table has $Q_2(P_{A2})$ rows:

$$Q_2(P_{A2}) = 2^{R_0+R_V}. \tag{3.19}$$

There is $Q_2(P_{A2}) = 16$ in the case of P_{A2} Mealy FSM S_1. The table of EMBer2 is represented by Table 3.7.

Table 3.7 Table of EMBer2 for P_{A2} Mealy FSM S_1

$K(a_m)$	V	Y		$K(a_m)$	V	Y	
$T_1T_2T_3$	v_1	$y_1y_2y_3y_4y_5y_6y_7$	q	$T_1T_2T_3$	v_1	$y_1y_2y_3y_4y_5y_6y_7$	h
000	0	000000	1	100	0	110000	9
000	1	000000	2	100	1	010100	10
001	0	000000	3	101	0	000010	11
001	1	000000	4	101	1	000001	12
010	0	110000	5	110	0	000000	13
010	1	110000	6	110	1	000000	14
011	0	001000	7	111	0	000000	15
011	1	001000	8	111	1	000000	16

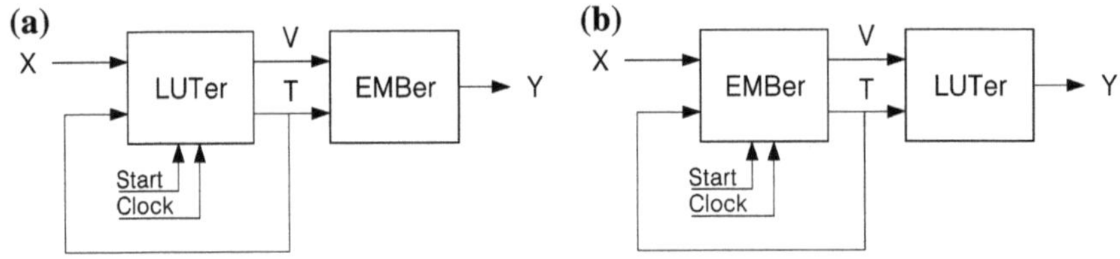

Fig. 3.8 Heterogeneous structural diagrams of P_A Mealy FSM

Table 3.8 Structural diagrams of P_AY Mealy FSMs

Block	$P_{A1}Y$	$P_{A2}Y$	$P_{A3}Y$	$P_{A4}Y$	$P_{A5}Y$	$P_{A6}Y$	$P_{A7}Y$	$P_{A8}Y$
BIMF	L_1	L_1	L_1	B_1	L_1	B_1	B_1	B_1
BCTY	L_2	L_2	B_1	L_1	B_2	B_2	L_1	B_2
BMOY	L_3	B_1	L_2	L_2	B_3	L_3	B_2	B_3

The rows 1 and 2 correspond to state a_1 with $Y(a_1) = \emptyset$. The rows 3 and 4 correspond to the code 001 which is not used. The same is true for rows 13–16. The rows 5 and 6 correspond to state a_2. There is the CMO Y_2 under transition into a_2. Different CMOs are generated under transitions in the state a_4. If $v_1 = 0$, then it is Y_2. If $v_1 = 1$, then it is Y_4.

Now let us discuss the heterogeneous approach for implementing the P_A Mealy FSM. It is possible to implement the BIMF as LUTer and BMO as EMBer. It leads to P_{A3} Mealy FSM (Fig. 3.8a). If the BIMF is implemented as EMBer and the BMO as LUTer, it leads to P_{A4} Mealy FSM (Fig. 3.8b).

It is very easy to work out the design methods for both P_{A3} and P_{A4} FSMs. It can be done on the base of the method for P_{A1} Mealy FSM. We do not discuss those methods in this book.

It is possible to make the following conclusion from the previous discussion. If there are K_B blocks in the structural diagram of an FSM, then it is possible

$$n_{SD} = 2^{K_B} \tag{3.20}$$

different combinations of LUTers and EMBers in its structural diagram.

Let us discuss different approaches for implementing P_AY Mealy FSMs. It follows from Fig. 3.2a that there is $K_B = 3$. So, it is possible 8 different combinations for FPGA-based P_AY Mealy FSMs. They are represented by Table 3.8. The symbol L_i stands for LUTer i (i = 1, 2, 3), whereas the symbol B_i stands for EMBer i (i = 1, 2, 3).

There are the following steps in design methods for $P_{Ai}Y$ Mealy FSMs ($i = \overline{1, 8}$):

1. Marking the initial GSA Γ and creating the set A.
2. Constructing the structure table of P Mealy FSM.
3. Finding the collections of microoperations $Y_t \in Y$.

Table 3.9 Transformed ST of PY Mealy FSM S_1

a_m	$K(a_m)$	a_s	$K(a_s)$	X_h	Y_t	$C(Y_t)$	Z_h	Φ_h	h
a_1	000	a_2	010	x_1	y_2	001	z_3	D_2	1
		a_2	011	$\bar{x_1}$	y_3	010	z_2	D_2D_3	2
a_2	010	a_2	010	x_2	y_2	001	z_3	D_2	3
		a_3	011	$\bar{x_2}x_3$	y_3	010	z_2	D_2D_3	4
		a_4	100	$\bar{x_2}\bar{x_3}$	y_3	010	z_2	D_2D_3	5
a_3	011	a_4	100	x_1	y_4	011	z_2z_3	D_2	6
		a_5	101	$\bar{x_1}$	y_5	100	z_1	D_2D_3	7
a_4	100	a_5	101	1	y_6	101	z_1z_3	D_1D_3	8
a_5	101	a_2	010	x_2x_3	y_2	001	z_3	D_2	9
		a_3	011	$x_2\bar{x_3}$	y_3	010	z_2	D_2D_3	10
		a_5	101	$\bar{x_2}x_4$	y_6	101	z_1z_3	D_2D_3	11
		a_1	000	$\bar{x_2}\bar{x_4}$	–	000	–	–	12

4. Constructing the set of identifiers $I = \{I_1, \ldots, I_K\}$.
5. Representing CMOs by pairs $\langle I_k, a_s \rangle$.
6. Encoding of collections of microoperations.
7. Constructing the transformed ST of PY Mealy FSM.
8. Constructing the table of BIMF.
9. Constructing the table of BCTY.
10. Constructing the table of BMOY.
11. Transformation of tables for implementing LUTers or EMBers.

Let us discuss an example of synthesis for $P_{A6}Y$ Mealy FSM S_1. Table 3.1 represents the P Mealy FSM S_1. Let us point out that steps 1–6 are already executed. Remind that there are $T_0 = 6$ of different CMOs in the discussed case: $Y_1 = \emptyset$, $Y_2 = \{y_1, y_2\}$, $Y_3\{y_3\}$, $Y_4 = \{y_2, y_5\}$, $Y_5 = \{y_6\}$ and $Y_6 = \{y_3, y_7\}$.

Using (1.32), we can find $R_Z = 3$. It gives the set $Z = \{z_1, z_2, z_3\}$. Let us encode the CMOs $Y_t \subseteq Y$ in the trivial way: $C(Y_1) = 000,\ldots,C(Y_6) = 101$.

The transformed ST of PY Mealy FSM is constructed on the base of its ST. To do it, the column Y_h is replaced by columns Y_t, $C(Y_t)$ and Z_h. The column Z_h includes the variables $z_r \in Z$ equal 1 in the code $C(Y_t)$ from h-th row of the table ($h = \overline{1, H_0}$). It is Table 3.9 in the discussed case.

Basing on Table 3.8, it is possible to derive the structural diagram shown in Fig. 3.9. So, in the $P_{A6}Y$ Mealy FSM, the BIMF is implemented as EMBer1, the BCTY as EMBer2, and the BMOY as LUTer1.

Tables of BIMF are the same for equivalent P_A and $P_{Ai}Y$ Mealy FSMs. So, Table 3.2 represents the table of BIMF for $P_{A6}Y$ Mealy FSM S_1.

There are the following columns in the table of BCTY: a_s, $K(a_s)$, I_h, $K(I_h)$, Y_h, $C(Y_h)$, Z_h, h. This table contains H_{BMO} rows determined by (3.9). Table 3.10 represents the table of BCTY in the discussed case. It is based on Table 3.3.

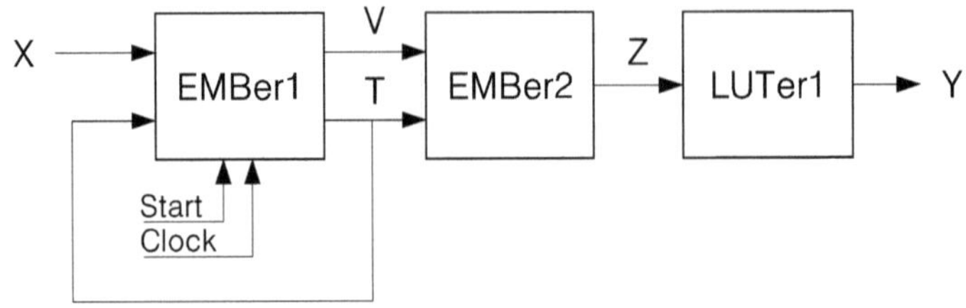

Fig. 3.9 Structural diagram of $P_{A6}Y$ Mealy FSM

Table 3.10 Table of BCTY for $P_{A6}Y$ Mealy FSM S_1

a_s	$K(a_s)$	I_h	$K(I_h)$	Y_h	$C(Y_h)$	Z_h	h
a_1	000	–	–	y_1	000	–	1
a_2	010	–	–	y_2	001	z_3	2
a_3	011	–	–	y_3	010	z_2	3
a_4	100	I_1	0	y_2	001	z_3	4
a_4	100	I_2	1	y_4	011	z_2z_3	5
a_5	101	I_1	0	y_5	100	z_1	6
a_5	101	I_2	1	y_6	101	z_2z_3	7

This table is used for constructing the system (3.3). There are the following equations derived from Table 3.10:

$$\begin{aligned} z_1 &= T_1T_3; \\ z_2 &= T_2T_3 \vee T_1\bar{T}_2\bar{T}_3v_1; \\ z_3 &= T_2\bar{T}_3 \vee T_1. \end{aligned} \tag{3.21}$$

We use codes 110 and 111 to optimize the equations (3.21).

There are the following columns in table of BMOY: $Y_t, C(Y_t), Y_h, h$. It is Table 3.11 in the discussed case.

This table is used to derive the system (1.40). For example, it is possible to find the equations $y_2 = \bar{z}_1z_3$ or $y_3 = \bar{z}_1z_2\bar{z}_3 \vee z_1\bar{z}_2z_3$ from Table 3.11.

As it is pointed before, the tables of BIMF are the same for P_A ans $P_{Ai}Y$ Mealy FSMs ($i = \overline{1,6}$). It means that Table 3.6 represents the table of EMBer1 for $P_{A6}Y$ Mealy FSM S_1.

Table 3.11 Table of BMOY for $P_{A6}Y$ Mealy FSM S_1

Y_t	$C(Y_t)$	Y_h	h	Y_t	$C(Y_t)$	Y_h	h
Y_1	000	–	1	Y_4	011	y_2y_5	4
Y_2	001	y_1y_2	2	Y_5	100	y_6	5
Y_3	010	y_3	3	Y_6	101	y_3y_7	6

Table 3.12 Table of EMBer2 for $P_{A6}Y$ Mealy FSM S_1

$K(a_s)$	V	Z	q	$K(a_s)$	V	Z	q
$T_1T_2T_3$	v_1	$z_1z_2z_3$		$T_1T_2T_3$	v_1	$z_1z_2z_3$	
000	0	000	1	100	0	001	9
000	1	000	2	100	1	011	10
001	0	000	3	101	0	100	11
001	1	000	4	101	1	101	12
010	0	001	5	110	0	000	13
010	1	001	6	110	1	000	14
011	0	010	7	111	0	000	15
011	1	010	8	111	1	000	16

Table 3.13 Table of LUTer2 for $P_{A6}Y$ Mealy FSM S_1

$C(Y_t)$	Y_t	t	$C(Y_t)$	Y_t	t
$z_1z_2z_3$	$y_1y_2y_3y_4y_5y_6y_7$		$z_1z_2z_3$	$y_1y_2y_3y_4y_5y_6y_7$	
000	000000	1	100	000010	5
001	110000	2	101	000001	6
010	001000	3	110	000000	7
011	010100	4	111	000000	8

Table 3.14 Structural diagrams of P_AD Mealy FSMs

Block	$P_{A1}D$	$P_{A2}D$	$P_{A3}D$	$P_{A4}D$
BIMF	L_1	L_1	B_1	B_1
BCTD	L_2	B_1	L_1	B_2
BMOD	L_3	L_2	L_2	L_1

There are the following columns in the table of EMBer2: $K(a_s)$, V (they determine the address of a cell), Z, q. It is Table 3.12 in the discussed case.

Table 3.12 is constructed on the base of Table 3.10. For example, the cells with $q = 5$ and $q = 6$ correspond to the row $h = 2$ of Table 3.10; the cell with $q = 9$ corresponds to $h = 4$, and so on.

There are the following columns in the table of LUTer1: $C(Y_t)$, Y_t, t. In the discussed case, it has $T_0 = 8$ rows (Table 3.13).

This table is constructed on the base ot Table 3.11. It determines N of tables for LUTs forming the circuit of the LUTer1.

There is no sense in using EMBs for implementing the circuit of BMOD. So, there are four different structures of P_AD Mealy FSMs (Table 3.14).

There are the following steps in design methods for $P_{Ai}D$ Mealy FSMs ($i = \overline{1,4}$):

1. Marking the initial GSA Γ and creating the set A.
2. Constructing the structure table of P Mealy FSM.
3. Finding the collections of microoperations $Y_t \in Y$.
4. Constructing the set of identifiers $I = \{I_1, \ldots, I_K\}$.
5. Representing CMOs by pairs $\langle I_K, a_s \rangle$.
6. Encoding of the fields of compatible microoperations.
7. Constructing the transformed ST of PD Mealy FSM.
8. Constructing the table of BIMF.
9. Constructing the table of BCTD.
10. Constructing the table of BMOD.
11. Transformation of tables for implementing LUTers and EMBers.
12. Implementing the FSM logic circuit.

We do not discuss this approach in this Chapter. Let us point out that it is possible using RLC for each of the discussed FSMs. It gives $MP_{Ai}(i = \overline{1,4})$, $MP_{Ai}Y(i = \overline{1,8})$ and $MP_{Ai}D(i = \overline{1,4})$ models of Mealy FSM based on the transformation of $a_m \in A$ into $Y_t \in Y$. We discuss some examples a bit later in this chapter.

3.3 Synthesis of FPGA-Based Mealy FSMs with Transformation of CMOs

Let us start from P_Y Mealy FSM (Fig. 3.4a) There are four different structural diagrams of this FSM (Table 3.15).

There are the following steps in the design methods for P_{Yi} Mealy FSMs ($i = \overline{1,4}$):

1. Marking the initial GSA Γ and creating the set A.
2. Constructing the structure table of P Mealy FSM.
3. Finding the collections of microoperations $Y_t \in Y$.
4. Constructing the set of identifiers $I = \{I_1, \ldots, I_K\}$.
5. Representing the states $a_m \in A$ by pairs $\langle I_k, Y_t \rangle$.
6. Constructing the table of BIMF.
7. Constructing the table of BMO.
8. Transformation of tables for implementing LUTers and EMBers.
9. Implementing the FSM logic circuit.

Table 3.15 Structural diagrams of P_{Yi} Mealy FSMs

Block	$P_{Y1}D$	$P_{Y2}D$	$P_{Y3}D$	$P_{Y4}D$
BIMF	L_1	L_1	B_1	B_1
BMO	L_2	B_1	L_1	B_2

Table 3.16 Structure table of Mealy FSM S_2

a_m	$K(a_m)$	a_s	$K(a_s)$	X_h	Y_t	Φ_h	h
a_1	000	a_2	001	x_1	y_1y_2	D_3	1
		a_3	010	$\bar{x}_1$	y_3	D_2	2
a_2	001	a_2	001	x_2	y_1y_2	D_3	3
		a_3	010	$\bar{x}_2x_3$	y_1y_4	D_2	4
		a_4	011	$\bar{x}_2\bar{x}_3$	y_1y_2	D_2D_3	5
a_3	010	a_4	011	x_1	y_3	D_2D_3	6
		a_5	100	$\bar{x}_1$	y_1y_4	D_1	7
a_4	011	a_5	100	1	y_4y_5	D_1	8
a_5	100	a_2	001	x_2x_3	y_3	D_3	9
		a_3	010	$x_2\bar{x}_3$	y_6	D_2	10
		a_6	101	$\bar{x}_2x_4$	y_5y_7	D_1D_3	11
		a_1	000	$\bar{x}_2\bar{x}_4$	–	–	12
a_6	101	a_3	010	1	y_6	D_2	13

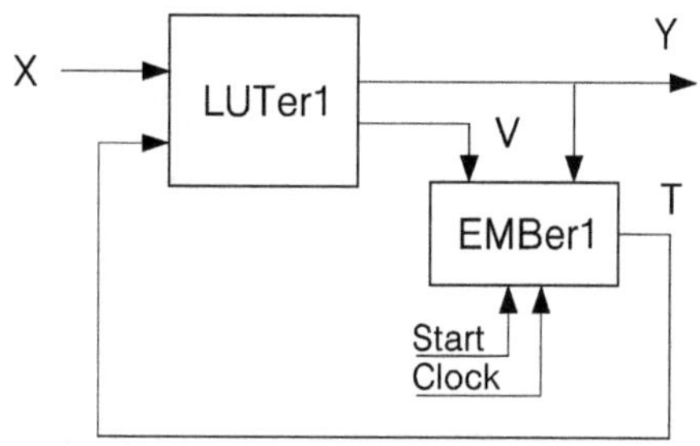

Fig. 3.10 Structural diagram of P_{Y3} Mealy FSM

Let us discuss an example of design for P_{Y3} Mealy FSM S_2. Table 3.16 represents the Mealy FSM S_2. It is the structural diagram of P_{Y3} Mealy FSM shown in Fig. 3.10.

The following sets can be derived from Table 3.16: $A = \{a_1, \ldots, a_6\}$, $X = \{x_1, \ldots, x_4\}$, $Y = \{y_1, \ldots, y_7\}$, $\Phi = \{D_1, D_2, D_3\}$, $T = \{T_1, T_2, T_3\}$. It gives the values: $M = 6, L = 4, N = 7, R_0 = 3$. There are $T_0 = 7$ of CMOs in Table 3.16: $Y_1 = \emptyset$, $Y_2 = \{y_1, y_2\}$, $Y_3 = \{y_3\}$, $Y_4 = \{y_1, y_4\}$, $Y_5 = \{y_4, y_5\}$, $Y_6 = \{y_6\}$, $Y_7 = \{y_5, y_7\}$. So, there are executed three steps of the design.

Let us find the set I. The CMO Y_1 determines only the state a_1. The CMO Y_2 determines the states a_2 and a_4. The CMO Y_3 determines the states a_2, a_3, a_4. The CMO Y_5 determines the state a_5. The CMO Y_6 determines the state a_3. The CMO Y_7 determines the state a_6. Analysis of these connections gives $K = 3$. So, there is the set $I = \{I_1, I_2, I_3\}$.

Table 3.17 Table of BIMF for P_{Y3} Mealy FSM S_2

Y_t	I_k	$K(I_k)$	a_s	$K(a_s)$	Φ_h	h
$\emptyset$	–	–	a_1	000	–	1
y_1y_2	I_1	00	a_2	001	D_3	2
y_1y_2	I_2	01	a_4	011	D_2D_3	3
y_3	I_1	00	a_2	001	D_2D_3	4
y_3	I_2	01	a_3	010	D_3	5
y_3	I_3	10	a_4	011	D_2	6
y_1y_4	I_1	00	a_3	010	D_2D_3	7
y_1y_4	I_2	01	a_5	100	D_2	8
y_4y_5	–	–	a_5	011	D_1	9
y_6	–	–	a_3	100	D_2	10
y_5y_7	–	–	a_6	101	D_1D_3	11

There are the following formulae for states $a_m \in A$:

$$\begin{aligned}
a_1 &= \langle \emptyset, Y_1 \rangle;\\
a_2 &= \langle I_1, Y_2 \rangle \vee \langle I_1, Y_3 \rangle;\\
a_3 &= \langle I_2, Y_3 \rangle \vee \emptyset, Y_6 \rangle;\\
a_4 &= \langle I_2, Y_2 \rangle \vee \langle I_2, Y_2 \rangle \vee \langle I_1, Y_4 \rangle;\\
a_5 &= \langle I_2, Y_4, \rangle \vee \langle \emptyset, Y_5 \rangle;\\
a_6 &= \langle \emptyset, Y_7 \rangle.
\end{aligned} \tag{3.22}$$

There is $K = 3$ in the discussed case. It gives $R_V = 2$ and the set $V = \{v_1, v_2\}$. Let us encode the identifiers $I_k \in I$ in the following way: $K(I_1) = 00$, $K(I_2) = 01$ and $K(I_3) = 10$.

There are H_{BIMF} rows in the table of BIMF determined by (3.12). In the discussed case, there is $H_{BIMF} = 11$. It follows from the system (3.22), for example. There are the following columns in the table of BIMF: $Y_t, I_k, K(I_k), a_s, K(a_s), \Phi_h, h$. It is Table 3.17 in the discussed case.

It is possible to derive the system (3.5) from the table of BIMF. It is the following system in the discussed case (after minimizing):

$$\begin{aligned}
D_1 &= y_1y_y\bar{v_1}v_2 \vee y_4y_5 \vee y_5y_7;\\
D_2 &= y_1y_2\bar{v_1}v_2 \vee y_3 \vee y_1y_4\bar{v_1}\bar{v_2} \vee y_6;\\
D_3 &= y_1y_2 \vee y_3\bar{v_2} \vee y_5y_7.
\end{aligned} \tag{3.23}$$

There are the following columns in the table of BMO: $a_m, K(a_m), X_h, V_h$, Y_h, h. It is constructed on the base of ST and a system similar to the system (3.22). It is Table 3.18 in the discussed case.

Table 3.18 Table of BMO for P_{Y3} Mealy FSM S_2

a_m	$K(a_m)$	X_h	V_h	Y_h	h
a_1	000	x_1	–	y_1y_2	1
		$\bar{x_1}$	v_2	y_3	2
a_2	001	x_2	–	y_1y_2	3
		$\bar{x_2}x_3$	–	y_1y_4	4
		$\bar{x_2}\bar{x_3}$	v_2	y_1y_2	5
a_3	010	x_1	v_1	y_3	6
		$\bar{x_1}$	v_2	y_1y_4	7
a_4	011	1	–	y_4y_5	8
a_5	100	x_2x_3	–	y_3	9
		$x_2\bar{x_3}$	–	y_6	10
		$\bar{x_2}x_4$	–	y_5y_7	11
		$\bar{x_2}\bar{x_4}$	–	–	12
a_6	101	1	–	y_6	13

This table is used for deriving the systems (1.3) and (3.1). For example, the following equations can be found from Table 3.18:

$$\begin{aligned} v_1 &= \bar{T_1}T_2\bar{T_3}x_1; \\ v_2 &= \bar{T_1}\bar{T_2}\bar{T_3}\bar{x_1} \vee \bar{T_1}\bar{T_2}T_3\bar{x_1}\bar{x_3} \vee \bar{T_1}T_2\bar{T_3}\bar{x_1}. \end{aligned} \tag{3.24}$$

These equations can be minimized.

The table of BMO should be transformed into the table of LUTer1. The table of LUTer1 corresponds to $N + R_V$ tables for circuits implementing functions $v_r \in V$ and $y_n \in Y$. The table of BIMF should be transformed into the table of EMBer1. Both transformations are executed in the same way as it was done in the previous cases.

Using the encoding of CMOs $Y_t \subseteq Y$ leads to $P_{Yi}Y$ models of Mealy FSMs $(i = \overline{1,8})$. The possible models are shown in Table 3.19.

This table is based on Fig. 3.4b. There are the following steps in the design methods for $P_{Yi}Y$ Mealy FSMs $(i = \overline{1,8})$:

1. Marking the initial GSA Γ and creating the set A.
2. Constructing the structure table of P Mealy FSM.

Table 3.19 Structural diagrams of $P_{Yi}Y$ Mealy FSMs

Blocks	$P_{Y1}Y$	$P_{Y2}Y$	$P_{Y3}Y$	$P_{Y4}Y$	$P_{Y5}Y$	$P_{Y6}Y$	$P_{Y7}Y$	$P_{Y8}Y$
BMOY	L_1	L_1	L_1	L_1	B_1	B_1	B_1	B_1
BIMF	L_2	L_2	B_2	B_2	L_2	L_2	B_2	B_2
BMO	L_3	B_3	L_3	B_3	L_3	B_3	L_3	B_3

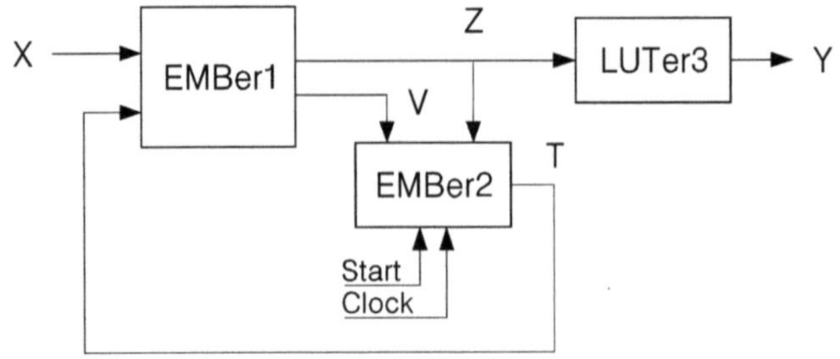

Fig. 3.11 Structural diagram of $P_{Y7}Y$ Mealy FSM

3. Finding the collections of microoperations $Y_t \subseteq Y$.
4. Constructing the set of identifiers $I = \{I_1, \ldots, I_K\}$.
5. Executing the encoding of CMOs $Y_t \subseteq Y$.
6. Representing the states $a_m \in A$ by pairs $\langle I_K, Y_t \rangle$.
7. Constructing the transformed ST of PY Mealy FSM.
8. Constructing the table of BIMF.
9. Constructing the table of BMOY.
10. Constructing the table of BMO.
11. Transformation of tables for implementing LUTers and EMBers.
12. Implementing the FSM logic circuit.

Let us discuss an example of design for $P_{Y7}Y$ Mealy FSM S_2 (Table 3.16). The corresponding structural diagram is shown in Fig. 3.11.

In $P_{Y7}Y$ FSM, the EMBer1 implements the systems (2.30) and (3.1), the EMBer2 the system (3.5), and the LUTer3 the system (1.40).

We have executed the steps 1–4 and 6 in the previous example. There is $T_0 = 7$. Using (1.32), we can find the value $R_Z = 3$ and the set $Z = \{z_1, z_2, z_3\}$. Let us encode the CMOs $Y_t \subseteq Y$ in the trivial way: $C(Y_1) = 000, \ldots, C(Y_7) = 111$.

The transformed ST is constructed on the base of an initial ST. To do it , it is necessary to replace the CMOs by their symbols and to add the columns $C(Y_h)$ and Z_h. There are variables $z_r = 1$ in the code $C(Y_h)$ in the row h of the transformed ST ($h = \overline{1, H_0}$). In the discussed case, it is Table 3.20.

There are the following columns in the table of BIMF: $C(Y_t)$, $K(I_k)$ (they determine the address of a cell), A_m, $K(a_m)$, Φ_h (it is the content of a cell), h. This table is constructed in a way similar to the table of BIMF for P_{Y3} FSM discussed before. For $P_{Y7}Y$ Mealy FSM S_2, this table is represented by Table 3.21.

Comparison shows that Tables 3.17 and 3.21 are practically identical. To construct Table 3.21, it is necessary to replace the collections $Y_t \subseteq Y$ by their codes.

The table of BIMF is used to derive the system (3.6). For example, the following equation can be derived from Table 3.21 (after minimization): $D_1 = T_2T_3v_2 \vee T_1\bar{T}_3$. To get it, we used the code 11 for $K(I_k)$ and the law of expansion for codes 100 and 110 of $C(Y_t)$.

Table 3.20 Transformed ST of PY Mealy FSM S_2

a_m	$K(a_m)$	a_s	$K(a_s)$	X_h	Y_t	$C(Y_t)$	Z_h	Φ_h	h
a_1	000	a_2	001	x_1	Y_2	001	z_3	D_3	1
		a_3	010	$\bar{x_1}$	Y_3	010	z_2	D_2	2
a_2	001	a_2	001	x_2	Y_2	001	z_3	D_3	3
		a_3	010	$\bar{x_2}x_3$	Y_4	011	z_2z_3	D_2	4
		a_4	011	$\bar{x_2}\bar{x_3}$	Y_2	001	z_3	D_2D_3	5
a_3	010	a_4	011	x_1	Y_3	010	z_2	D_2D_3	6
		a_5	100	$\bar{x_1}$	Y_4	011	z_2z_3	D_1	7
a_4	011	a_5	100	1	Y_5	100	z_1	D_1	8
a_5	100	a_2	001	x_2x_3	Y_3	010	z_2	D_3	9
		a_3	010	$x_2\bar{x_3}$	Y_6	101	z_1z_3	D_2	10
		a_6	101	$\bar{x_2}x_4$	Y_7	110	z_1z_2	D_1D_3	11
		a_1	000	$\bar{x_2}\bar{x_4}$	Y_1	000	–	–	11
a_6	101	a_3	010	1	Y_6	101	z_1z_3	D_2	12

Table 3.21 Table of BIMF for $P_{Y7}Y$ Mealy FSM S_2

$C(Y_t)$	$K(I_k)$	a_m	$K(a_m)$	Φ_h	h
000	*	a_1	000	–	1
001	00	a_2	001	D_3	2
001	01	a_4	011	D_2D_3	3
010	00	a_2	001	D_3	4
010	01	a_3	010	D_2	5
010	10	a_4	011	D_2D_3	6
011	01	a_3	010	D_2	7
011	00	a_5	100	D_1	8
100	01	a_5	100	D_1	9
101	*	a_3	010	D_2	10
110	*	a_6	101	D_1D_3	11

To construct the table of BMOY, it is necessary to eliminate the columns a_s, $K(a_s)$, Φ_h from the transformed ST of PY Mealy FSM. They are replaced by the columns I_k, $K(I_k)$ and V_h. For example, the state a_2 (h = 1 in Table 3.20) is determined by the pair $\langle Y_2, I_1\rangle$ (it follows, for example, from Table 3.21). So, it should be I_1 in the column I_k, the code 00 in the column $K(I_k)$ and no variables in the column V_h (h = 1 in the table of BMO). Acting in the same way, we construct Table 3.22.

Table of BIMF is used for constructing the systems (2.30) and (3.1). For example, the following equations can be derived from Table 3.22:

Table 3.22 Table of BMOY for $P_{Y7}Y$ Mealy FSM S_2

a_m	$K(a_m)$	X_h	Y_h	$C(Y_h)$	Z_h	I_k	$K(I_k)$	V_h	h
a_1	000	x_1	Y_2	001	z_3	I_1	00	–	1
		$\bar{x_1}$	Y_3	010	z_2	I_2	01	v_2	2
a_2	001	x_2	Y_2	001	z_3	I_1	00	–	3
		$\bar{x_2}x_3$	Y_4	011	z_2z_3	I_1	00	–	4
		$\bar{x_2}\bar{x_3}$	Y_2	001	z_3	I_2	01	v_2	5
a_3	010	x_1	Y_3	010	z_2	I_3	10	v_1	6
		$\bar{x_1}$	Y_4	011	z_2z_3	I_2	01	v_2	7
a_4	011	1	Y_5	100	z_1	–	–	–	8
a_5	100	x_2x_3	Y_3	010	z_2	I_1	00	–	9
		$x_2\bar{x_3}$	Y_6	101	z_1z_3	–	–	–	10
		$\bar{x_2}x_4$	Y_7	110	z_1z_2	–	–	–	11
		$\bar{x_2}\bar{x_4}$	Y_1	000	–	–	–	–	12
a_6	101	1	Y_6	101	z_1z_3	–	–	–	13

Table 3.23 Table of LUTer3 for $P_{Y7}Y$ Mealy FSM S_2

$C(Y_t)$	Y	t	$C(Y_t)$	Y	t
$z_1z_2z_3$	$y_1y_2y_3y_4y_5y_6y_7$		$z_1z_2z_3$	$y_1y_2y_3y_4y_5y_6y_7$	
000	0000000	1	100	0001100	5
001	1100000	2	101	0000010	6
010	0010000	3	110	0000101	7
011	1001000	4	111	0000000	8

$$\begin{aligned}
z_1 &= \bar{T_1}T_2T_3 \lor T_1\bar{T_2}\bar{T_3}x_2\bar{x_3} \lor T_1\bar{T_2}\bar{T_3}\bar{x_2}x_4;\\
z_2 &= \bar{T_1}\bar{T_3}\bar{x_1} \lor \bar{T_1}T_2\bar{T_3} \lor \bar{T_1}\bar{T_2}T_3\bar{x_2}x_3 \lor T_1\bar{T_2}\bar{T_3}(x_2x_3 \lor \bar{x_2}x_3);\\
z_3 &= \bar{T_1}\bar{T_2}\bar{T_3}x_1 \lor \bar{T_1}\bar{T_2}T_3 \lor \bar{T_1}T_2\bar{T_3}\bar{x_1} \lor T_1\bar{T_2}\bar{T_3}x_2\bar{x_3} \lor T_1\bar{T_2}T_3;\\
v_1 &= T_2\bar{T_3}x_1;\\
v_2 &= \bar{T_1}\bar{T_3}\bar{x_1} \lor \bar{T_1}\bar{T_2}T_3\bar{x_2}\bar{x_3}
\end{aligned} \tag{3.25}$$

There are the following columns in the table of LUTer3: $C(Y_t)$, Y, t. It is Table 3.23 in the discussed case.

This table determines N circuits implementing the functions $y_n \in Y$. These functions are represented in the form (1.40). If the following condition takes place

$$S \geq R_Z, \tag{3.26}$$

then there are exactly N of LUTs in the circuit of LUTer3.

There is no sense in using EMBs for implementing microoperations in the case of encoding of fields of compatible microoperations. Because of it, the block BMO

Table 3.24 Structural diagrams of $P_{Y_i}D$ Mealy FSMs

Blocks	$P_{Y1}D$	$P_{Y2}D$	$P_{Y3}D$	$P_{Y4}D$
BMOD	L_1	B_1	L_1	B_1
BIMF	L_2	L_1	B_1	B_2
BMO	L_3	L_2	L_2	L_1

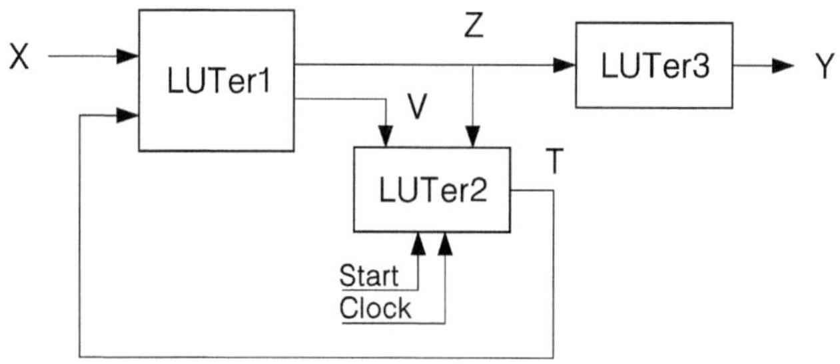

Fig. 3.12 Structural diagram of $P_{Y1}D$ Mealy FSM

Fig. 3.4c is implemented as LUTer. So, there are 4 different structures of FPGA-based P_YD Mealy FSMs (Table 3.24).

For example, Fig. 3.12 represents the structural diagram of $P_{Y1}D$ Mealy FSM.

To develop design methods for $P_{Y_i}D$ FSMs ($i = \overline{1,4}$), it is enough to modify the design methods for $P_{Y_i}Y$ FSMs ($i = \overline{1,8}$). There are two differences. The point 5 is the executing the encoding of fields of compatible microoperations. The point 5 is the constructing the table of BMOD. We do not discuss these methods in that section.

3.4 Replacement of Logical Conditions in Mealy FSMs with OCT

It is possible to use the RLC for optimizing any model of Mealy FSM with OCT. It results in adding the BRLC as it is shown, for example, in Fig. 3.5. It has sense to implement the block BRLC as LUTer.

Let us, for example, discuss the design methods for $MP_{Y7}Y$ Mealy FSM (Fig. 3.13).

In $MP_{Y7}Y$ Mealy FSM, the LUTer1 executes the RLC. It implements the system (1.47). The EMBer1 implements the system (3.4) and the system

$$Z = Z(T, P). \tag{3.27}$$

The EMBer2 implements the system (3.6). The LUTer3 implements the system (1.40).

There are the following steps in the design method for $MP_{Y7}P$ Mealy FSM:

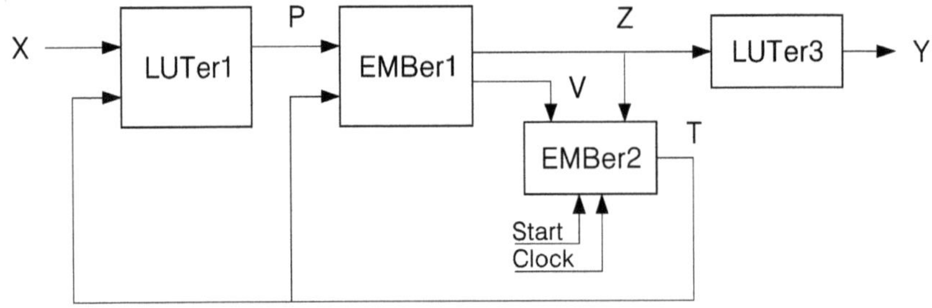

Fig. 3.13 Structural diagram of $MP_{Y7}Y$ Mealy FSM

1. Marking the initial GSA Γ and creating the set A.
2. Constructing the ST of Mealy FSM.
3. Finding the collections of microoperations $Y_t \subseteq Y$.
4. Constructing the set of identifiers I.
5. Executing the maximal encoding of CMOs $Y_t \subseteq Y$.
6. Representing the states $a_m \in A$ by pairs $\langle I_k, Y_t \rangle$.
7. Executing the replacement of logical conditions.
8. Constructing transformed ST of MPY Mealy FSM.
9. Constructing the table of BIMF.
10. Constructing the table of BMOY.
11. Constructing the table of BMO.
12. Transformation the tables for representing the LUTers and EMBers.
13. Implementing the FSM logic circuit.

We already have executed the points 1–6 of this method for $P_{Y7}Y$ Mealy FSM S_2. Let us continue this example for the $MP_{Y7}Y$ Mealy FSM S_2.

There are the following sets $X(a_m)$ in Table 3.16: $X(a_1) = \{x_1\}, X(a_2) = \{x_2, x_3\}, X(a_3) = \{x_1\}, X(a_4) = X(a_6) = \emptyset, X(a_5) = \{x_2, x_3, x_4\}$. It gives $G = 3$ and $P = \{p_1, p_2, p_3\}$. Let us replace the logical conditions in the following manner (Table 3.25).

Table 3.25 Table of RLC for MP Mealy FSM S_2

a_m	a_1	a_2	a_3	a_4	a_5	a_6
p_1	x_1	x_2	x_1	–	x_2	–
p_2	–	x_3	–	–	x_3	–
p_3	–	–	–	–	x_4	–

Table 3.26 Transformed ST of MPY Mealy FSM S_2

a_m	$K(a_m)$	a_s	$K(a_s)$	P_h	Y_t	$C(Y_t)$	Z_h	Φ_h	h
a_1	000	a_2	001	p_1	Y_2	001	z_3	D_3	1
		a_3	010	$\bar{p_1}$	Y_3	010	z_2	D_2	2
a_2	001	a_2	001	p_1	Y_2	001	z_3	D_3	3
		a_3	010	$\bar{p_2}x_3$	Y_4	011	z_2z_3	D_2	4
		a_4	011	$\bar{p_2}\bar{x_3}$	Y_2	001	z_3	D_2D_3	5
a_3	010	a_4	011	p_1	Y_3	010	z_2	D_2D_3	6
		a_5	100	$\bar{p_1}$	Y_4	011	z_2z_3	D_1	7
a_4	011	a_5	100	1	Y_5	100	z_1	D_1	8
a_5	100	a_2	001	p_1x_3	Y_3	010	z_2	D_3	9
		a_3	010	$p_1\bar{x_3}$	Y_6	101	z_1z_3	D_2	10
		a_6	101	$\bar{p_1}x_4$	Y_7	110	z_1z_2	D_1D_3	11
		a_1	000	$\bar{p_1}\bar{x_4}$	Y_1	000	–	–	11
a_6	101	a_3	010	1	Y_6	101	z_1z_3	D_2	12

There are the following functions (1.47) derived from Table 3.25:

$$\begin{aligned} p_1 &= \bar{T}_1\bar{T}_3x_1 \lor \bar{T}_1\bar{T}_2T_3x_2 \lor T_1\bar{T}_2\bar{T}_3x_2;\\ p_2 &= x_3;\\ p_3 &= x_4. \end{aligned} \quad (3.28)$$

Analysis of (3.28) shows that it is necessary only to implement a circuit for the function p_1.

The transformed ST is represented by Table 3.26 in the case of MPY Mealy FSM S_2. It is similar to Table 3.20, but the column X_h is replaced by column P_h. We do not replace the conditions $x_3, x_4 \in X$.

There are the same columns in tables of BIMF for $P_{Y_7}Y$ and $MP_{Y_7}Y$ Mealy FSM S_2. The same is true for the table of BMO. In the discussed case, it is the same as Table 3.22.

The table of BMOY is constructed in the same way as Table 3.22. But it is necessary to replace the column X_h by the column P_h. It gives Table 3.27.

This table is used for constructing the functions (3.4) and (3.27). For example, the following functions can be derived from Table 3.27:

$$\begin{aligned} z_1 &= T_2T_3 \lor T_1\bar{T}_2\bar{T}_3(p_1\bar{x_3} \lor T_1x_4) \lor T_1T3;\\ v_1 &= T_2\bar{T}_3P_1. \end{aligned} \quad (3.29)$$

Acting in the same manner, it is possible to develop design methods for all proposed models of Mealy FSMs with OCT and RLC. We leave this task to a reader.

Table 3.27 Table of BMOY for $MP_{Y7}Y$ Mealy FSM S_2

a_m	$K(a_m)$	P_h	Y_h	$C(Y_h)$	Z_h	I_k	$K(I_k)$	V_h	h
a_1	000	p_1	Y_2	001	z_3	I_1	00	–	1
		$\bar{p_1}$	Y_3	010	z_2	I_2	01	v_2	2
a_2	001	p_2	Y_2	001	z_3	I_1	00	–	3
		$\bar{p_2}x_3$	Y_4	011	z_2z_3	I_1	00	–	4
		$\bar{p_2}\bar{x_3}$	Y_2	001	z_3	I_2	01	v_2	5
a_3	010	p_1	Y_3	010	z_2	I_3	10	v_1	6
		$\bar{p_1}$	Y_4	011	z_2z_3	I_2	01	v_2	7
a_4	011	1	Y_5	100	z_1	–	–	–	8
a_5	100	p_1x_3	Y_3	010	z_2	I_1	00	–	9
		$p_1\bar{x_3}$	Y_6	101	z_1z_3	–	–	–	10
		$\bar{p_1}x_4$	Y_7	110	z_1z_2	–	–	–	11
		$\bar{p_1}\bar{x_4}$	Y_1	000	–	–	–	–	12
a_6	101	1	Y_6	101	z_1z_3	–	–	–	13

3.5 Analysis of Proposed Methods

Using the OCT leads to increasing for the number of functions which can be implemented by EMBs. Let us find conditions allowing the application of EMBs. Modern EMBs have a lot of possible configurations corresponding to the following expression:

$$V \cdot t_F = V_0. \tag{3.30}$$

In (3.30), V_0 is a number of memory cells for $t_F = 1$.

Let us start from P_A Mealy FSM (Table 3.1). It is possible to use EMBs till the following conditions take places:

$$2^{L+R_0} \leq V_0; \tag{3.31}$$

$$2^{R_0+R_V} \leq V_0. \tag{3.32}$$

If the condition (3.31) takes place, then the block BIMF can be implemented as EMBer. If the condition (3.32) takes place, then the BMO can be implemented as EMBer.

In the best case, each block is implemented as a single EMB. The best case is determined for BIMF by the condition(3.33) and for BMO by (3.34):

$$2^{L+R_0}(R_0 + R_V) \leq V_0; \tag{3.33}$$

$$2^{R_0+R_V} \cdot N \leq V_0. \tag{3.34}$$

If the condition (3.31) takes place and the condition (3.33) is violated, then it is necessary $n(BIMF, P_A)$ of blocks for implementing the circuit of BIMF:

$$n(BIMF, P_A) = \left\lceil \frac{R_0 + R_V}{t_F} \right\rceil. \tag{3.35}$$

If the condition (3.32) takes place and the condition (3.34) is violated, then it is necessary $n(BMO, P_A)$ of blocks for implementing the circuit of BMO for P_A Mealy FSM:

$$n(BMO, P_A) = \left\lceil \frac{N}{t_F} \right\rceil. \tag{3.36}$$

In the case of P_AY Mealy FSM, the corresponding conditions for BIMF are determined by (3.31), (3.33) and (3.35). To use EMBs for implementing the circuit of BCTY (Fig. 3.2a), it is necessary to satisfy the following condition:

$$2^{R_V+R_0} \cdot R_Z \leq V_0; \tag{3.37}$$

If this condition is violated, but the condition (3.32) takes place, then it is necessary $n(BCTY, P_AY)$ of blocks for implementing the circuit of BCTY:

$$n(BCTY, P_AY) = \left\lceil \frac{R_Z}{t_F} \right\rceil. \tag{3.38}$$

EMBs can be used for implementing the circuit of BMOY, if the following condition takes place:

$$2^{R_Z} \leq V_0. \tag{3.39}$$

In this case, the number of EMBs is determined by the following equation:

$$n(BMOY, P_AY) = \left\lceil \frac{N}{t_F} \right\rceil. \tag{3.40}$$

As we can see, expressions (3.36) and (3.40) are the same. It is true for the block BMOD (Fig. 3.2b).

If the model of P_AD Mealy FSM is used, then there are R_D elements in the set Z:

$$R_D - \sum_{k=1}^{K} R_k. \tag{3.41}$$

In (3.41), the symbol K stands for the number of classes of compatible MOs. The symbol R_k stands for the number of bits in the codes for a field $k (k = \overline{1, K})$. It is determined as (1.35).

As a rule, the following relation is true:

$$R_D > R_Z. \tag{3.42}$$

Because of it, it is better to implement the BMOD as a LUTer.

It is possible to use EMBs in the BCTD of P_AD Mealy FSM, if the following condition takes place:

$$2^{R_0+R_V} \cdot R_D \leq V_0. \tag{3.43}$$

The number of EMBs in the circuit of BCTD is determined as

$$n(BCTD, P_AY) = \left\lceil \frac{R_D}{t_F} \right\rceil. \tag{3.44}$$

Using the RLC leads to adding the BRLC in the structures of P_A Mealy FSMs. This block is implemented as LUTer. In this case, there is a change in conditions for the BIMF. For example, there are new conditions (3.45) and (3.46) corresponding to the conditions (3.31) and (3.33), representing as

$$2^{G+R_0} \leq V_0; \tag{3.45}$$

$$2^{G+R_0}(R_0 + R_V) \leq V_0. \tag{3.46}$$

EMBs can be used for implementing the circuit of BMO for P_Y Mealy FSM (Fig. 3.4a). The condition (3.31) shows that EMBs can be used for implementing the BMO. The following condition shows that the circuit of BMO is implemented as a single EMB:

$$2^{L+R_0}(R_V + N) \leq V_0. \tag{3.47}$$

If the condition (3.31) is true and the condition (3.47) is violated, then it is necessary $n(BMO, P_Y)$ of blocks in the BMO:

$$n(BMO, P_Y) = \left\lceil \frac{R_V + N}{t_F} \right\rceil. \tag{3.48}$$

It is possible to use EMBs in the circuit of BIMF of PY FSM, if the following condition takes place:

$$2^{R_V+N} \leq V_0. \tag{3.49}$$

In this case, the number of EMBs is determined as

$$n(BIMF, P_Y) = \left\lceil \frac{R_0}{t_F} \right\rceil. \tag{3.50}$$

EMBs can be used in the BMOY (Fig. 3.4b), if the condition (3.31) takes place. It is enough to use a single EMB, if the following condition takes place:

$$2^{L+R_0}(R_V + R_Z) \leq V_0. \tag{3.51}$$

If this condition is violated, then it is necessary $n(BMOY, P_Y Y)$ blocks:

$$n(BMOY, P_Y Y) = \left\lceil \frac{R_V + R_Z}{t_F} \right\rceil. \tag{3.52}$$

The condition (3.35) determines the using of EMBs in the circuit of BMO for $P_Y Y$ FSM. The expression (3.40) can be used for finding the number of blocks in BMO.

EMBs can be used for implementing the BIMF of $P_Y Y$ FSM, if the following condition takes place:

$$2^{R_Z+R_V} \leq V_0. \tag{3.53}$$

The expression (3.50) can be used to find the number of EMBs in this circuit.

Acting in the same way, it is possible to find the conditions for using EMBs in the circuits of BMOD, BMO and BIMF of $P_Y D$ Mealy FSMs, respectively:

$$2^{R_0+L} \cdot (R_D + R_V) \leq V_0; \tag{3.54}$$

$$2^{R_D} \cdot N \leq V_0; \tag{3.55}$$

$$2^{R_D+R_V} \cdot R_0 \leq V_0. \tag{3.56}$$

If conditions (3.54)–(3.56) take places, then each block of $P_Y D$ Mealy FSM is implemented as a single EMB. If the conditions are violated, then numbers of EMB can be calculated as:

$$n(BMOD, P_Y D) = \left\lceil \frac{R_V + R_D}{t_F} \right\rceil; \tag{3.57}$$

$$n(BMO, P_Y D) = \left\lceil \frac{N}{t_F} \right\rceil; \tag{3.58}$$

$$n(BIMF, P_Y D) = \left\lceil \frac{R_0}{t_F} \right\rceil. \tag{3.59}$$

It is possible to use the RLC in all models of PY, P_YY and P_YD FSM. As it is pointed before, the BRLC is implemented as a LUTer. It is change for conditions determining the application of EMBs in BMO, BMOY and BMOD. Now, there are the following conditions:

$$2^{G+R_0} \cdot (R_V + N) \leq V_0; \tag{3.60}$$

$$2^{G+R_0} \cdot (R_V + R_Z) \leq V_0; \tag{3.61}$$

$$2^{G+R_0} \cdot (R_V + R_D) \leq V_0. \tag{3.62}$$

If the conditions are violated, then the following formulae determine the number of EMBs in the circuits of BMO, BMOY and BMOD respectively:

$$n(BMO, MP_Y) = \left\lceil \frac{R_V + N}{t_F} \right\rceil; \tag{3.63}$$

$$n(BMOY, MP_YY) = \left\lceil \frac{R_V + R_Z}{t_F} \right\rceil; \tag{3.64}$$

$$n(BMOD, MP_YD) = \left\lceil \frac{R_V + R_D}{t_F} \right\rceil. \tag{3.65}$$

It is very important to choose a model of FSM having, for example, minimal amount of hardware. It is possible to make a preliminary choice of a model with minimal amount of EMBs using the formulae (3.31)–(3.65).

Table 3.28 Characteristics of Mealy FSM with OCT

№	FSM	BRLC	BIMF	BCTA	BMOA	BMO
1	P_A	–	$V(T,X)$; $\Phi(T,X)$	–	–	$Y(V,T)$
2	P_AY	–	$V(T,X)$; $\Phi(T,X)$	$Z(T,V)$	$Y(Z)$	–
3	P_AD	–	$V(T,X)$; $\Phi(T,X)$	$Z(T,V)$	$Y(Z)$	–
4	MP_A	$P(T,X)$	$V(T,P)$; $\Phi(T,P)$	–	–	$Y(V,T)$
5	MP_AY	$P(T,X)$	$V(T,P)$; $\Phi(T,P)$	$Z(T,V)$	$Y(Z)$	–
6	MP_AD	$P(T,X)$	$V(T,P)$; $\Phi(T,P)$	$Z(T,V)$	$Y(Z)$	–
7	P_Y	–	$\Phi(V,Y)$	–	–	–
8	P_YY	–	$\Phi(V,Z)$	–	$Z(T,X)$; $V(T,X)$	–
9	P_YD	–	$\Phi(V,Z)$	–	$Z(T,X)$; $V(T,X)$	–
10	MP_Y	$P(T,X)$	$\Phi(V,Y)$	–	–	–
11	MP_YY	$P(T,X)$	$\Phi(V,Z)$	–	$Z(P,X)$; $V(P,X)$	–
12	MP_YD	$P(T,X)$	$\Phi(V,Z)$	–	$Z(P,X)$; $V(P,X)$	–

There are 12 basic models of Mealy FSMs with OCT. They are represented by Table 3.28. We use the abbreviations BCTA and BMOA to show the blocks BCTY, BCTD, BMOY and BMOD.

References

1. A. Barkalov, A. Barkalov Jr, Synthesis of Finite-State Machines with transformation of the object's codes. In *Proceedings of the International Conference TCSET'2004* pp. 61–64, Lviv, Ukraina, 2004. Lviv Polytechnic National University, Lviv, Publishing House of Lviv Polytechnic
2. A. Barkalov, A. Barkalov Jr, Design of Mealy Finite-State Machines with the transformation of object codes. Int. J. Appl. Math. Comput. Sci. **15**(1), 151–158 (2005)
3. A. Barkalov, L. Titarenko, *Logic Synthesis for FSM-based Control Units*, vol. 53. Lecture Notes in Electrical Engineering (Springer, Berlin, 2009)
4. A. Barkalov, L. Titarenko, A. Barkalov Jr, Moore FSM synthesis with coding of compatible microoperations fields. In *Proceedings of IEEE East-West Design & Test Symposium—EWDTS'07*, pp. 644–646, Yerevan, Armenia, Kharkov, 2007. Kharkov National University of Radioelectronics
5. A. Barkalov, L. Titarenko, A. Barkalov Jr, Structural decomposition as a tool for the optimization of an FPGA-based implementation of a mealy FSM. Cybern. Syst. Anal. **48**(2), 313–322 (2012)
6. A. Barkalov, A. Węgrzyn, A. Barkalov Jr, Synthesis of control units with transformation of the codes of objects. In *Proceedings of the IXth International Conference CADSM 2007 (The Experience of Designing and Application of CAD Systems in Microelectronics)*, pp. 260–261, Lviv - Polyana, Ukraine, 2007. Lviv Polytechnic National University, Lviv, Publishing House of Lviv Polytechnic National University
7. E. McCluskey, *Logic Design Principles* (Prentice Hall, Englewood Cliffs, 1986)

Chapter 4
Object Codes Transformation for Moore FSMs

4.1 Principle of OCT for Moore FSMs

There are the same basic principles for object transformation in the case of Mealy and Moore FSMs. In both cases, it is necessary to find one-to-one correspondences $A \rightarrow Y$ (P_A Moore FSM) or $Y \rightarrow A$ (P_Y Moore FSM). Let us start from P_A Moore FSMs.

Due to dependence (1.5), the P Moore FSM can be viewed as P_A Moore FSM. So, there are no additional blocks in P_A Moore FSMs in comparison with P Moore FSMs.

In the case of P_AY Moore FSM, it is necessary to encode the collections $Y_t \subseteq Y$ by codes $C(Y_t)$ having R_Z bits, where R_Z is determined by (1.22). It leads to the following structural diagram (Fig. 4.1).

In P_AY Moore FSM, the BIMF implements the functions (1.4), whereas BMOY the functions (1.40). A block of state codes transformer (BSCT) executes transformation of state codes $K(a_m)$ into codes of CMOs $C(Y_t)$. De facto, it implements the principle of OCT based on the system

$$Z = Z(T). \tag{4.1}$$

It is possible to use the encoding of the fields of compatible MOs together with the principle of OCT. It leads to P_AD Moore FSM (Fig. 4.2).

Two methods can be used for reducing hardware in BIMF [5]: the optimal state assignment and the transformation of state codes $K(a_m)$ in the codes of classes of PES $K(B_i)$. In the first case, there is no change in structural diagrams. But it is quite possible that some state variables are insignificant for functions Φ. Let us denote the corresponding models as $P_{0A}Y$ and $P_{0A}D$. In these FSMs, the BIMF implements functions

$$\Phi = \Phi(T', X). \tag{4.2}$$

In (4.2), there is the relation $T' \subseteq T$.

A. Barkalov et al., *Logic Synthesis for FPGA-Based Finite State Machines*,
Studies in Systems, Decision and Control 38, DOI 10.1007/978-3-319-24202-6_4

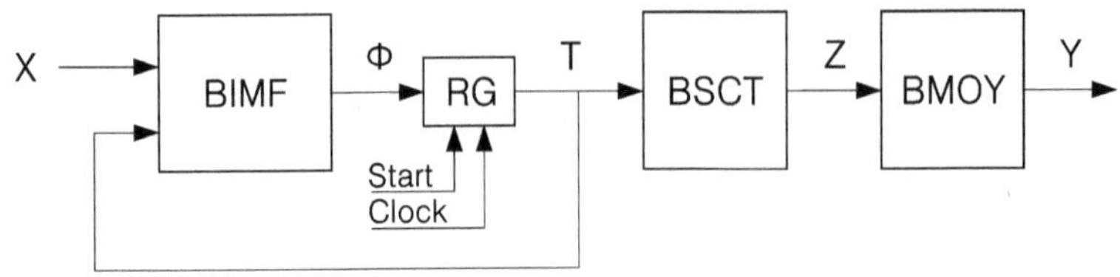

Fig. 4.1 Structural diagram of P_AY Moore FSM

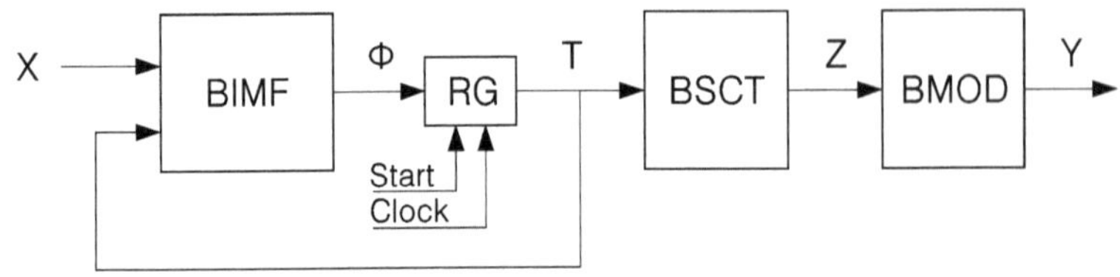

Fig. 4.2 Structural diagram of P_AD Moore FSM

Let us encode classes $B_i \in \Pi_A$ by binary codes $K(B_i)$ having R_A bits:

$$R_A = \lceil log_2 I \rceil. \tag{4.3}$$

Remind, that there is $I = |\Pi_A|$. Let us use variables $\tau_r \in \tau$ for encoding of the class $B_i \in \Pi_A$. Obviously, there is $|\tau| = R_A$. Let us find the system

$$\tau = \tau(T). \tag{4.4}$$

The system (4.4) shows the correspondence among the codes $K(a_m)$ and $K(B_i)$. To implement it, it is necessary to use the BSCT. Let us use the letter "E" to show using of this principle in a Moore FSM. There are the structural diagrams of $P_{EA}Y$ and $P_{EA}D$ Moore FSMs shown in Fig. 4.3.

In both cases, the BIMF implements the system

$$\Phi = \Phi(\tau, X). \tag{4.5}$$

whereas the BSCT the systems (4.1) and (4.4).

It is possible to double the number of models adding the BRLC in the all discussed models. There are four different models of Moore FSMs shown in Fig. 4.4. In the case of MP_AY and MP_AD models, the BRLC implements the system (1.47) and the BIMF the system (1.49). In the case of $MP_{EA}Y$ and $MP_{EA}D$ models, the BRLC implements the system

$$P = P(\tau, X). \tag{4.6}$$

The BIMF implements the system (4.5). We discuss all these methods a bit later. Of course, there are models of $MP_{0A}Y$ and $MP_{0A}D$ Moore FSMs.

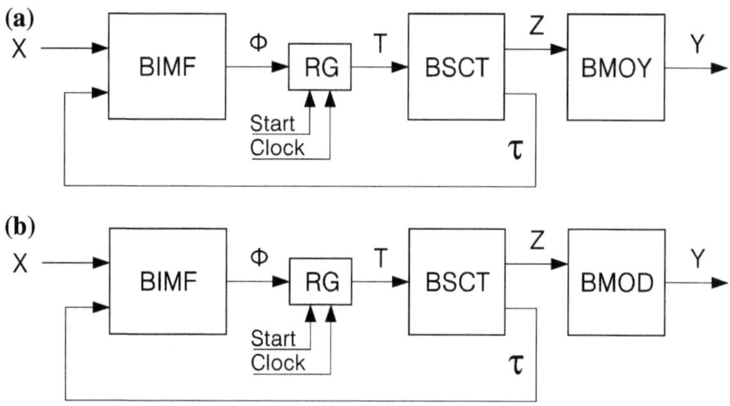

Fig. 4.3 Structural diagrams of $P_{EA}Y$ **a** and $P_{EA}D$ **b** Moore FSMs

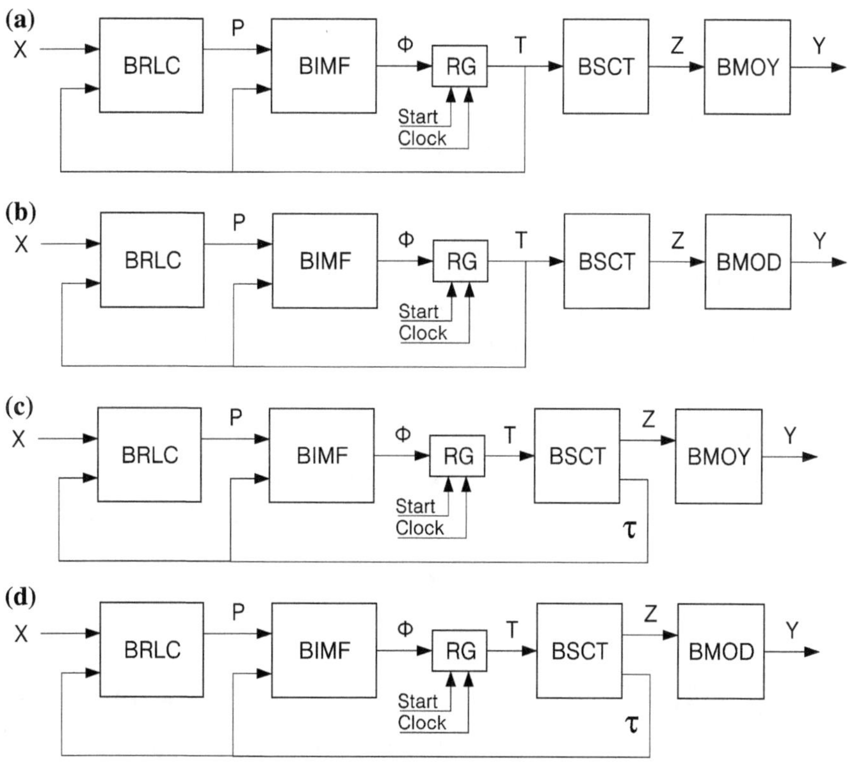

Fig. 4.4 Structural diagrams of MP_AY (**a**), MP_AD (**b**), $MP_{EA}Y$ (**c**) and $MP_{EA}D$ (**d**) Moore FSMs

It follows from Figs. 4.1, 4.2, 4.3 and 4.4 that there are no identifiers in the P_A Mealy FSM and its modifications. It is connected with the following relation:

$$M_1 \geq T_0. \tag{4.7}$$

On the other hand, there is a need for identifiers for P_Y Moore FSMs. They are based on the correspondence $Y \rightarrow A$. In this case, the CMOs $Y_t \subseteq Y$ are the primary objects and the states $a_m \in A$ are the secondary objects.

Three structures of P_Y Moore FSMs are possible. It is P_Y FSM based on the unitary encoding of MOs. In this case, they use N bits for representing N microoperations [9]. We can name this approach "N-hot" encoding. There is the structural diagram of P_Y Moore FSM shown in Fig. 4.5.

In P_Y FSM, the BMO implements the system (1.5) and the BIMF the system (3.5). A block of identifiers (BI) implements the system (3.1).

Let us encode the CMOs $Y_t \subset Y$ by binary codes having R_Z bits, where R_Z is determined by (1.32). This approach is a maximal encoding of collections of micropoperations [4, 9]. Combining this approach with OCT leads to $P_Y Y$ Moore FSM Fig. 4.6.

In $P_Y Y$ FSM, the BMO implements the system of functions $z_r \in Z$ used for the maximal encoding of CMOs. It is represented as

$$Z = Z(T). \tag{4.8}$$

The BIMF implements the system (3.6), whereas the BI the system (3.1).

Using the principle of encoding of the fields of compatible microoperations [2, 17] together with OCT leads to $P_Y D$ Moore FSM (Fig. 4.7). In this model, the BMOD

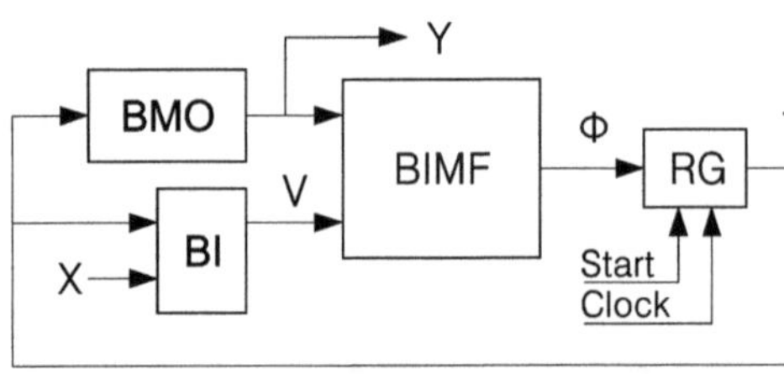

Fig. 4.5 Structural diagram of P_Y Moore FSM

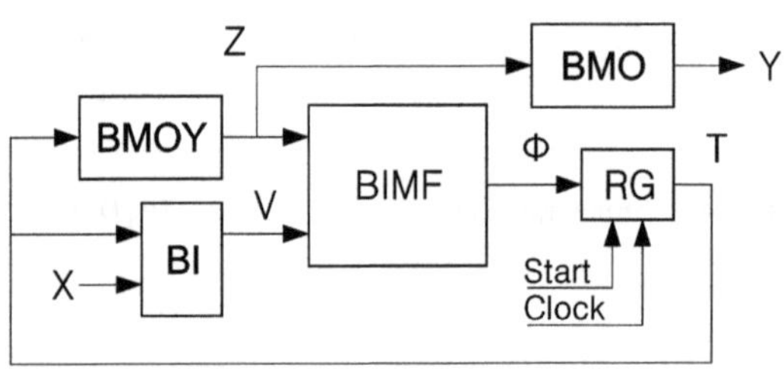

Fig. 4.6 Structural diagram of $P_Y Y$ Moore FSM

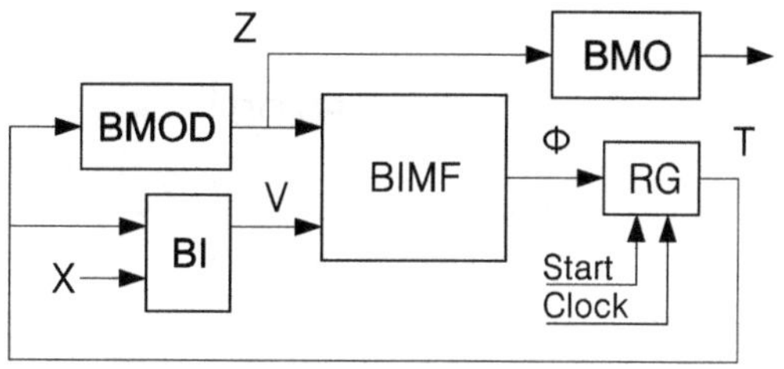

Fig. 4.7 Structural diagram of $P_Y D$ Moore FSM

implements R_D of functions (4.8). Other blocks execute the same functions as it is for $P_Y Y$ Moore FSM.

It is possible to reduce the hardware amount in the circuit of BI taking into account the existence of PES [5, 9, 18]. There are two approaches based on PES, discussed beforehand.

Using the optimal state encoding leads to three models of P_Y Moore FSMs, namely, to P_{0Y}, $P_{0Y}Y$ and $P_{0Y}D$ models. They have the same structural diagrams as their counterparts without the optimal state encoding. There is the only difference. It is possible that functions V will be depended on some part of the set T:

$$V = V(T', X). \tag{4.9}$$

Three models are based on the transformation of state codes into the codes of classes of PES. They are the models P_{EY} (Fig. 4.8), $P_{EY}Y$ (Fig. 4.9) and $P_{EY}D$ (Fig. 4.10).

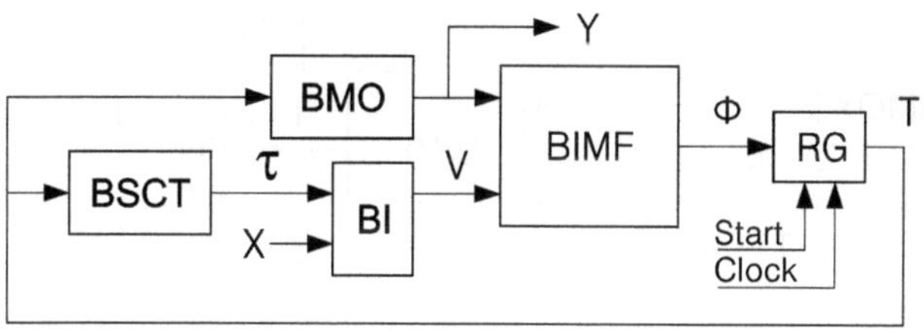

Fig. 4.8 Structural diagram of P_{EY} Moore FSM

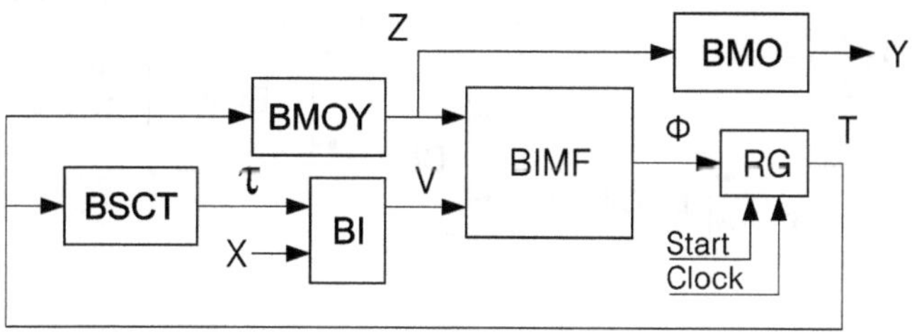

Fig. 4.9 Structural diagram of $P_{EY}Y$ Moore FSM

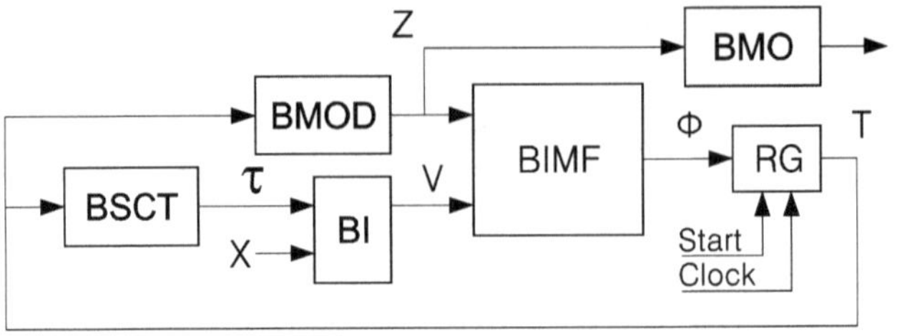

Fig. 4.10 Structural diagram of $P_{EY}D$ Moore FSM

There are the same functions of BMO and BIMF as for their counterparts in P_Y Moore FSM (Fig. 4.5). The BSCT implements the system (4.4). It executes the transformation of state codes $K(a_m)$ into the class codes $K(B_i)$. The BI implements the system (4.10):

$$V = V(\tau, X). \quad (4.10)$$

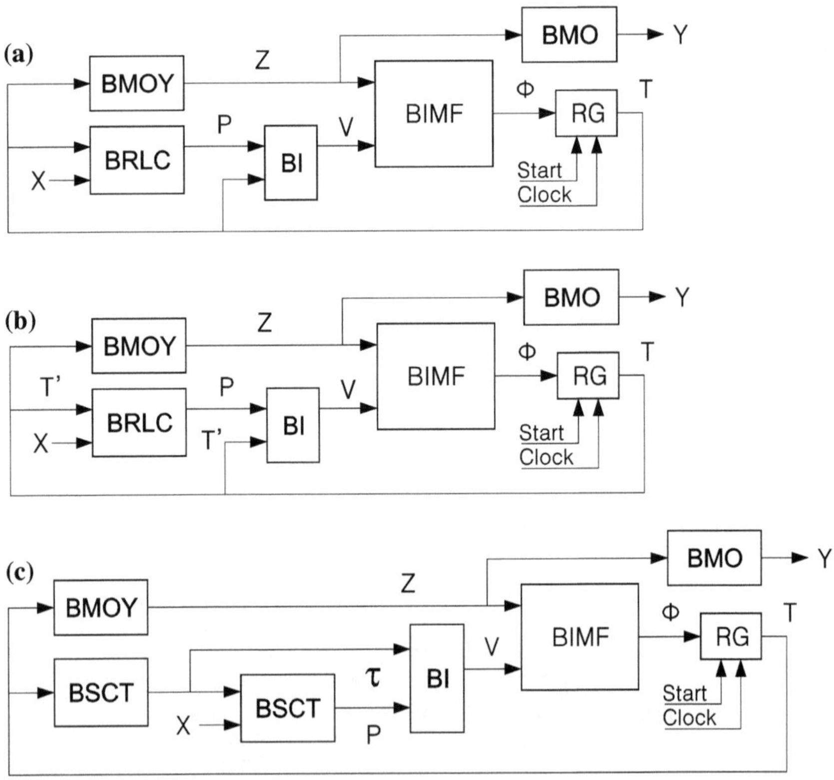

Fig. 4.11 Structural diagrams of MP_YY (**a**), $MP_{OY}Y$ **b** and $MP_{EY}Y$ **c** Moore FSMs

The functions of different blocks can be obtained from analysis of Figs. 4.6 and 4.8. There are the same functions of BMOY, BMO and BIMF of $P_{EY}Y$ FSM as for their counterparts from P_YY FSM. There are the same functions of BSCT and BI as for their counterparts from P_YY Moore FSM.

In $P_{EY}D$ FSM, the BMOD implements the system $Z = Z(T)$, the BMO the system $Y = Y(Z)$, the BIMF the system $\Phi = \Phi(V, Z)$. These systems are represented in the same way as for P_YD Moore FSM. The BSCT implements the system $\tau = \tau(T)$, the BI implements the system $V = V(\tau, X)$. These functions are represented as in the P_{EY} Moore FSM.

It is possible further hardware reduction in the circuit of BI. It is based on using the principle of replacement of logical conditions [3]. It is reduced to introduction of the block BRLC into a structural diagram of any FSM. For example, there are structural diagrams of MP_{YY}, $MP_{OY}Y$, $MP_{EY}Y$ Moore FSMs shown in Fig. 4.11.

All these models differ in functions implemented by the blocks BRLC and BI. There are functions (1.47) and (3.4) in the case of MP_{YY} FSMs. There are functions $P = P(T', X)$ and $V = V(P, T')$ in the case of $MP_{OY}Y$ FSMs. There are functions $P = P(\tau, X)$ and $V = V(\tau, P)$ in the case of $MP_{EY}Y$ FSMs.

It is possible to generate all other models based on the RLC. We leave this task to our reader. There are characteristics of possible models of FSMs shown in Table 4.1. As follows from Table 4.1, there are 18 different models of Moore FSM based on the relation $Y \to A$.

Table 4.1 Characteristics of Moore FSM with $Y \to A$

№	FSM	BRLC	BI	BIMF	BMOA	BMO	BSCT
1	P_Y	–	$V(T, X)$	$\Phi(V, Y)$	–	$Y(T)$	–
2	P_{OY}	–	$V(T', X)$	$\Phi(V, Y)$	–	$Y(T)$	–
3	P_{EY}	–	$V(\tau, X)$	$\Phi(V, Y)$	–	$Y(T)$	$\tau(T)$
4	P_YY	–	$V(T, X)$	$\Phi(V, Y)$	$Z(T)$	$Y(Z)$	–
5	$P_{OY}Y$	–	$V(T', X)$	$\Phi(V, Z)$	$Z(T)$	$Y(Z)$	–
6	$P_{EY}Y$	–	$V(\tau, X)$	$\Phi(V, Z)$	$Z(T)$	$Y(Z)$	$\tau(T)$
7	P_YD	–	$V(T, X)$	$\Phi(V, Z)$	$Z(T)$	$Y(Z)$	–
8	$P_{OY}D$	–	$V(T', X)$	$\Phi(V, Z)$	$Z(T)$	$Y(Z)$	–
9	$P_{EY}D$	–	$V(\tau, X)$	$\Phi(V, Z)$	$Z(T)$	$Y(Z)$	$\tau(T)$
10	MP_Y	$P(T, X)$	$V(T, P)$	$\Phi(V, Y)$	–	$Y(T)$	–
11	MP_{OY}	$P(T, X)$	$V(T', P)$	$\Phi(V, Y)$	–	$Y(T)$	–
12	MP_{EY}	$P(\tau, X)$	$V(\tau, P)$	$\Phi(V, Y)$	–	$Y(T)$	$\tau(T)$
13	MP_YY	$P(T, X)$	$V(T, P)$	$\Phi(V, Z)$	$Z(T)$	$Y(Z)$	–
14	$MP_{OY}Y$	$P(T, X)$	$V(T', P)$	$\Phi(V, Z)$	$Z(T)$	$Y(Z)$	–
15	$MP_{EY}Y$	$P(\tau, X)$	$V(\tau, P)$	$\Phi(V, Z)$	$Z(T)$	$Y(Z)$	$\tau(T)$
16	MP_YD	$P(T, X)$	$V(T, P)$	$\Phi(V, Z)$	$Z(T)$	$Y(Z)$	–
17	$MP_{OY}D$	$P(T, X)$	$V(T', P)$	$\Phi(V, Z)$	$Z(T)$	$Y(Z)$	–
18	$MP_{EOY}D$	$P(\tau, X)$	$V(\tau, P)$	$\Phi(V, Z)$	$Z(T)$	$Y(Z)$	$\tau(T)$

Table 4.2 Characteristics of Moore FSM with $A \to Y$

№	FSM	BRLC	BIMF	BSCT	BMOA
1	$P_A Y$	–	$\Phi(T, X)$	$Z(T)$	$Y(Z)$
2	$P_{OA} Y$	–	$\Phi(T, X)$	$Z(T)$	$Y(Z)$
3	$P_{EA} Y$	–	$\Phi(\tau, X)$	$Z(T); \tau(T)$	$Y(Z)$
4	$P_A D$	–	$\Phi(T, X)$	$Z(T)$	$Y(Z)$
5	$P_{OA} D$	–	$\Phi(T, X)$	$Z(T)$	$Y(Z)$
6	$P_{EA} D$	–	$\Phi(\tau, X)$	$Z(T); \tau(T)$	$Y(Z)$
7	MP_{AY}	$P(T, X)$	$\Phi(T, P)$	$Z(T)$	$Y(Z)$
8	$MP_{OA} Y$	$P(T, X)$	$\Phi(T, P)$	$Z(T)$	$Y(Z)$
9	$MP_{EA} Y$	$P(\tau, X)$	$\Phi(\tau, P)$	$Z(T); \tau(T)$	$Y(Z)$
10	MP_{AD}	$P(T, X)$	$\Phi(T, P)$	$Z(T)$	$Y(Z)$
11	$MP_{OA} D$	$P(T, X)$	$\Phi(T, P)$	$Z(T)$	$Y(Z)$
12	$MP_{EA} D$	$P(\tau, X)$	$\Phi(\tau, P)$	$Z(T); \tau(T)$	$Y(Z)$

It is possible to create the table with different structures of Moore FSMs based on the relation $A \to Y$. It is Table 4.2. As follows from Table 4.2, there are 12 different models of Moore FSM based on the transformation of states codes into collections of microoperations. In both tables, we used the abbreviation BMOA to determine either BMOY or BMOD. As follows from Tables 4.1 and 4.2, there are 30 different models of Moore FSMs with OCT. Now, let us discuss how to implement these models using FPGAs.

4.2 Synthesis of FPGA-Based Moore FSMs with Transformations of States

Let us start from $P_A Y$ Moore FSM (Fig. 4.1). It includes the following blocks: BIMF, BSCT and BMOY. Using (3.20), the number $n_{SD} = 8$ can be found. These possible structures are shown in Table 4.3.

Let us discuss the structural diagram of $P_{A6} Y$ Moore FSM Fig. 4.12. In this FSM, the EMBer1 implements the system $\Phi = \Phi(T, X)$, the EMBer2 the system $Z =$

Table 4.3 Structural diagrams of $P_{Ai} Y$ Moore FSMs

Block	$P_{A1} Y$	$P_{A2} Y$	$P_{A3} Y$	$P_{A4} Y$	$P_{A5} Y$	$P_{A6} Y$	$P_{A7} Y$	$P_{A8} Y$
BIMF	L_1	L_1	L_1	B_1	L_1	B_1	B_1	B_1
BSCT	L_2	L_2	B_1	L_1	B_2	B_2	L_1	B_2
BMOY	L_3	B_1	L_2	L_2	B_3	L_3	B_2	B_3

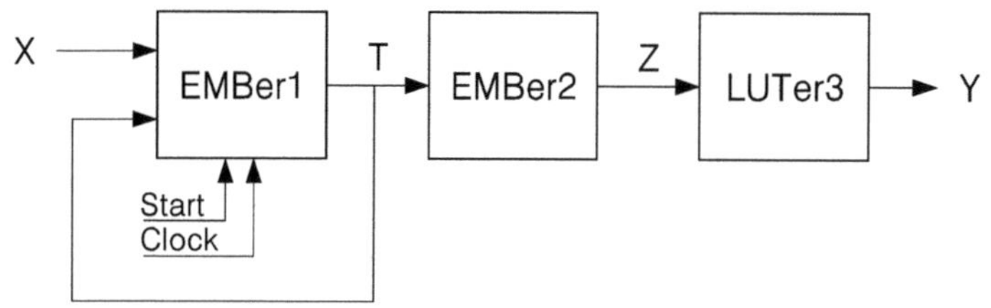

Fig. 4.12 Structural diagram of $P_{A6}Y$ Moore FSM

$Z(T)$, and LUTer3 the system $Y = Y(Z)$. We used synchronized EMBs in EMBer1, so there is no register in the circuit shown in Fig. 4.12.

There are the following steps in the design method for $P_{A6}Y$ Moore FSM:

1. Marking the initial GSA Γ and creating the set A.
2. Executing the state assignment.
3. Constructing the ST of P Moore FSM.
4. Executing the encoding of CMOs $Y_t \subseteq Y$.
5. Constructing the table of EMBer1.
6. Constructing the table of EMBer2.
7. Constructing the table of LUTer3.
8. Implementing the FSM logic circuit.

Let us discuss an example of design for Moore FSM $P_{A6}Y(\Gamma_5)$, where the GSA Γ_5 is shown in Fig. 4.13. The following sets can be found from this GSA: $X = \{x_1, \ldots, x_4\}$, $Y = \{y_1, \ldots, y_7\}$, $A = \{a_1, \ldots, a_9\}$. We use rules [1] for marking the states. So, there is M=9. It gives $R_1 = 4$ and two sets: $T = \{T_1, \ldots, T_4\}$ and $\Phi = \{D_1, \ldots, D_4\}$. Let us encode the states in the trivial way: $K(a_1) = 0000$, $K(a_2) = 0001, \ldots, K(a_9) = 1000$. Now, it is possible to execute the point 3 of the method (Table 4.4). There is the column $Y(a_m)$ showing the CMO generated in the state $a_m \in A$.

There are the following CMOs in the column $Y(a_m)$ of Table 4.4: $Y_1 = \emptyset$, $Y_2 = \{y_1, y_2\}$, $Y_3 = \{y_3\}$, $Y_4 = \{y_2, y_4\}$, $Y_5 = \{y_3, y_5\}$, $Y_6 = \{y_6, y_7\}$, $Y_7 = \{y_3, y_6\}$. So, there is $T_0 = 7$. It gives the value $R_Z = 3$. Let us encode the collections $Y_t \subseteq Y$ in the following way (Fig. 4.14).

This Karnaugh map can be used for deriving the equations of the system (1.40). We discuss this step a bit later.

There are the following columns in the table of EMBer1: $K(a_m)$, X (it is an address of a cell), T (it is a content of the cell), q (it is a number of a cell). The number of rows is determined by the following formula:

$$Q_1(P_{A6}Y) = 2^{R_1+L}. \tag{4.11}$$

It is necessary $H(a_m)$ cells to represent the transitions from a state $a_m \in A$. This number is determined by (3.18).

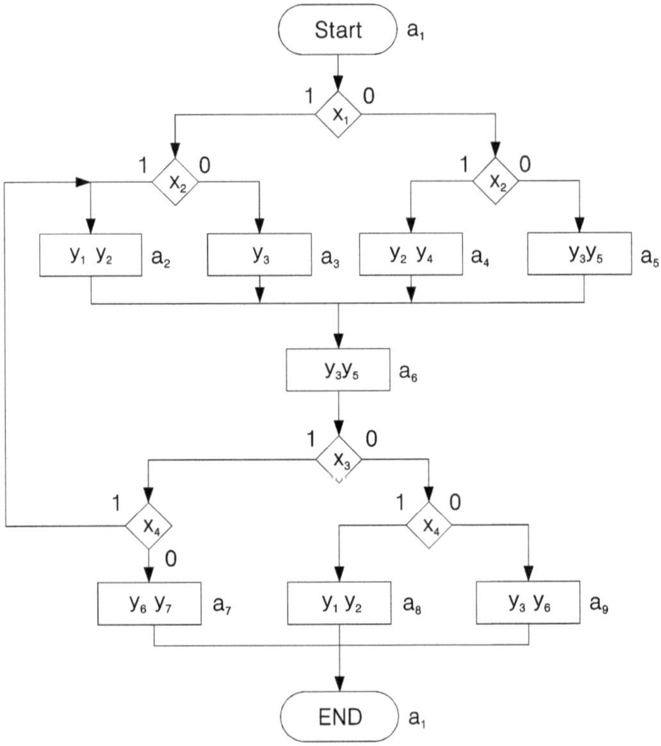

Fig. 4.13 Initial GSA Γ_5

For example, there are the following characteristics of EMBer1 for the discussed example: $Q_1(P_{A6}Y) = 256$ and $H(a_m) = 16$. There is a part of the table of EMBer1 represented by Table 4.5.

Table 4.5 shows transitions from the state $a_1 \in A$. It is added the column h to show the connection between the Tables 4.4 and 4.5.

There are the following columns in the table of EMBer2: $K(a_m)$ (it is an address of a cell), Z (it is a content of the cell), m. It is constructed on the base of the column $Y(a_m)$ of the ST. In the discussed case, it is the Table 4.6.

To fill the column Z, we use the codes $C(Y_t)$ from Fig. 4.14.

There are the following columns in the table of LUTer3: Z (it is an address of a cell), Y (it is a content of the cell), t. In the discussed case, it is Table 4.7.

This table is constructed on the base of the column $Y(a_m)$ from Table 4.4 and the codes from Table 4.14. It corresponds to N circuits for implementing the functions $Y(Z)$. These functions can be derived from Table 4.14:

Table 4.4 Structure table of Moore FSM $P_{A6}Y(\Gamma_5)$

a_m	$K(a_m)$	a_s	$K(a_s)$	X_h	Φ_h	$Y(a_m)$	h
a_1	0000	a_2	0001	x_1x_2	D_4	$\emptyset$	1
		a_3	0010	$x_1\bar{x}_2$	D_3		2
		a_4	0011	$\bar{x}_1x_2$	D_3D_4		3
		a_5	0100	$\bar{x}_1\bar{x}_2$	D_2		4
a_2	0001	a_6	0101	1	D_2D_4	y_1y_2	5
a_3	0010	a_6	0101	1	D_2D_4	y_3	6
a_4	0011	a_6	0101	1	D_2D_4	y_2y_4	7
a_5	0100	a_6	0101	1	D_2D_4	y_3y_5	8
a_6	0101	a_2	0001	x_3x_4	D_4	y_3y_5	9
		a_7	0110	$x_3\bar{x}_4$	D_2D_3		10
		a_8	0111	$\bar{x}_3x_4$	$D_2D_3D_4$		11
		a_9	1000	$\bar{x}_3\bar{x}_4$	D_1		12
a_7	0110	a_1	0000	1	–	y_6y_7	13
a_8	0111	a_1	0000	1	–	y_1y_2	14
a_9	1000	a_1	0000	1	–	y_3y_6	15

Fig. 4.14 Codes of collections of microoperations for Moore FSM $P_{A6}Y(\Gamma_5)$

T_3 \ T_1T_2	00	01	11	10
0	Y_2	Y_3	Y_5	Y_1
1	Y_4	*	Y_7	Y_6

$$\begin{aligned}
y_1 &= Y_2 = \bar{z}_1\bar{z}_2\bar{z}_3;\\
y_2 &= Y_2 \vee Y_4 = \bar{z}_2\bar{z}_3;\\
y_3 &= Y_3 \vee Y_5 \vee Y_7 = z_3;\\
y_4 &= Y_4 = z_1\bar{z}_2;\\
y_5 &= Y_5 = \bar{z}_1z_2z_3;\\
y_6 &= Y_7 = z_1z_3;\\
y_7 &= Y_6 = z_1z_2\bar{z}_3.
\end{aligned} \tag{4.12}$$

Now let us discuss $P_{0Ai}Y$ Moore FSMs ($i = \overline{1,8}$). These models can be represented by the table similar to Table 4.3. It is necessary to add the subscript "0" in the names of models. Let us discuss an example of design for $P_{0A6}Y(\Gamma_5)$ Moore FSM. There are the following steps in the design method for $P_{0A6}Y$ Moore FSMs:

1. Marking the initial GSA Γ and creating the set A.
2. Constructing the partition $\Pi_A = \{B_1, \ldots, B_I\}$.
3. Executing the optimal state encoding.
4. Constructing the transformed table of P_0 Moore FSM.
5. Executing the encoding of CMOs $Y_t \subseteq Y$.
6. Constructing the table of EMBer1.

Table 4.5 Part of table of EMBer1 for Moore FSM $P_{A6}Y(\Gamma_5)$

$K(a_m)$	X	T	q	h
$T_1T_2T_3T_4$	$x_1x_2x_3x_4$	$T_1T_2T_3T_4$		
0000	0000	0100	1	4
0000	0001	0100	2	4
0000	0010	0100	3	4
0000	0011	0100	4	4
0000	0100	0011	5	3
0000	0101	0011	6	3
0000	0110	0011	7	3
0000	0111	0011	8	3
0000	1000	0010	9	2
0000	1001	0010	10	2
0000	1010	0010	11	2
0000	1011	0010	12	2
0000	1100	0001	13	1
0000	1101	0001	14	1
0000	1110	0001	15	1
0000	1111	0001	16	1

Table 4.6 Table of EMBer2 for Moore FSM $P_{A6}Y(\Gamma_5)$

$K(a_m)$	Z	m	$K(a_m)$	Z	m
$T_1T_2T_3T_4$	$z_1z_2z_3$		$T_1T_2T_3T_4$	$z_1z_2z_3$	
0000	010	1	1000	111	9
0001	000	2	1001	000	10
0010	001	3	1010	000	11
0011	100	4	1011	000	12
0100	011	5	1100	000	13
0101	011	6	1101	000	14
0110	110	7	1110	000	15
0111	000	8	1111	000	16

7. Constructing the table of EMBer2.
8. Constructing the table of LUTer3.
9. Implementing the FSM logic circuit.

There is the following set $\Pi_A = \{B_1, \ldots, B_4\}$ in the discussed case with $B_1 = \{a_1\}$, $B_2 = \{a_2, \ldots, a_5\}$, $B_3 = \{a_6\}$, $B_4 = \{a_7, a_8, a_9\}$. The codes of states $a_m \in B_4$ can be treated as "don't care", because these states are connected only with the state a_1 (Fig. 4.13). Let us encode the states in the way shown in Fig. 4.15.

Table 4.7 Table of LUTer3 for Moore FSM $P_{A6}Y(\Gamma_5)$

Z	Y	t	Z	Y	t
$z_1z_2z_3$	$y_1y_2y_3y_4y_5y_6y_7$		$z_1z_2z_3$	$y_1y_2y_3y_4y_5y_6y_7$	
000	1100000	2	100	0101000	4
001	0010000	3	101	0000000	*
010	0000000	1	110	0000011	6
011	0010100	5	111	0101001	7

Fig. 4.15 State codes for Moore FSM $P_{0A6}Y(\Gamma_5)$

T_1T_2 \ T_3T_4	00	01	11	10
00	a_1	a_2	*	a_6
01	*	a_3	*	a_7
11	*	a_4	*	a_8
10	*	a_5	*	a_9

There are the following class codes derived from the Karnaugh map (Fig. 4.15): $K(B_1) = **00$, $K(B_2) = ***1$, $K(B_3) = **1*$. It gives the set $T' = \{T_3, T_4\}$.

They use the system of generalised formulae of transitions (GFT) [9] to construct the transformed ST of P_0 Moore FSM. It is the following system in the discussed case:

$$\begin{aligned} &B_1 \rightarrow x_1x_2a_2 \vee x_1\bar{x}_2a_3 \vee \bar{x}_1x_2a_4 \vee \bar{x}_1\bar{x}_2a_5;\\ &B_2 \rightarrow a_6;\\ &B_3 \rightarrow x_3x_4a_2 \vee x_3\bar{x}_4a_7 \vee \bar{x}_3x_4a_8 \vee \bar{x}_3\bar{x}_4a_9. \end{aligned} \tag{4.13}$$

There are the following columns in the transformed ST: $B_i, K(B_i), a_s, K(a_s), X_h, \Phi_h, h$. It is constructed using the system of GFT. In the discussed case, it is Table 4.8 having $H(P_{0A6}Y(\Gamma_5)) = 9$ rows. Let us point out that this number is the

Table 4.8 Transformed ST of Moore FSM $P_{0A6}Y(\Gamma_5)$

B_i	$K(B_i)$	a_s	$K(a_s)$	X_h	Φ_h	h
B_1	**00	a_2	0001	x_1x_2	D_4	1
		a_3	0101	$x_1\bar{x}_2$	D_2D_4	2
		a_4	1101	$\bar{x}_1x_2$	$D_1D_3D_4$	3
		a_5	1001	$\bar{x}_1\bar{x}_2$	D_1D_4	4
B_2	***1	a_6	0010	1	D_3	5
B_3	**1*	a_2	0001	x_3x_4	D_4	6
		a_7	0110	$x_3\bar{x}_4$	D_2D_3	7
		a_8	1110	$\bar{x}_3x_4$	$D_1D_2D_3$	8
		a_9	1010	$\bar{x}_3\bar{x}_4$	D_1D_3	9

same as $H_0(\Gamma_5)$. The number of transitions are determined by the number of terms in the system (4.13). Comparison of Tables 4.4 and 4.8 shows that there is 60 % of compression for the transformed ST (9 rows instead of 15).

This table is used for deriving the system (4.2). Let us use the symbol R_T for the following relation:

$$R_T = |T'|. \tag{4.14}$$

Let us point out that there is the following relation:

$$R_T \geq R_0. \tag{4.15}$$

In the discussed case, there is $R_0 = 2$ and, therefore, $R_T = R_0$. It is the best possible outcome of the optimal state assignment.

After minimizing, the following system can be derived from Tabel 4.8:

$$\begin{aligned} D_1 &= \bar{T_3}\bar{T_4}\bar{x_1} \vee T_3\bar{x_3}; \\ D_2 &= \bar{T_3}\bar{T_4}x_1\bar{x_2} \vee T_3x_3\bar{x_4} \vee T_3\bar{x_3}x_4; \\ D_3 &= \bar{T_3}\bar{T_4}\bar{x_1}x_2 \vee T_4 \vee T_3x_3\bar{x_4} \vee T_3\bar{x_3}; \\ D_4 &= \bar{T_3}\bar{T_4} \vee T_3x_3x_4. \end{aligned} \tag{4.16}$$

This system is used in the case of using LUTer for implementing the circuit of BIMf.

The step 5 is already executed in the step 4 of the previous example. So, let us use the codes from Fig. 4.14.

There are the following columns in the table of EMBer1: $K(B_i)$, X (it is an address of a cell), T (it is a content of the cell), q, h (to show the correspondence between the transformed ST and the table of EMBer1). The number of required cells is determined as:

$$Q_1(P_{0A6}Y) = 2^{R_T+L}. \tag{4.17}$$

The expression (3.18) determines the number of cells used for representing the transitions for the class $B_i \in \Pi_A$.

There is $Q_1(P_{0A6}Y) = 64$ and $H(a_m) = 16$. There are the transitions for the class B_1 represented by Table 4.9.

The tables of EMBer2 and LUTer3 are constructed in the same way as their counterparts for $P_{A6}Y$ Moore FSMs. So, these steps are omitted. Let us point out that it should be used the codes from Fig. 4.15 in the table of EMBer3. The tables of LUTer3 are the same for $P_{A6}Y(\Gamma_5)$ and $P_{0A6}Y(\Gamma_5)$.

Let us discuss the FPGA-based implementation of $P_{EA}Y$ Moore FSMs (Fig. 4.3a). Obviously, there are 8 different models of this FSM. They corresponds to the models from Table 4.3. So, there are the models $P_{EAi}Y$ where $i = \overline{1,8}$.

Let us discuss an example of design for Moore FSM $P_{EA6}Y(\Gamma_5)$. Its design method is similar to the one for $P_{0A6}Y$ FSM. There are the following differences: (1) it can be the arbitrary state assignment in the point 3; (2) it should be executed the class assignment before constructing the transformed ST.

Table 4.9 Part of table of EMBer1 for Moore FSM $P_{0A6}Y(\Gamma_5)$

$K(B_i)$	X	T	q	h
T_3T_4	$x_1x_2x_3x_4$	$T_1T_2T_3T_4$		
00	0000	1001	1	4
00	0001	1001	2	4
00	0010	1001	3	4
00	0011	1001	4	4
00	0100	1011	5	3
00	0101	1011	6	3
00	0110	1011	7	3
00	0111	1011	8	3
00	1000	0101	9	2
00	1001	0101	10	2
00	1010	0101	11	2
00	1011	0101	12	2
00	1100	0001	13	1
00	1101	0001	14	1
00	1110	0001	15	1
00	1111	0001	16	1

So, let us leave the state codes from Table 4.4 in the case of $P_{EA6}Y(\Gamma_5)$. There are the same classes $B_i \in \Pi_A$ for $P_{EA6}Y(\Gamma_5)$: $B_1 = \{a_1\}, B_2 = \{a_2, \ldots, a_5\}, B_3 = \{a_6\}$ and $B_4 = \{a_7, a_8, a_9\}$. There are four classes of PES(I=4). It is necessary R_A of variables $\tau_r \in \tau$ for class encoding:

$$R_A = \lceil log_2 I \rceil. \tag{4.18}$$

In the discussed case, there is $R_A = 2$. It determines the set $\tau = \{\tau_1, \tau_2\}$.

Let us encode the classes $B_i \in \Pi_A$ in the following manner: $K(B_1) = 00$, $K(B_2) = 01$, $K(B_3) = 10$ and $K(B_4) = 11$. As in the previous case, the code $K(B_4)$ is treated as "don't care".

They use a system of GFT for constructing the transformed ST of $P_{EA6}Y$ Moore FSM. It is the system (4.13) in the discussed case. There are the same columns in transformed STs for $P_{0A6}Y$ and $P_{EA6}Y$ FSMs. But there are the class codes determined by variables $\tau_r \in \tau$ in the column $K(B_i)$ for $P_{EA6}Y$ Moore FSM. It is Table 4.10 in the discussed case. We use the input assignment 11 to simplify the codes $K(B_2)$ and $K(B_3)$. Now, they are equal to *1 and 1* respectively. Of course, it is important only if LUTs are used for implementing the system (4.5).

They use the transformed ST to get the system (4.5). It is the following system in the discussed case:

Table 4.10 Transformed ST of Moore FSM $P_{EA6}Y(\Gamma_5)$

B_i	$K(B_i)$	a_s	$K(a_s)$	X_h	Φ_h	h
B_1	00	a_2	0001	x_1x_2	D_4	1
		a_3	0010	$x_1\bar{x}_2$	D_3	2
		a_4	0011	$\bar{x}_1x_2$	D_3D_4	3
		a_5	0100	$\bar{x}_1\bar{x}_2$	D_2	4
B_2	$*1$	a_6	0101	1	D_2D_4	5
B_3	$1*$	a_2	0001	x_3x_4	D_4	6
		a_7	0010	$x_3\bar{x}_4$	D_2D_3	7
		a_8	0111	$\bar{x}_3x_4$	$D_2D_3D_4$	8
		a_9	1000	$\bar{x}_3\bar{x}_4$	D_1	9

$$
\begin{aligned}
D_1 &= \tau_1\bar{x}_3\bar{x}_4;\\
D_2 &= \bar{\tau}_1\bar{\tau}_2\bar{x}_1\bar{x}_2 \vee \tau_2 \vee \tau_1x_3\bar{x}_4 \vee \tau_1\bar{x}_3x_4;\\
D_3 &= \bar{\tau}_1\bar{\tau}_2x_1\bar{x}_2 \vee \bar{\tau}_1\bar{\tau}_2\bar{x}_1x_2 \vee \tau_1x_3\bar{x}_4 \vee \tau_1\bar{x}_3x_4;\\
D_4 &= \bar{\tau}_1\bar{\tau}_2x_2 \vee \tau_2 \vee \tau_1x_4.
\end{aligned}
\tag{4.19}
$$

There are the same columns in the tables of EMBer1 for $P_{OA6}Y$ and $P_{EA6}Y$ Moore FSMs. There are $Q_1(P_{EA6}Y)$ cells in the EMBer1:

$$
Q_1(P_{EA6}Y) = 2^{R_A+L}. \tag{4.20}
$$

As in all cases, they use the formula (3.18) to find the number of cells representing the transitions from the state $a_m \in A$. There are $Q_1(P_{EA6}Y(\Gamma_5)) = 64$ and $H(a_m) = 16$ in the discussed case.

We hope that there is no need in constructing all tables in the discussed case. Moreover, the tables of EMBer2 and LUTer3 are the same for both $P_{A6}Y(\Gamma_5)$ and $P_{EA6}Y(\Gamma_5)$ FSMs. We discuss the models based on the RLC in Sect. 4.4.

4.3 Synthesis of FPGA-Based Moore FSMs with Transformation of CMOs

Let us discuss the design methods based on using the encoding of the fields of compatible microoperations (FCO). There are many approaches devoted to solution of this problem [2, 10, 12–17]. There are the following sets Y^k in the case of GSA $\Gamma_5 : Y^1 = \{y_1, y_3, y_4\}$, $Y^2 = \{y_2, y_5, y_6\}$ and $Y^3 = \{y_7\}$. So, there are $R_1 = R_2 = 2$ and $R_3 = 1$ (See (1.35)). It determines the set $Z = \{z_1, \ldots, z_5\}$. The variables z_1, z_2 encode the field Y^1, the variables z_3, z_4 the field Y^2 and the variable z_5 the field Y^3 (Table 4.11).

Table 4.11 Codes of FCO for GSA Γ_5

Y^1	$K(Y_n)^1$	Y^2	$K(Y_n)^2$	Y^3	$K(y_n)^3$
	$z_1 z_2$		$z_3 z_4$		z_5
$\emptyset$	00	$\emptyset$	00	$\emptyset$	0
y_1	01	y_2	01	y_7	1
y_3	10	y_5	10		
y_5	11	y_6	11		

Let us start from $P_Y D$ Moore FSM (Fig. 4.7). The LUTs are used for implementing the circuit of BMO. So, there are 8 different models of FPGA-based Moore FSMs witch transformation of FCOs (Table 4.12).

Design methods are practically the same for all these models. All methods include the step of encoding of FCO. Let us discuss the design method for $P_{Y6}D$ Moore FSM. It includes the following steps:

1. Marking the initial GSA Γ and creating the set A.
2. Executing the state assignment.
3. Constructing the ST of P Moore FSM.
4. Executing the encoding of FCO.
5. Constructing the set of identifiers $I = \{I_1, \ldots, I_K\}$.
6. Representing the states $a_m \in A$ by the pairs $\langle I_k, Y_t \rangle$.
7. Executing the encoding of identifiers $I_k \in I$.
8. Constructing the table of EMBer1.
9. Constructing the table of EMBer2.
10. Constructing the table of LUTer3.
11. Constructing the system of equations for BMO.
12. Implementing the FSM logic circuit.

There is the structural diagram of $P_{Y6}D$ Moore FSM shown in Fig. 4.16. The LUTer represents the BMO.

Let us discuss an example for design of Moore FSM $P_{Y6}D(\Gamma_5)$. There are already executed the steps 1–4 for the given example. There is the structure table of Moore FSM $P(\Gamma_5)$ represented by Table 4.4. There are the codes of FCO shown in in Table 4.11.

Table 4.12 Structural diagram of $P_{Yi}D$ FSMs

Block	$P_{Y1}D$	$P_{Y2}D$	$P_{Y3}D$	$P_{Y4}D$	$P_{Y5}D$	$P_{Y6}D$	$P_{Y7}D$	$P_{Y8}D$
BIMF	L_1	L_1	L_1	B_1	L_1	B_1	B_1	B_1
BMOD	L_2	L_2	B_1	L_1	B_2	B_2	L_1	B_2
BI	L_3	B_1	L_2	L_2	B_3	L_3	B_2	B_3

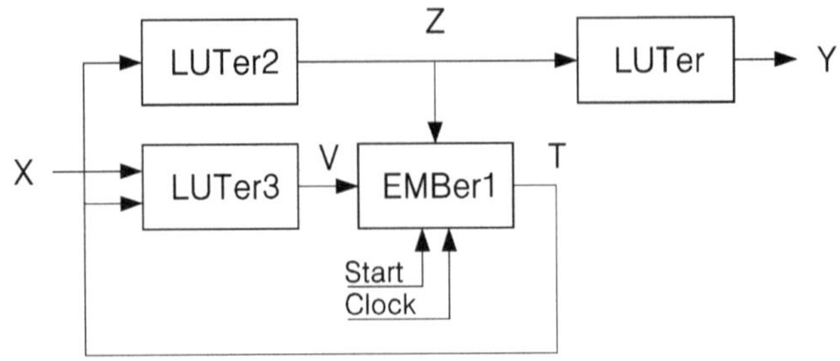

Fig. 4.16 Structural diagram of $P_{Y6}D$ Moore FSM

Let us find the sets $A(Y_t)$ in the discussed case. It can be done on the base of Table 4.4. There are the following sets: $A(y_1) = \{a_1\}, A(y_2) = \{a_2, a_8\}, A(y_3) = \{a_3\}, A(y_4) = \{a_4\}, A(y_5) = \{a_5, a_6\}, A(y_6) = \{a_7\}, A(y_7) = \{a_9\}$. It gives $K = 2$ and $I = \{I_1, I_2\}$.

Let us represent the states $a_m \in A$ by the pairs $\langle I_k, Y_t \rangle$. Using Table 4.4, the following formulae can be obtained:

$$\begin{aligned} a_1 &= \langle \emptyset, y_1 \rangle; \\ a_2 &= \langle I_1, y_2 \rangle; \\ a_3 &= \langle \emptyset, y_3 \rangle; \\ a_4 &= \langle \emptyset, y_4 \rangle; \\ a_5 &= \langle I_1, y_5 \rangle; \\ a_6 &= \langle I_2, y_5 \rangle; \\ a_7 &= \langle \emptyset, y_6 \rangle; \\ a_8 &= \langle I_2, y_2 \rangle; \\ a_9 &= \langle \emptyset, y_7 \rangle; \end{aligned} \tag{4.21}$$

Using (3.7), we can get $R_V = 1$. It gives the set $V = \{v_1\}$. Let us encode the identifiers in the following way: $K(I_1) = 0, K(I_2) = 1$.

The table of EMBer1 (table of BIMF) is constructed from the ST and the system of pairs $\langle I_k, Y_t \rangle$. There are the following columns in the table of EMBer1: Z, V, T, q, m. The first two columns are determined by the column T. The column m is added to show the correspondence between the rows of the table and states $a_m \in A$. This table includes $Q_1(P_{Y6}D)$ lines:

$$Q_1(P_{Y6}D) = 2^{R_D + R_V}. \tag{4.22}$$

In the discussed case, there is $Q_1(PY6D(\Gamma_5)) = 64$.

Table 4.13 contains only 9 rows corresponding to the states $a_m \in A$. There are the decimal equivalents of corresponding addresses in the column q.

Table 4.13 Part of table of EMBer1 for Moore FSM $P_{Y6}D(\Gamma_5)$

Z	V	T	q	m
$z_1z_2z_3z_4z_5$	v_1	$T_1T_2T_3T_4$		
00000	*	0000	1	1
01010	0	0001	21	2
10000	*	0010	33	3
11010	*	0011	53	4
10100	0	0100	41	5
10100	1	0101	42	6
00111	*	0110	15	7
01010	1	0111	22	8
10110	*	1000	45	9

To find the code $C(Y_t)$, the following formula is used:

$$C(Y_t) = K(y_n)_t^1 * K(y_n)_t^2 * \cdots * K(y_n)_t^K. \tag{4.23}$$

The symbol $K(y_n)_t^k$ stands for the code of MO $y_n \in Y_t$ used in the k-th field of $C(Y_t)$. For example, let us analyse the CMO $Y_2 = \{y_1, y_2\}$.Using Table 4.11, it can be found that $C(Y_2) = 01010$.

It is necessary $H(Y_t)$ rows to represent the pair $\langle I_k, Y_t\rangle$ where

$$H(Y_t) = 2^{R_V}. \tag{4.24}$$

So, each symbol "*" in Table 4.13 determines 2 cells of an EMB. We show the decimal equivalent of the first cell's address in the column q of Table 4.13. Let us point out that these equivalents are incremented starting from 1.

There are the following columns in the table of EMBer2 (table of BMOD): T, Z, m. It is Table 4.14 in the discussed case.

Table 4.14 Table of EMBer2 for Moore FSM $P_{Y6}D(\Gamma_5)$

T	Z	m	T	Z	m
$T_1T_2T_3T_4$	$z_1z_2z_3z_4z_5$		$T_1T_2T_3T_4$	$z_1z_2z_3z_4z_5$	
0000	00000	1	0101	10100	6
0001	01010	2	0110	00111	7
0010	10000	3	0111	01010	8
0011	11010	4	1000	10110	9
0100	10100	5	1001	00000	10

We do not show the rows 11–16 because there are no state codes starting from 1001. These rows contain only zeros in the column Z.

There are the following columns in the table of LUTer3 (table of BI): T, X, V, q. It is necessary to find the Eq. (3.2) to construct this table. Let us form the transformed ST of Moore FSM $P_{A6}D(\Gamma_5)$ (Table 4.15).

This table is constructed on the base of Table 4.4 and system (4.20). For example, there is the state $a_s = a_2$ in the first row of Table 4.4. It is determined as $\langle I_1, Y_2 \rangle$. There is 0 in the columns V_h of the first row.

There is $a_s = a_6$ in the row 5 of Table 4.4. There is $a_6 = \langle I_2, Y_5 \rangle$. So, there is 1 in the column V_h for the row 5.If there is the pair $\langle \emptyset, Y_t \rangle$, then we sign it as "$*$" in the column V_h.

To optimize the function $v_1 = v_1(T, X)$, it is possible to use the "don't care" state codes (1001–1111) and the term $A_m X_h$ where there is $V_h = *$. After minimizing, the following equation can be obtained from Table 4.15:

$$v_1 = T_2 x_3 \vee T_3 \vee T_4. \quad (4.25)$$

So, there are the variables T_2, T_3, T_4 in the column T and the variable x_3 in the column X of the table of LUTer3. It is Table 4.16.

The rows 10, 12, 14, 16 correspond to the term $T_2 x_3$ of (4.25); the rows 5–8 and 13–16 to T_3; the rows 3, 4, 7, 8, 11, 12, 15, 16 to T_4.

Table 4.15 Transformed ST for Moore FSM $P_{Y6}D(\Gamma_5)$

a_m	$K(a_m)$	X_h	V_h	h
a_1	0000	$x_1 x_2$	0	1
		$x_1 \bar{x}_2$	$*$	2
		$\bar{x}_1 x_2$	$*$	3
		$\bar{x}_1 \bar{x}_2$	0	4
a_2	0001	1	1	5
a_3	0010	1	1	6
a_4	0011	1	1	7
a_5	0100	1	1	8
a_6	0101	$x_3 x_4$	0	9
		$x_3 \bar{x}_4$	$*$	10
		$\bar{x}_3 x_4$	1	11
		$\bar{x}_3 \bar{x}_4$	$*$	12
a_7	0110	1	$*$	13
a_8	0111	1	$*$	14
a_9	1000	1	$*$	15

Table 4.16 Table of LUTer3 for Moore FSM $P_{Y6}D(\Gamma_5)$

T	X	v_1	q	T	X	v_1	q
$T_2T_3T_4$	x_3			$T_2T_3T_4$	x_3		
000	0	0	1	100	0	0	9
000	1	0	2	100	1	1	10
001	0	1	3	101	0	1	11
001	1	1	4	101	1	1	12
010	0	1	5	110	0	1	13
010	1	1	6	110	1	1	14
011	0	1	7	111	0	1	15
011	1	1	8	111	1	1	16

To form the table of LUTer (table of BMO), it is necessary to construct the system $Y(Z)$. In the discussed example, it can be found from Table 4.11:

$$\begin{aligned} y_1 &= \bar{z}_1 z_2; \\ y_2 &= \bar{z}_3 z_4; \\ y_3 &= z_1 \bar{z}_2; \\ y_4 &= z_1 z_2; \\ y_5 &= z_3 \bar{z}_4; \\ y_6 &= z_3 z_4; \\ y_7 &= z_5. \end{aligned} \tag{4.26}$$

To implement these equations, it is enough 6 of LUTs having $S = 2$. The corresponding tables of LUTs can be found either from Table 4.11 or from the system (4.26).

Let us discuss the models of $P_{0Y_i}D$ Moore FSMs ($i = \overline{1,8}$). These models are the same as the ones shown in Table 4.12. But they are based on the optimal state assignment [5]. Let us discuss the $P_{0Y4}D$ Moore FSM (Fig. 4.17).

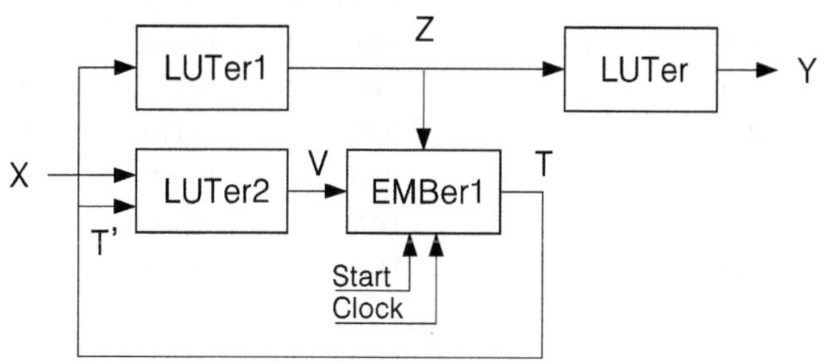

Fig. 4.17 Structural diagram of $P_{0Y4}D$ Moore FSM

In this FSM model, the LUTer1 corresponds to BMOD. It implements the system $Z = Z(T)$. The LUTer2 corresponds to BI and implements the system $V = V(T', X)$. Both EMBer1 and LUTer execute the same functions as their counterparts in $P_{Y6}D$ FSM (Fig. 4.16).

To design the circuit of $P_{OY4}D$ Moore FSM, it is necessary to change the point 2 and add some new steps in the design method discussed for the $P_{Y6}D$ FSM. Now, step 2 is "executing the optimal state assignment". It is necessary to add the following steps:

1a. Constructing the partition $\Pi_A = \{B_1, \ldots, B_I\}$.
7a. Constructing the transformed ST of P_0 Moore FSM.

Let us discuss an example of design for Moore FSM $P_{OY4}D(\Gamma_5)$. There is the GSA Γ_5 shown in Fig. 4.13. The states $a_m \in A$ are encoded in the optimal way shown in the Karnaugh map (Fig. 4.15). Using the system (4.13) the transformed ST of Moore FSM is constructed. There are the codes of FCO for the discussed case shown in Table 4.11.

There are the following columns in the table of LUTer1: T, Z, m. It is similar to Table 4.14, but there are the state codes from Fig. 4.15. In the discussed case, the LUTer1 is represented by Table 4.17.

There is a sign "*" in the column m if there is no state code as the one shown in the corresponding row of the column Z. This table determines $R_D = 5$ of tables for LUTer implementing the functions $z_r \in Z$. These LUTs implement the following functions:

$$\begin{aligned}
z_1 &= T_1\bar{T}_2 \vee T_2\bar{T}_3;\\
z_2 &= T_1T_2 \vee \bar{T}_1\bar{T}_2T_4 \vee \bar{T}_1T_3;\\
z_3 &= T_1\bar{T}_2 \vee \bar{T}_1T_2\bar{T}_4;\\
z_4 &= \bar{T}_1\bar{T}_2 \vee \bar{T}_2T_3 \vee T_1T_2;\\
z_5 &= \bar{T}_1T_2\bar{T}_4.
\end{aligned} \tag{4.27}$$

Table 4.17 Table of LUTer1 for Moore FSM $P_{OY4}D(\Gamma_5)$

T	Z	m	T	Z	m
$T_1T_2T_3T_4$	$z_1z_2z_3z_4z_5$		$T_1T_2T_3T_4$	$z_1z_2z_3z_4z_5$	
0000	00000	1	1000	00000	*
0001	01010	2	1001	10100	5
0010	01010	6	1010	10110	9
0011	00000	*	1011	00000	*
0100	00000	*	1100	00000	*
0101	10000	3	1101	11010	4
0110	01101	7	1110	01010	8
0111	00000	*	1111	00000	*

Analysis of (4.27) shows that only a LUT for z_2 requires 4 inputs. To implement other functions $Z(T)$, it is necessary to use LUTs having $S = 3$.

It is necessary to form the transformed ST of Moore FSM $P_{OY4}D(\Gamma_5)$ to find the function v_1. It is Table 4.18 which is similar to Table 4.15. There is the following equation for the function v_1:

$$v_1 = T_4 \vee T_3\bar{x}_3. \tag{4.28}$$

This equation determines Table 4.19.

Comparison of Tables 4.16 and 4.19 shows that using the optimal state assignment results in the decreasing for the number of inputs in LUTs implementing the system $Z = Z(T', X)$.

Tables of EMBer1 differ only in the columns T for equivalent Moore FSMs $P_{Y_i}D$ and $P_{OY_i}D$. So, there are the codes from Fig. 4.15 in the table of EMBer1 (Table 4.20).

Tables of LUTer are the same for all equivalent $P_{OY_i}D$ Moore FSMs. In the discussed case, the LUTer is represented by the system (4.26).

Table 4.18 Transformed ST for Moore FSM $P_{OY4}D(\Gamma_5)$

B_i	$K(B_i)$	X_h	V_h	h
B_1	**00	x_1x_2	0	1
		$x_1\bar{x}_2$	*	2
		$\bar{x}_1x_2$	*	3
		$\bar{x}_1\bar{x}_2$	0	4
B_2	***1	1	1	5
B_3	**1*	x_3x_4	0	6
		$x_3\bar{x}_4$	*	7
		$\bar{x}_3x_4$	1	8
		$\bar{x}_3\bar{x}_4$	*	9

Table 4.19 Table of LUTer2 for Moore FSM $P_{OY4}D(\Gamma_5)$

T	X	v_1	q	T	X	v_1	q
T_3T_4	x_3			T_3T_4	x_3		
00	0	0	1	10	0	1	5
00	1	0	2	10	1	0	6
01	0	1	3	11	0	1	7
01	1	1	4	11	1	1	8

Table 4.20 Part of table of EMBer1 for Moore FSM $P_{0Y4}D(\Gamma_5)$

Z	V	T	q	m
$z_1z_2z_3z_4z_5$	v_1	$T_1T_2T_3T_4$		
00000	*	0000	1	1
01010	0	0001	21	2
10000	*	0101	33	3
11010	*	1101	53	4
10100	0	1001	41	5
10100	1	0010	42	6
00111	*	0110	15	7
01010	1	1110	22	8
10110	*	1010	45	9

4.4 Replacement of Logical Conditions in Moore FSMs with OCT

The method of RLC is an universal method. It can be used to improve the hardware amount in any model of FSM. Let us discuss, for example, $MP_{EYi}D$ Moore FSMs. They are based on the following model (Fig. 4.18).

There is the same nature in the functions $Z(T)$ and $\tau(T)$. So, it is possible to combine the blocks BMOD and BSCT in a single block BMT. It gives the circuit having 5 blocks and the RG. As in the previous cases, a LUTer is used to implement the circuit of BMO. Therefore, there are 16 possible models of $MP_{EYi}D$ Moore FSMs (Table 4.21).

There are the following steps in design method for a $MP_{EYi}D$ Moore FSM:

1. Constructing the set of states A.
2. Executing the encoding of states $a_m \in A$.
3. Constructing the ST of P Moore FSM.
4. Constructing the partition $\Pi_A = \{B_1, \ldots, B_I\}$.
5. Executing the encoding of classes $B_i \in \Pi_A$.

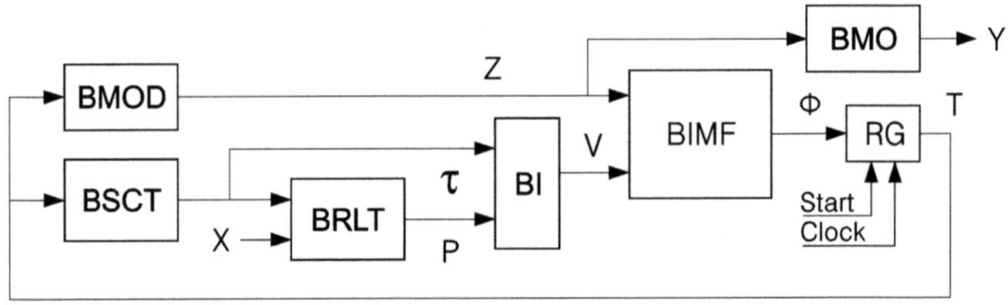

Fig. 4.18 Structural diagram of $MP_{EY}D$ Moore FSM

Table 4.21 Structural diagrams of $MP_{EYi}D$ Moore FSMs

Block	BMT	BRLC	BI	BIMF
1	L_1	L_2	L_3	L_4
2	L_1	L_2	L_3	B_4
3	L_1	L_2	B_3	L_4
4	L_1	L_2	B_3	B_4
5	L_1	B_2	L_3	L_4
6	L_1	B_2	L_3	B_4
7	L_1	B_2	B_3	L_4
8	L_1	B_2	B_3	B_4
9	B_1	L_2	L_3	L_4
10	B_1	L_2	L_3	B_4
11	B_1	L_2	B_3	L_4
12	B_1	L_2	B_3	B_4
13	B_1	B_2	L_3	L_4
14	B_1	B_2	L_3	B_4
15	B_1	B_2	B_3	L_4
16	B_1	B_2	B_3	B_4

6. Executing the encoding of FCO.
7. Executing the replacement of logical conditions.
8. Constructing the transformed ST of MP Moore FSM.
9. Constructing the table of BMT.
10. Constructing the table of BRLC.
11. Constructing the table of BI.
12. Constructing the table of BIMF.
13. Constructing the tables of LUter2 and EMBers.
14. Implementing the FSM logic circuit.

Let us discuss an example of design for $MP_{EY1}D$ Moore FSM on the base of the GSA Γ_5 (Fig. 4.13). Let us use the results obtained from previous examples for the steps 1–6. So, there is the ST of $P(\Gamma_5)$ Moore FSM (Table 4.4), there is the transformed ST of $P_E(\Gamma_5)$ Moore FSM (Table 4.10), the codes of FCO are shown in Table 4.11.

Let the symbol $X(B_i)$ stand for the set of logical conditions determining transitions from states $a_m \in B_i$. There are the following sets in the discussed case: $X(B_1) = \{x_1, x_2\}, X(B_2) = X(B_4) = \emptyset, X(B_3) = \{x_3, x_4\}$. So, there is $G = 2$ and $P = \{p_1, p_2\}$. Let us replace the logical conditions as it is shown in Table 4.22.

It is possible to find the equations of the system $P = P(\tau, X)$ from Table 4.22:

$$\begin{aligned} p_1 &= B_1x_1 \vee B_3x_3; \\ p_2 &= B_1x_2 \vee B_3x_3. \end{aligned} \tag{4.29}$$

Table 4.22 Table of RLC for $MP(\Gamma_5)$

B_i	B_1	B_2	B_3	B_4
p_1	x_1	–	x_3	–
p_2	x_2	–	x_4	–

There are the following class codes: $K(B_1) = 00$, $K(B_2) = 01$, $K(B_3) = 10$ and $K(B_4) = 11$. The codes of B_2 and B_4 are treated as "don't care". So, the following system can be obtained on the base of (4.29):

$$\begin{aligned} p_1 &= \bar{\tau}_1 x_1 \vee \tau_1 x_3; \\ p_2 &= \bar{\tau}_1 x_2 \vee \tau_1 x_4. \end{aligned} \tag{4.30}$$

They use Tables 4.10 and 4.22 to construct the transformed ST of Moore FSM $MP(\Gamma_5)$ (Table 4.23).

This table is used to find functions $\Phi = \Phi(\tau, P)$. There are the following minimized functions derived from Table 4.23:

$$\begin{aligned} D_1 &= \tau_1 \bar{p}_1 \bar{p}_2; \\ D_2 &= \bar{\tau}_1 \bar{\tau}_2 \bar{p}_1 \bar{p}_2 \vee \tau_2 \vee \tau_1 p_1 \bar{p}_2 \vee \tau_1 \bar{p}_1 p_2; \\ D_3 &= \bar{\tau}_1 \bar{\tau}_2 p_1 \bar{p}_2 \vee \bar{\tau}_1 \bar{\tau}_2 \bar{p}_1 p_2 \vee \tau_1 p_1 \bar{p}_2 \vee \tau_1 \bar{p}_1 p_2; \\ D_4 &= \bar{\tau}_1 \bar{\tau}_2 p_2 \vee \tau_2 \vee \tau_1 p_2. \end{aligned} \tag{4.31}$$

There are the following columns in the table of BMT: a_m, $K(a_m)$, B_i, $K(B_i)$, τ_m, Y_t, $C(Y_t)$, Z_m, m. It is Table 4.24 in the discussed example. We use here the codes $C(Y_t)$ from Table 4.13.

The table of BMT is used for deriving the functions $\tau = \tau(T)$ and $Z = Z(T)$. For example, the following minimized functions can derived from Table 4.24:

Table 4.23 Transformed ST of Moore FSM $MP(\Gamma_5)$

B_i	$K(B_i)$	a_s	$K(a_s)$	P_h	Φ_h	h
B_1	00	a_2	0001	$p_1 p_2$	D_4	1
		a_3	0010	$p_1 \bar{p}_2$	D_3	2
		a_4	0011	$\bar{p}_1 p_2$	$D_3 D_4$	3
		a_5	0100	$\bar{p}_1 \bar{p}_2$	D_2	4
B_2	$*1$	a_6	0101	1	$D_2 D_4$	5
B_3	$1*$	a_2	0001	$p_1 p_2$	D_4	6
		a_7	0010	$p_1 \bar{p}_2$	$D_2 D_3$	7
		a_8	0111	$\bar{p}_1 p_2$	$D_2 D_3 D_4$	8
		a_9	1000	$\bar{p}_1 \bar{p}_2$	D_1	9

Table 4.24 Table of BMT for Moore FSM $MP_{EY1}D(\Gamma_5)$

a_m	$K(a_m)$	B_i	$K(B_i)$	τ_m	Y_h	$C(Y_t)$	Z_m	m
a_1	0000	B_1	00	–	y_1	00000	–	1
a_2	0001	B_2	01	τ_2	y_2	01010	z_2z_4	2
a_3	0010	B_2	01	τ_2	y_3	10000	z_1	3
a_4	0011	B_2	01	τ_2	y_4	11010	$z_1z_2z_4$	4
a_5	0100	B_2	01	τ_2	y_5	10100	z_1z_3	5
a_6	0101	B_3	10	τ_1	y_5	10100	z_2z_3	6
a_7	0110	B_4	11	$\tau_1\tau_2$	y_6	00111	$z_3z_4z_5$	7
a_8	0111	B_4	11	$\tau_1\tau_2$	y_2	01010	z_2z_4	8
a_9	1000	B_4	11	$\tau_1\tau_2$	y_7	10110	$z_1z_3z_4$	9

$$
\begin{aligned}
\tau_1 &= T_1 \vee T_2T_4 \vee T_2T_3;\\
\tau_2 &= T_1 \vee T_3 \vee \bar{T}_2T_4;\\
z_1 &= T_1 \vee T_3 \vee \bar{T}_2T_4;\\
z_2 &= T_3T_4 \vee \bar{T}_2T_4;\\
z_3 &= T_1 \vee T_2\bar{T}_3 \vee T_2\bar{T}_4;\\
z_4 &= T_1 \vee \bar{T}_2T_4 \vee T_2T_3;\\
z_5 &= T_2T_3\bar{T}_4.
\end{aligned}
\tag{4.32}
$$

To find the functions $V(\tau, P)$, it is necessary to construct the table of BI. It is similar to Table 4.18 in the way of constructing. There are the following columns in the table of BI for $MP_{EY1}D$ Moore FSM: B_i, $K(B_i)$, P_h, V_h, h. In the discussed case, it is Table 4.25.

There is the following minimized equation derived from Table 4.25:

$$v_1 = \tau_2 \vee \tau_1\bar{p}_1. \tag{4.33}$$

Using the Eq. (4.33), the table of LUTer3 can be constructed (Table 4.26).

Table 4.25 Table of BI for Moore FSM $MP_{EY1}D(\Gamma_5)$

B_i	$K(B_i)$	P_h	V_h	h
B_1	00	p_1p_2	0	1
		$p_1\bar{p}_2$	*	2
		$\bar{p}_1p_2$	*	3
		$\bar{p}_1\bar{p}_2$	0	4
B_2	*1	1	1	5
B_3	1*	p_1p_2	0	6
		$p_1\bar{p}_2$	*	7
		$\bar{p}_1p_2$	1	8
		$\bar{p}_1\bar{p}_2$	*	9

Table 4.26 Table of LUTer3 for Moore FSM $MP_{EY1}D(\Gamma_5)$

τ	P	v_1	q	τ	P	v_1	q
$\tau_1\tau_2$	P_1		$\tau_1\tau_2$	P_1			
00	0	0	1	10	0	0	5
01	1	0	2	10	1	1	6
10	0	1	3	11	0	1	7
11	1	1	4	11	1	1	8

The table of BIMF includes the following columns: Z_h, V_h, Φ_h, h. It is Table 4.27 in the discussed case.

This table is used to derive the system $\Phi(Z, V)$. After minimizing each of equations, it is possible to create the tables of LUTs implementing each of input memory functions $D_r \in \Phi$.

Acting in the same way, it is possible to develop design methods for all Moore FSMs based on both OCT and RLC.Now, let us discuss the ways of improving the circuits for blocks BRLC and BMOD.

It is possible to change the number of functions $P = P(T, X)$ or $P = P(\tau, X)$. To do it, it is necessary to transform an initial GSA Γ [9]. Let us discuss the following example.

Let it be the following subgraph Γ (Fig. 4.19a). It presents the transitions from the state $a_3 \in A$.

Let it be $M_1 = 19$. Let $|X(a_3)| = max(|X(a_1)|, \ldots, |X(a_13)|)$. In this case, there is $G = 3$ and $P = \{p_1, p_2, p_3\}$. The value of G can be diminished by introducing additional states (additional operator vertices) [9]. Let us add the vertex marked by the state a_{14} (Fig. 4.19b). Now, there are $X(a_3) = \{x_1, x_3\}$ and $X(a_{14}) = \{x_2\}$. It leads to $G = 2$. Let us point out that adding the state a_{14} does not change the value of $R_1 = 4$. It is very important because there is no change in the required parameters of either LUTs or EMBs.

Table 4.27 Table of BIMF for Moore FSM $MP_{EY1}D(\Gamma_5)$

Z_h	V_h	Φ_h	h
$\bar{z}_1\bar{z}_2\bar{z}_3\bar{z}_4\bar{z}_5$	*	–	1
$\bar{z}_1 z_2\bar{z}_3 z_4\bar{z}_5$	$\bar{v}_1$	D_4	2
$z_1\bar{z}_2\bar{z}_3\bar{z}_4\bar{z}_5$	*	D_3	3
$z_1 z_2\bar{z}_3 z_4\bar{z}_5$	*	D_3D_4	4
$z_1\bar{z}_2 z_3\bar{z}_4\bar{z}_5$	$\bar{v}_1$	D_2	5
$z_1\bar{z}_2 z_3\bar{z}_4\bar{z}_5$	v_1	D_2D_4	6
$\bar{z}_1\bar{z}_2 z_3 z_4 z_5$	*	D_2D_3	7
$\bar{z}_1 z_2\bar{z}_3 z_4\bar{z}_5$	1	$D_2D_3D_4$	8
$z_1\bar{z}_2 z_3 z_4\bar{z}_5$	*	D_1	9

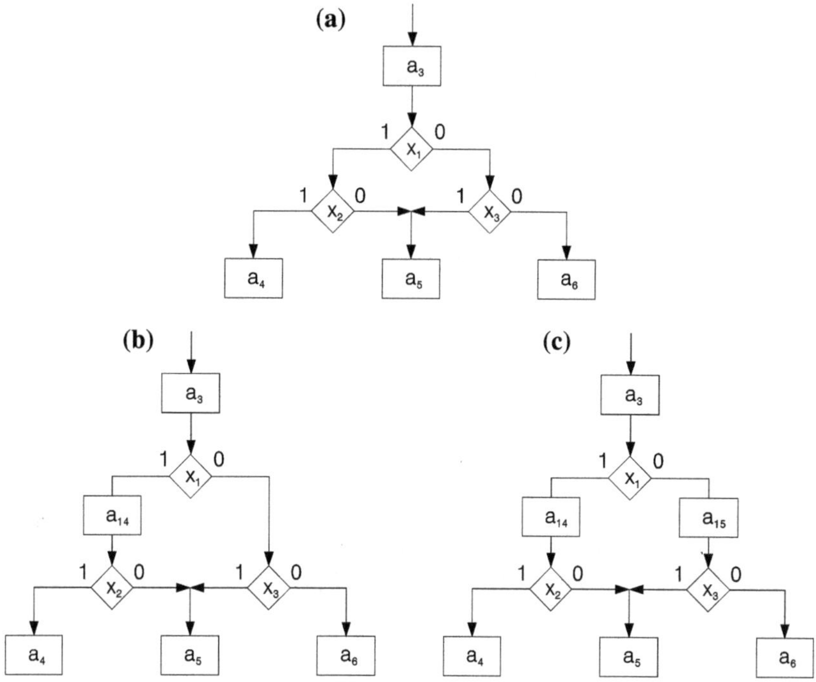

Fig. 4.19 The subgraph of GSA Γ

Let us introduce the state a_{15} (Fig. 4.19c). Now, there is $G = 1$. At the same time, there is no change for the value of R_1. So, the introducing the additional states can decrease the value of G. In turn, it allows using either LUTs or EMBs with smaller values of inputs. Obviously, such a transformation has sense till there is no changing for the value of R_1.

It is possible to change the value of R_D using so called verticalization of initial GSA Γ [1, 6–8, 11]. Let us discuss the subgraph of a GSA Γ_i (Fig. 4.20a). Let it be $M_1 = 21$ for the GSA Γ_i.

Let the CMO $Y(a_6)$ determine the number K of the FCO. Obviously, there is $K = 3$. Let us divide the state a_6 by two states (Fig. 4.20b). Now, there are two CMOs: $Y(a_6) = \{y_1, y_2\}$ and $Y(a_{22}) = \{y_0, y_3\}$. Now, there is $K = 2$. The additional variable y_0 shows that all microoperations are generated.

The MOs are loaded into a special register of microoperations (RMO). There is no execution of MOs after the transition into the state a_6 (Fig. 4.20b). Now, there are the MOs y_1 and y_2 into RMO. There are all MOs (y_1, y_2, y_3) in RMO after the transition into the state a_{22}. So, this method leads to the delay into a digital systems operation. Also, it is necessary to have an additional bit of RMO to keep the variable y_0.

The number of fields can be decreased up to 1. To do it, it is necessary to insert the states a_{22}, a_{23}, a_{24} into the initial GSA (Fig. 4.20c). In this case, the value of R_D is determined by the following expression:

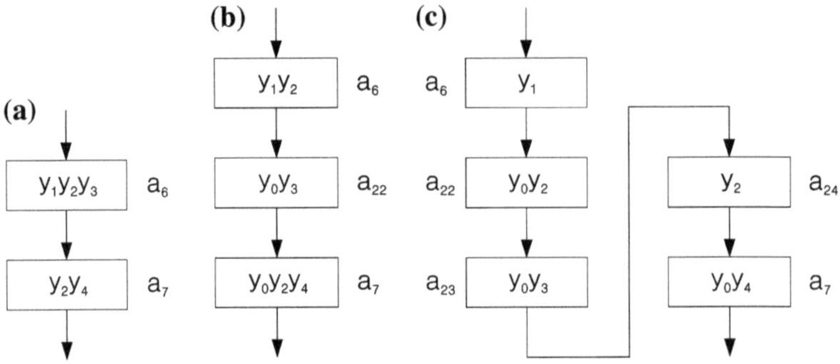

Fig. 4.20 The subgraph of GSA Γ_i

$$R_D = \lceil log_2 N \rceil + 1. \tag{4.34}$$

In (4.34), the one is added to take into account the additional variable y_0.

We do not discuss the last two approaches in this chapter. Let us point out that the using them increases significantly the amount of possible models of Moore FSMs with OCT.

References

1. M. Adamski, A. Barkalov, A. Bukowiec, Structures of mealy FSM logic circuits under implementation of verticalized flow-chart, in *Proceedings of the IEEE East-West Design and Test Workshop (EWDTW'05)*, Kharkov National University of Radioelectronics, Kharkov, 2005, pp. 70–74
2. T. Agerwala, Microprogram optimization: a survey. IEEE Trans. Comput. **25**(10), 962–973 (1976)
3. S. Baranov, *Logic Synthesis of Control Automata* (Kluwer Academic Publishers, 1994)
4. A. Barkalov, Multilevel PLA schemes for microprogram automata. Cybern. Syst. Anal. **31**(4), 489–495 (1995)
5. A. Barkalov, Principles of logic optimization for Moore microprogram automaton. Cybern. Syst. Anal. **34**(1), 54–60 (1998)
6. A. Barkalov, O. Beleckij, A. Nedal, Applying of optimization methods of Moore automaton for synthesis of compositional microprogram control unit. Autom. Control Comput. Sci. **33**(1), 44–52 (1999)
7. A. Barkalov, A. Bukowiec, Synthesis of Mealy Finite-States Machines for interpretation of verticalized flow-charts. Theor. Appl. Inf. **5**(5), 39–51 (2005)
8. A. Barkalov, L. Titarenko, Design of control units with programmable logic devices, in *Measurements*, methods, systems and design, ed. by J. Korbicz (Wydawnictwo Komunikacji i Łączności, Warsaw, Poland, 2007), pp. 371–391
9. A. Barkalov, L. Titarenko, *Logic Synthesis for FSM-Based Control Units*, vol. 53, Lecture notes in electrical engineering (Springer, Berlin, 2009)
10. A. Barkalov, M. Węgrzyn, *Design of Control Units With Programmable Logic* (University of Zielona Góra Press, 2006)
11. A. Bukowiec, *Synthesis of Finite State Machines for Programmable devices based on multi-level implementation*. Ph.D thesis, University of Zielona Góra, 2008

12. Y.C. Chu. *Computer Organization and Microprogramming* (Prentice Hall, 1972)
13. S. Dasgupta, The organization of microprogram stores. ACM Comput. Surv. **24**, 101–176 (1979)
14. M.J. Flynn, R.F. Rosin, Microprogramming: an introduction and a viewpoint. IEEE Trans. Comput. C **20**(7), 727–731 (1971)
15. S. Habib, *Microprogramming and Firmware Engineering Methods* (Wiley, New York, 1988)
16. E. Pugh, L. Johnson, J. Palmer, *IBM's 360 and Early 370 Systems* (MIT Press, Cambridge, MA, 1991)
17. S. Schwartz. An algorithm for minimizing read-only memories for machine control. IEEE 10th annual symposium on switching and automata theory, 1968, pp. 28–33
18. V. Sklyarov, I. Skliarova, A. Barkalov, L. Titarenko, *Synthesis and Optimization of FPGA-Based Systems*, vol. 294, Lecture notes in electrical engineering (Springer, Berlin, 2014)

Chapter 5
Distribution of Class Codes in Moore FSMs

5.1 The Case of CPLD-Based FSMs

The approach was proposed in [6]. There are design methods for different models of Moore FSMs presented in [7–10]. There are the results of investigation of this approach and other known methods in [11]. Let us discuss the CPLD-based approach of design.

The majority of CPLDs are based on PAL cells with programmable flip-flops [1, 2, 16]. Each cell can be viewed as q of s-input AND gates connected with OR gate. The number q determines the upper limit for product terms to be implemented by a single PAL. This value is rather small ($q \leq 8$) [1, 2, 16]. Some CPLDs include EMBs, for example, the Delta39K family from Cypress [12, 13]. It allows using PALers and EMBers in the logic design.

The proposed method [6] uses two peculiarities of CPLD. The first is a wide fan-in of macrocells (around 30). The second is existence of configurable EMBs. The first feature allows using more than a single source of codes of PES (rather, the class codes $K(B_i)$). The second feature results in implementing the system $Y(T)$ by EMBs. Let us discuss the model of CPLD-based P Moore FSM (Fig. 5.1).

A PALer is a network of PALs implementing the system of input memory functions $\Phi = \Phi(T, X)$. There is the register RG distributed among the cells producing the functions $D_r \in \Phi$. Because of it, there are signals Start and Clock connected with the PALer. As in the case of FPGA-based Moore FSMs, the EMBer implements the system of microoperations $Y = Y(T)$.

It is possible to diminish the number of PALs in the PALer using either the optimal state assignment (P_0) or the transformation of the state codes $K(a_m)$ into the codes of classes of PES $K(B_i)$. The second approach leads to the P_E Moore FSM (Fig. 5.2).

In this case, the PALer2 executes the transformation of state codes. It implements the system $\tau = \tau(T)$.

There are the same structural diagrams for P_0 and P Moore FSM. But it is implemented the system $\Phi = \Phi(T', X)$ in the case of P_0 Moore FSM.

A. Barkalov et al., *Logic Synthesis for FPGA-Based Finite State Machines*,
Studies in Systems, Decision and Control 38, DOI 10.1007/978-3-319-24202-6_5

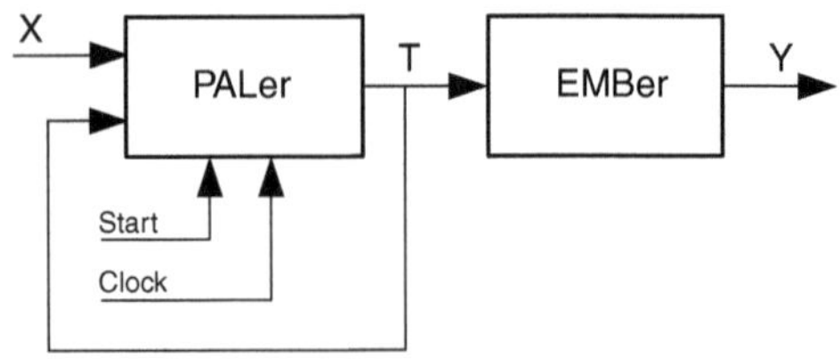

Fig. 5.1 Structural diagram of CPLD-based P Moore FSM

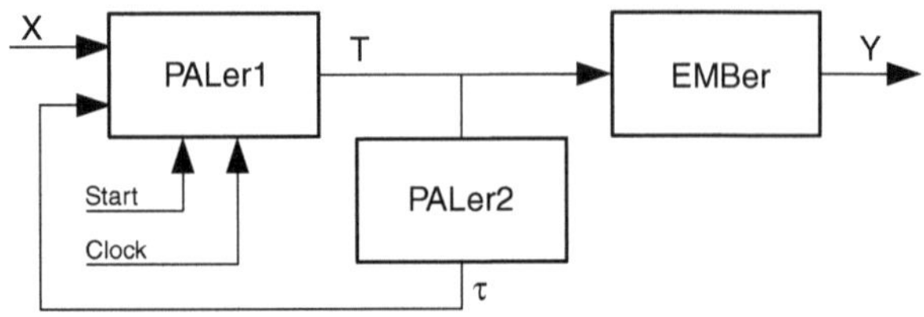

Fig. 5.2 Structural diagram of CPLD-based P_E Moore FSM

In the case of P_E Moore FSM, the PALer1 implements H_0 terms. This model is applied if the number of terms in $\Phi(T', X)$ exceed H_0 [3]. There is one very serious drawback of P_E FSM. It includes the PALer2 consuming some resources of a chip. The main goal of the approach from [3] is to decrease the number of PAL macrocells in the circuit of PALer2.

Let the partition Π_A be found for some GSA Γ marked by the states of Moore FSM. Let us encode the states $a_m \in A$ in the optimal way [5]. Let the symbol NI_i stand for the number of generalized intervals of R_1-dimensional Boolean space required for representing the code $K(B_i)$. Let us represent the set Π_A as $\Pi_B \cup \Pi_C$ where:

$$NI_i > 1 \rightarrow B_i \in \Pi_B; \tag{5.1}$$

$$NI_i = 1 \rightarrow B_i \in \Pi_C. \tag{5.2}$$

Obviously, it is necessary to transform only PES belonging to Π_B. Let it be $I_B = |\Pi_B|$. To encode the classes $B_i \in \Pi_B$, it is enough R_B bits:

$$R_B = \lceil \log_2 (I_B + 1) \rceil. \tag{5.3}$$

In (5.3), we add 1 to take into account the relation $B_i \notin \Pi_B$. Now, there are R_B elements in the set τ.

Let us use an EMB having t_F outputs for the case

$$2^{R_1} \leq V_O. \tag{5.4}$$

There are Δ_t unused outputs of EMBs forming the EMBer, where:

$$\Delta_t = \left\lceil \frac{N}{t_F} \right\rceil \cdot t_F - N. \tag{5.5}$$

Let $L(D_r)$ be the number of literals in the function $D_r \in \Phi$. Obviously, there is $L(D_r) \leq L + R_1 (r = \overline{1, R_1})$. Let the following condition take place:

$$L(D_r) + R_B \leq S. \tag{5.6}$$

Let the following condition take place:

$$\Delta_t \geq R_A. \tag{5.7}$$

In this case the state transformer can be implemented as a part of EMBer. It leads to P_{E1} Moore FSM shown in Fig. 5.3.

There is no block PALer2 in the P_{E1} FSM. The EMBer executes the functions of the PALer2. This solution requires the minimal number of PAL macrocells.

Let the condition (5.7) be violated but the following condition take place:

$$\Delta_t = 0. \tag{5.8}$$

In this case, it is possible to use the PALer1 as a source of codes $K(B_i)$ for $B_i \in \Pi_C$. The PALer2 generates the codes $K(B_i)$ for $B_i \in \Pi_B$. It leads to P_{E2} Moore FSM (Fig. 5.4).

The PALer1 implements the functions

$$\Phi = \Phi(T, \tau, X). \tag{5.9}$$

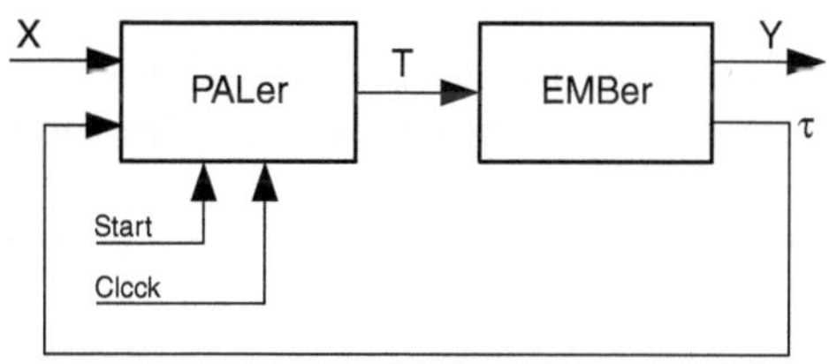

Fig. 5.3 Structural diagram of P_{E1} Moore FSM

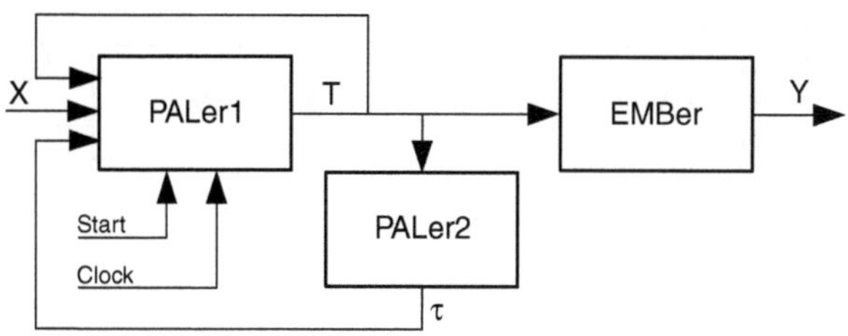

Fig. 5.4 Structural diagram of P_{E2} Moore FSM

There are the following terms in the functions (5.9):

$$F_h = \left(\bigwedge_{r=1}^{R_B} \tau_r^{e_{ir}}\right) \cdot X_h; \tag{5.10}$$

$$F_h = \left(\bigwedge_{r=1}^{R_B} \tau_r^0\right) \cdot \left(\bigwedge_{r=1}^{R_1} T_r^{e_{ir}}\right) \cdot X_h. \tag{5.11}$$

The terms (5.10) are used for classes $B_i \in \Pi_B$. Here $e_{ir} \in \{0, 1\}$ is a value of the rth bit of the code $K(B_i)$ from the hth row of the ST. The terms (5.11) are used for the classes $B_i \in \Pi_C$. The first part of (5.11) corresponds to the zero code used for the relation $B_i \in \Pi_B$. The second part of (5.11) corresponds to the code $K(B_i)$ of the class $B_i \in \Pi_C$ from the hth row of the ST. Here there is $e_{ir} \in \{0, 1, *\}$. Remind that the terms (5.11) can be used only if the condition (5.6) takes place.

Let the following condition take place:

$$R_B \leq \Delta_t < R_A. \tag{5.12}$$

In this case, the model of P_{E3} Moore FSM can be used (Fig. 5.5).

In this model, the EMBer executes the function of PALer2 from Fig. 5.4. The PALer implements functions (5.9) based on terms (5.10) and (5.11).

Let the following condition take place:

$$0 < \Delta_t < R_B. \tag{5.13}$$

In this case, the EMBer implements only some bits of the codes $K(B_i)$ for $B_i \in \Pi_B$. The rest of the bits are implemented by the PALer2. It leads to the P_{E4} Moore FSM (Fig. 5.6).

In this case, the set τ is represented as $\tau^1 \cup \tau^2$. The PALer2 implements the system

$$\tau^1 = \tau^2(T). \tag{5.14}$$

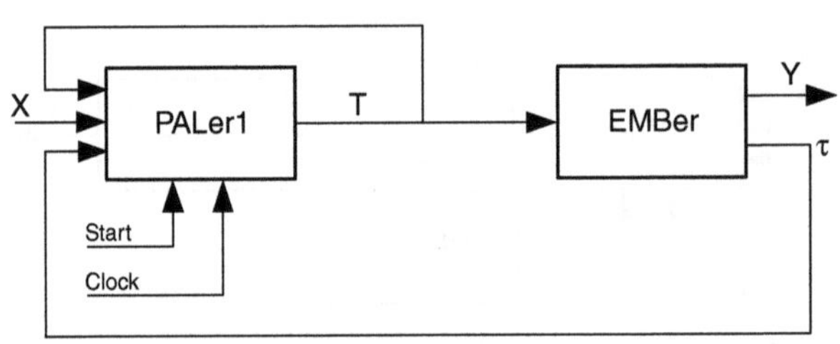

Fig. 5.5 Structural diagram of P_{E3} Moore FSM

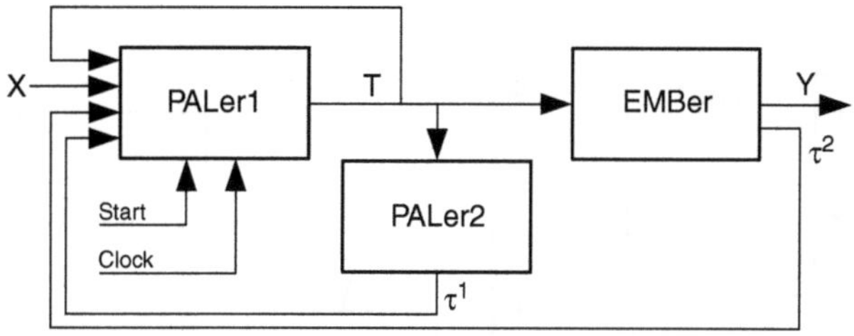

Fig. 5.6 Structural diagram of P_{E4} Moore FSM

The EMBer implements the functions $Y(T)$ and the system

$$\tau^2 = \tau^2(T). \tag{5.15}$$

The PALer1 implements the functions (5.9) based on the terms (5.10) and (5.11).

The discussed models are based on using more than a single source of codes $K(B_i)$. It is possible due to the wide fan-in of PAL macrocells. The investigation of this approach [15] shows that it allows saving around 18 % of hardware in comparison with equivalent P_0 and P_E Moore FSMs. Moreover, it gives a gain in performance. The resulting circuits are up to 24 % faster than their counterparts based on P_0 or P_E FSMs.

As a rule, it is possible to use the CPLD-based methods for hardware optimizing the FPGA-based solutions [14]. But the methods should be modified to take into account the specifics of FPGA. The most important specific of FPGA is a very limited amount of inputs per LUT [1, 16]. So, it is impossible to satisfy the condition (5.6). But it is worth trying to use more than one source of class codes in FPGA-based Moore FSMs.

5.2 Two Sources of Class Codes in FPGA-Based Moore FSMs

Let us start from the case when the condition (5.8) takes place. In this case, it is impossible to use EMBs as a source of class codes. In this case we propose the FPGA-based model of P_{E1} Moore FSM (Fig. 5.7).

Let us discuss this model. The set Φ is divided by the subsets Φ_1 and Φ_2. The set X is represented as $X' \cup X''$. In the common case, the following conditions take places:

$$\Phi_1 \cap \Phi_2 \neq \emptyset; \tag{5.16}$$

$$X' \cap X'' \neq \emptyset. \tag{5.17}$$

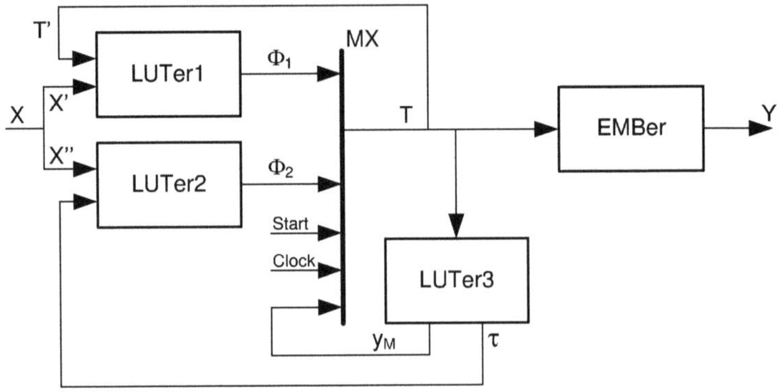

Fig. 5.7 Structural diagram of FPGA-based P_{E1} Moore FSM

The LUTer1 implements the functions

$$\Phi_1 = \Phi_1(T', X'). \tag{5.18}$$

The variables $T' \in T$ are generated by the multiplexer MX. The LUTer2 implements the functions

$$\Phi_2 = \Phi_2(\tau, X''). \tag{5.19}$$

The variables $\tau_r \in \tau$ are generated by the LUTer3. It implements the functions (4.4) and the function

$$y_M = y_M(T). \tag{5.20}$$

The function y_M is used to implement the following choice:

$$\Phi = y_M \Phi_1 \vee \bar{y_M} \Phi_2. \tag{5.21}$$

The function (5.21) is implemented by MX. The flip-flops of RG are distributed among the LUTs of MX. Because of it, there are the pulses Start and Clock connected with MX.

There are the following steps in the design method of P_{E1} Moore FSM:

1. Marking the initial GSA Γ and finding the set A.
2. Constructing the set $\Pi_A = \{B_1, \ldots, B_I\}$.
3. Executing the optimal state assignment.
4. Constructing the sets Π_B and Π_C.
5. Executing the encoding of classes $B_i \in \Pi_B$.
6. Constructing the table of LUTer1.
7. Constructing the table of LUTer2.
8. Constructing the table of LUTer3.

9. Constructing the table of MX.
10. Constructing the table of EMBer.
11. Implementing the FSM logic circuit.

Let us discuss an example of synthesis for the Moore FSM $P_{E2}(\Gamma_6)$. There is the GSA Γ_6 shown in Fig. 5.8. There are the following sets in the case of Moore FSM $P(\Gamma_6)$: $X = \{x_1, \dots, x_6\}$, $Y = \{y_1, \dots, y_8\}$, $A = \{a_1, \dots, a_{13}\}$. It is necessary $R_1 = 4$ of bits to encode the states $a_m \in A$. It gives the sets $\Phi = \{D_1, \dots, D_4\}$ and $T = \{T_1, \dots, T_4\}$.

Let us construct the partition Π_A of the states $a_m \in A$ by the classes of PES. There is the following partition $\Pi_A = \{B_1, \dots, B_7\}$, where $B_1 = \{a_1\}$, $B_2 = \{a_2, a_3\}$, $B_3 = \{a_4\}$, $B_4 = \{a_5, a_6, a_7\}$, $B_5 = \{a_8, a_9\}$, $B_6 = \{a_{10}\}$ and $B_7 = \{a_{11}, a_{12}, a_{13}\}$. So, there is $I = 7$.

Let us encode the state codes in the optimal way. One from the possible solutions is shown in Fig. 5.9. Let $I(B_i)$ be a set of generalized intervals of R_1-dimensional Boolean space covering the states $a_m \in B_i$.

There are the following sets $I(B_i)$ derived from the Karnaugh map (Fig. 5.9): $I(B_1) = \{0000\}$, $I(B_2) = \{0{*}01, {*}101\}$, $I(B_3) = \{001{*}\}$, $I(B_4) = \{010{*}, {*}{*}11\}$, $I(B_5) = \{1{*}10\}$, $I(B_6) = \{0110\}$, $I(B_7) = \{1{*}00, 100{*}\}$. Using these sets, it is possible to find the sets Π_B and Π_C: $\Pi_B = \{B_2, B_4, B_7\}$ and $\Pi_C = \{B_1, B_3, B_5, B_6\}$. So, it is necessary to use the transformation of state codes for the states $a_m \in B_2 \cup B_4 \cup B_7$.

Using (5.3), it is possible to find the value $R_B = 2$ determining the set $\tau = \{\tau_1, \tau_2\}$. Let it be the configuration 16×8 in the FPGA chip in use. So, it is enough only a single EMB to implement the system $Y(T)$. Using $t_F = 8$ and the expression (5.5), the value $\Delta_t = 0$ can be found. So, the model P_{E1} can be used for the FSM $P(\Gamma_6)$ in the discussed case.

Let us encode the classes $B_i \in \Pi_B$ as the following ones: $K(B_2) = 01, K(B_4) = 10$ and $K(B_7) = 11$. Let us use the code 00 to present the relation $B_i \notin \Pi_B$.

To construct the tables of LUTer1 and LUTer2, it is necessary to construct the system of GFT. It is the following system in the discussed case:

$$\begin{aligned}
&B_1 \to x_1 a_2 \vee \bar{x}_1 a_3;\\
&B_2 \to x_2 a_4 \vee \bar{x}_2 x_3 a_5 \vee \bar{x}_2 \bar{x}_3 a_6;\\
&B_3 \to x_3 a_6 \vee \bar{x}_3 x_4 a_8 \vee \bar{x}_3 \bar{x}_4 a_9;\\
&B_4 \to x_5 a_8 \vee \bar{x}_5 x_6 a_{11} \vee \bar{x}_5 \bar{x}_6 a_{13};\\
&B_5 \to a_{10};\\
&B_6 \to x_1 a_{12} \vee \bar{x}_1 a_1;\\
&B_7 \to x_4 a_1 \vee \bar{x}_4 x_5 a_{13} \vee \bar{x}_4 \bar{x}_5 x_6 a_7 \vee \bar{x}_4 \bar{x}_5 \bar{x}_6 a_5.
\end{aligned} \tag{5.22}$$

To construct the table of LUTer1, it is necessary to build the table of BIMF1. It is constructed on the base of the system of GFT. In the discussed case, it is necessary to extract from (5.22) the formulae for $B_i \in \Pi_B$. It is Table 5.1 in the case of Moore FSM $P_{E1}(\Gamma_6)$.

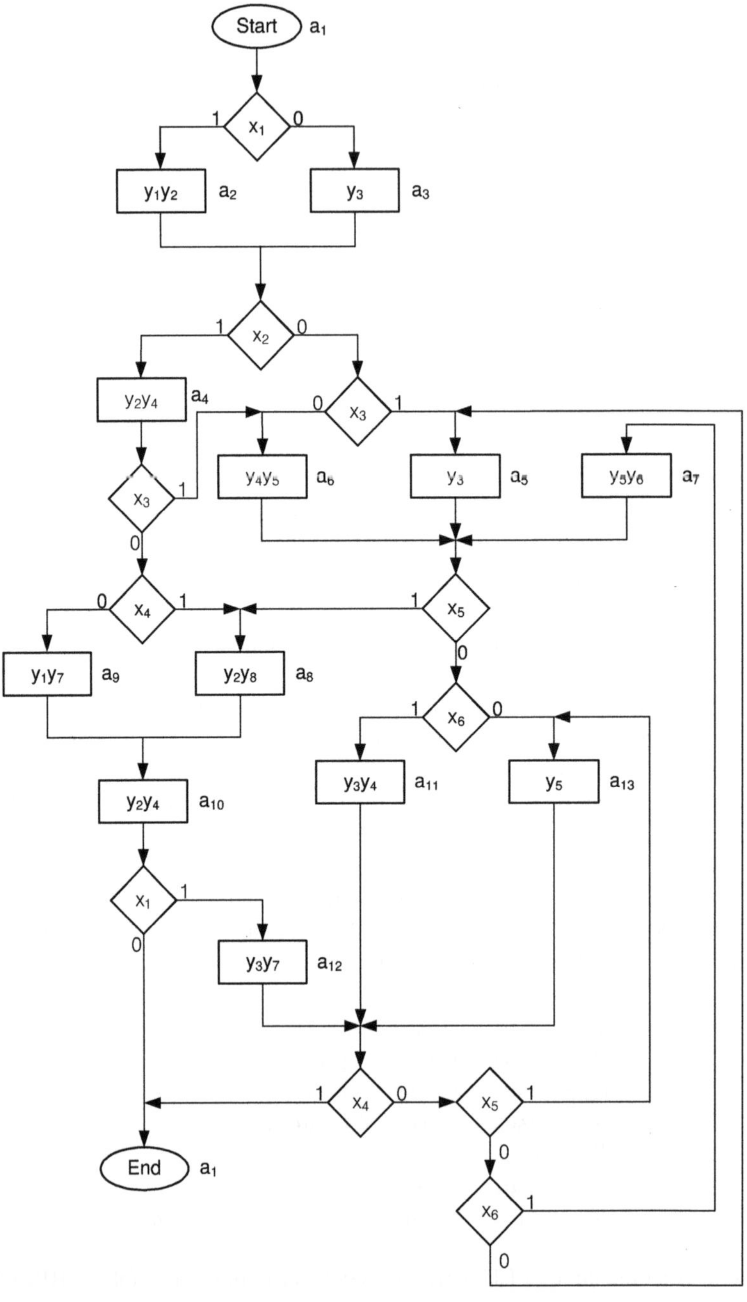

Fig. 5.8 Initial GSA Γ_6

Fig. 5.9 State codes of Moore FSM $P_0(\Gamma_6)$

T_1T_2 \ T_3T_4	00	01	11	10
00	a_1	a_2	*	a_4
01	a_5	*	a_6	a_{10}
11	a_{11}	a_3	a_7	a_8
10	a_{13}	a_{12}	*	a_9

Table 5.1 Table of BIMF1 for Moore FSM $P_{E1}(\Gamma_6)$

B_i	$K(B_i)$	a_s	$K(a_s)$	X_h	Φ_h	h
B_1	0000	a_2	0001	x_1	D_4	1
		a_3	1101	$\bar{x}_1$	$D_1D_2D_4$	2
B_3	001*	a_6	0111	x_3	$D_2D_3D_4$	3
		a_8	1110	$\bar{x}_3x_4$	$D_1D_2D_3$	4
		a_9	1010	$\bar{x}_3\bar{x}_4$	D_1D_3	5
B_5	1*10	a_{10}	0110	1	D_2D_3	6
B_6	0110	a_{12}	1001	x_1	D_1D_4	7
		a_2	0001	$\bar{x}_1$	D_4	8

This table is used to derive the equations for $D_r \in \Phi_1$. It is the following system in the discussed case:

$$
\begin{aligned}
D_1 &= \bar{T}_1\bar{T}_2\bar{T}_3\bar{T}_4\bar{x}_1 \vee \bar{T}_1\bar{T}_2T_3\bar{x}_3. \vee \bar{T}_1T_2T_3\bar{T}_4x_1;\\
D_2 &= \bar{T}_1\bar{T}_2\bar{T}_3\bar{T}_4\bar{x}_1 \vee \bar{T}_1\bar{T}_2T_3(x_3 \vee x_4) \vee T_1T_3\bar{T}_4;\\
D_3 &= \bar{T}_1\bar{T}_2T_3 \vee T_1T_2T_3;\\
D_4 &= \bar{T}_1\bar{T}_2\bar{T}_3\bar{T}_4 \vee \bar{T}_1\bar{T}_2T_3x_3 \vee \bar{T}_1T_2T_3\bar{T}_4;
\end{aligned}
\tag{5.23}
$$

The system (5.23) determines four tables for LUTs of the LUTer1. These tables are constructed in the way discussed in previous chapters.

To construct the table of LUTer2, it is necessary to use the system of GFT and construct the table of BIMF2. In the discussed case, it is constructed using formulae for $B_2, B_4, B_7 \in \Pi_C$ from (5.22). This table has 10 rows (Table 5.2).

This table is used for deriving the equations for system (5.19). The following equations can be found from Table 5.2:

$$
\begin{aligned}
D_1 &= \tau_1\bar{\tau}_2 \vee \tau_1\tau_2(\bar{x}_4x_5 \vee \bar{x}_4\bar{x}_5x_6);\\
D_2 &= \bar{\tau}_1\tau_2\bar{x}_2 \vee \tau_1\bar{\tau}_2(x_5 \vee x_6) \vee \tau_1\tau_2\bar{x}_4\bar{x}_5x_6;\\
D_3 &= \bar{\tau}_1\tau_2(x_2 \vee \bar{x}_3) \vee \tau_1\bar{\tau}_2x_5 \vee \tau_1\tau_2\bar{x}_4\bar{x}_5x_6;\\
D_4 &= \bar{\tau}_1\tau_2\bar{x}_2\bar{x}_3 \vee \tau_1\tau_2\bar{x}_4\bar{x}_5\bar{x}_6.
\end{aligned}
\tag{5.24}
$$

The system (5.24) determines $R_1 = 4$ of tables for LUTs implementing the circuit of LUTer2. We do not discuss this step.

Table 5.2 Table of BIMF2 for Moore FSM $P_{E1}(\Gamma_6)$

B_i	$K(B_i)$	a_s	$K(a_s)$	X_h	Φ_h	h
B_2	01	a_4	0010	x_2	D_3	1
		a_5	0100	$\bar{x}_2 x_3$	D_2	2
		a_6	0111	$\bar{x}_2 \bar{x}_3$	$D_2 D_3 D_4$	3
B_4	10	a_8	1110	x_5	$D_1 D_2 D_3$	4
		a_{11}	1100	$\bar{x}_5 x_6$	$D_1 D_2$	5
		a_{13}	1000	$\bar{x}_5 \bar{x}_6$	D_1	6
B_7	11	a_1	0000	x_4	–	7
		a_{13}	1000	$\bar{x}_4 x_5$	D_1	8
		a_7	1111	$\bar{x}_4 \bar{x}_5 x_6$	$D_1 D_2 D_3 D_4$	9
		a_5	0100	$\bar{x}_4 \bar{x}_5 \bar{x}_6$	D_2	10

To construct the table of LUTer3, it is necessary to find the system $\tau(T)$. It can be done using two Karnaugh maps shown in Fig. 5.10.

Let us explain the approach used for filling the cells of the maps. There are the signs "*" for unused state codes (0011, 0101 and 1011). There are the signs "*" for the state codes corresponding to states $a_m \in B_i$ where $B_i \in \Pi_C$ (0000, 0010, 0110, 1110 and 1010). Obviously, there are the following equations:

$$\begin{aligned} \tau_1 &= B_4 \vee B_7; \\ \tau_2 &= B_2 \vee B_7. \end{aligned} \tag{5.25}$$

It follows, for example, from Table 5.2.

So, there are 1's for states $a_m \in B_4 \cup B_7$ and 0's for states $a_m \in B_2$ (Fig. 5.10a). There are 1's for states $a_m \in B_2 \cup B_7$ and 0's for states $a_m \in B_4$ (Fig. 5.10b). It leads to the following equations:

$$\begin{aligned} \tau_1 &= \bar{T}_4 \vee T_3 \vee \bar{T}_2 \bar{T}_4; \\ \tau_2 &= T_1 \bar{T}_3 \vee \bar{T}_1 \bar{T}_2. \end{aligned} \tag{5.26}$$

To implement the system (5.26), it is enough 2 of LUTs having $S = 3$.

Fig. 5.10 Karnaugh maps for functions $\tau_1(a)$ and $\tau_2(b)$

(a)

T_1T_2 \ T_3T_4	00	01	11	10
00	*	0	*	*
01	1	*	1	*
11	1	0	1	*
10	1	1	*	*

(b)

T_1T_2 \ T_3T_4	00	01	11	10
00	*	1	*	*
01	0	*	0	*
11	1	1	0	*
10	1	1	*	*

It is necessary R_1 of LUTs having $S = 3$ to implement the system (5.21). There is the following equation implemented by rth LUT of MX:

$$D_r = y_M D_r^1 \vee \bar{y_M} D_r^2 \quad (r = \overline{1, R_1}). \tag{5.27}$$

There is the superscript 1 for $D_r \in \Phi_1$, whereas there is the superscript 2 for $D_r \in \Phi_2$.

The table of EMBer has the columns T, Y, m. It is constructed in a way we have discussed in previous chapters.

Now, let us discuss how to get the equation for y_M. There are two ways. Firstly, we can form an equation similar to the equations from (5.25):

$$y_M = B_2 \vee B_4 \vee B_7. \tag{5.28}$$

It gives the following minimized expression:

$$y_M = T_4 \vee T_2 \bar{T_3} \vee T_1 \bar{T_3}. \tag{5.29}$$

This equation is implemented by the LUTer3.

Secondly, the class codes can be used to get the equation for y_M. It is the following equation:

$$y_M = \tau_1 \vee \tau_2. \tag{5.30}$$

This equation is implemented by the LUTer2. It results in the P_{E2} Moore FSM (Fig. 5.11).

Let us discuss the case when the following conditions take places:

$$\Delta_t = R_B + 1; \tag{5.31}$$

$$R_B + 1 < R_A. \tag{5.32}$$

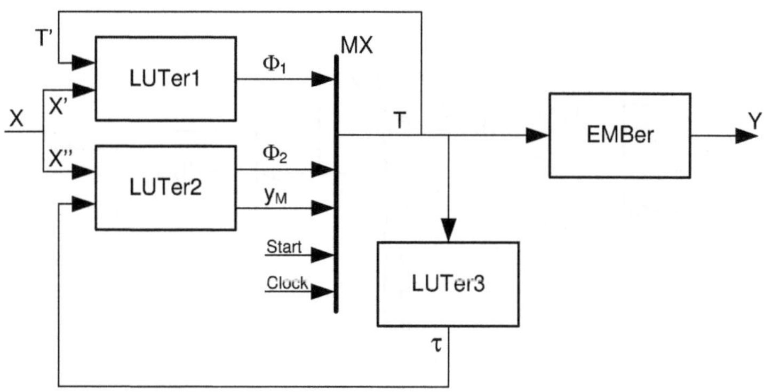

Fig. 5.11 Structural diagram of FPGA-based P_{E2} Moore FSM

The condition (5.32) shows that the EMBer can be used for generating the functions $\tau(T)$. The condition (5.32) shows that there are no enough outputs of EMBer to implement the transformation for all classes $B_i \in \Pi_A$.

In this case, the model of P_{E3} Moore FSM (Fig. 5.12) can be used. Here, the EMBer implements the functions $Y(T)$, $\tau(T)$ and $y_M(T)$.

There are the same functions of blocks LUTer1 and LUTer2 as they are for P_{E1} Moore FSM. The EMBer executes the functions of LUTer3. Design methods are the same for P_{E1} and P_{E3} FSMs. But there is no step 8 in the case of P_{E3} FSM.

Let us discuss how to construct the table of EMBer for the Moore FSM $P_{E3}(\Gamma_6)$. It contains the following columns: T (an address of a cell), Y, τ, y_M (the content of the cell), m. In the discussed case, it is Table 5.3. We use here the state codes from Fig. 5.9. The column m contains either the subscript of a corresponding state $a_m \in A$ or the sign "*" for "don't care" input assignments.

There is $y_M = 1$ for the cells corresponding to $a_m \in B_2 \cup B_4 \cup B_7$. There is the code 01 for states $a_2, a_3 \in B_2$, the code 10 for $a_5, a_6, a_7 \in B_4$ and the code 11 for $a_{11}, a_{12}, a_{13} \in B_7$ (the column τ of Table 5.3). There are the codes of CMOs taken from the GSA Γ_6 (the column Y of Table 5.3).

Let the following condition take place:

$$\Delta_t = R_B. \tag{5.33}$$

In this case, the function y_M is represented as

$$y_M = y_M(\tau). \tag{5.34}$$

It results in the P_{E4} Moore FSM shown in Fig. 5.13. In this model, the LUTer2 implements the function (5.34).

There are the same steps in design of FSMs P_{E3} and P_{E4}. But there are two peculiarities in the later case: (1) the LUTer2 implements the function (5.34) and (2) there is no the column y_M in the table of EMBer. In the discussed case, the equation (5.30) is implemented by the LUTer2.

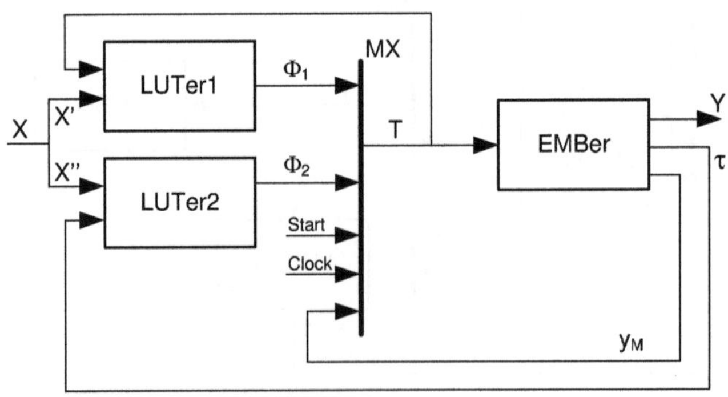

Fig. 5.12 Structural diagram of P_{E3} Moore FSM

Table 5.3 Table of EMBer for Moore FSM $P_{E3}(\Gamma_6)$

T	Y	τ	y_M	m
$T_1T_2T_3T_4$	$y_1y_2y_3y_4y_5y_6y_7y_8$	$\tau_1\tau_2$		
0000	00000000	00	0	1
0001	11000000	01	1	2
0010	01010000	00	0	4
0011	00000000	00	0	*
0100	00100000	10	1	5
0101	00000000	00	0	*
0110	01010000	00	0	10
0111	00011000	10	1	6
1000	00001000	11	1	13
1001	00100010	11	1	12
1010	10000010	00	0	9
1011	00000000	00	0	*
1100	00110000	11	1	11
1101	00100000	01	1	3
1110	01000001	00	0	8
1111	00001100	10	1	7

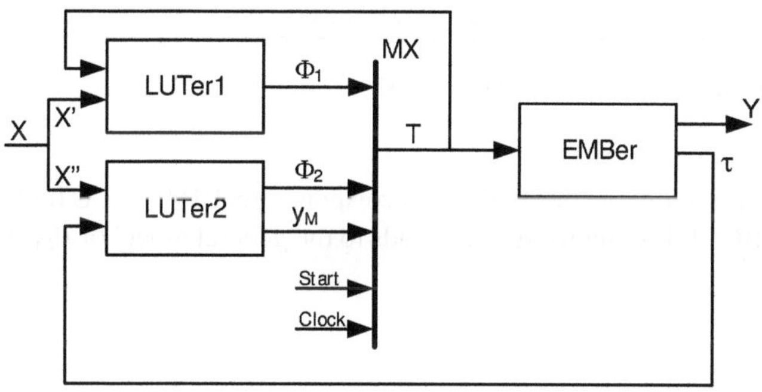

Fig. 5.13 Structural diagram of P_{E4} Moore FSM

5.3 Three Sources of Class Codes in FPGA-Based Moore FSMs

Let the condition (5.13) take place. In this case, we propose the model of P_{E5} Moore FSM (Fig. 5.14).

The LUTer3 plays the same role as the PALer2 (Fig. 5.6). The design methods are the same for FSMs P_{E1} and P_{E5}. As in the previous case, the LUTer2 can implement the function y_M. It leads to P_{E6} Moore FSM (Fig. 5.15).

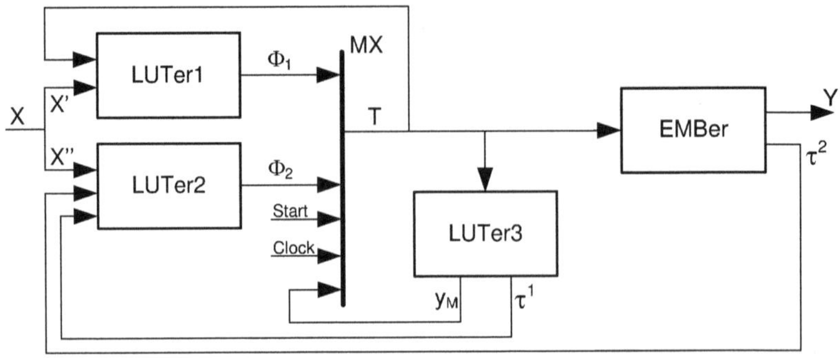

Fig. 5.14 Structural diagram of P_{E5} Moore FSM

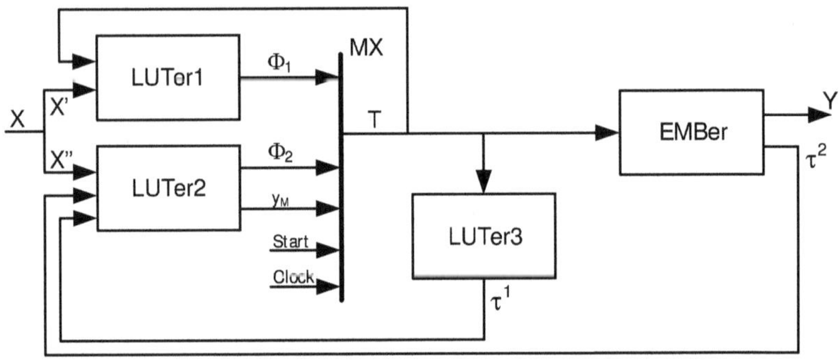

Fig. 5.15 Structural diagram of P_{E6} Moore FSM

In the common case, the LUTer1 corresponds to BIMF1, LUTer2 to BIMF2, LUTer3 to BCST, EMBer to BMO. It leads to the general model of P_{E6} Moore FSM (Fig. 5.16).

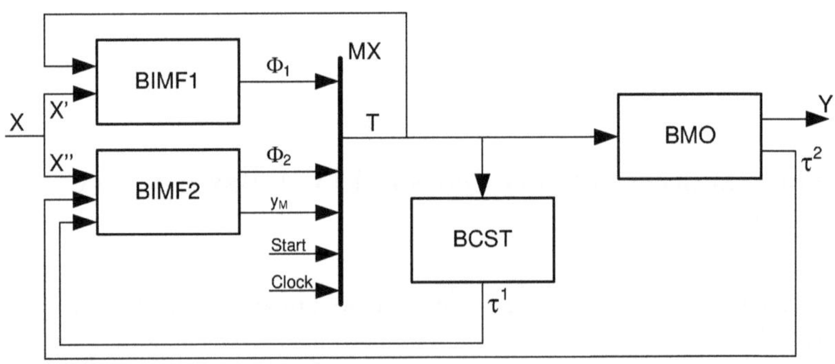

Fig. 5.16 General structural diagram of P_{E6} Moore FSM

As in previous cases, different elements can be used for implementing the circuits of different FSM's blocks. Analysis of Fig. 5.16 shows that the BMO should be implemented as EMBer, whereas BCST as LUTer, So, there are four different modifications of P_{E6} Moore FSM. Let us show the vectors $\langle BIMF1, BIMF2\rangle$ for different modifications.

The P_{E61} Moore FSM corresponds to the vector $\langle L_1, L_2\rangle$, the P_{E62} Moore FSM to $\langle L_1, B_2\rangle$, the P_{E63} Moore FSM to $\langle B_1, L_2\rangle$ and the P_{E64} Moore FSM to $\langle B_1, B_2\rangle$. All these models are designed using the same general method. But there are different approaches for designing the circuits for BIMF1 and BIMF2.

Obviously, there are four different modifications for all discussed P_{Ei} FSMs ($i = \overline{1, 6}$). It gives 24 basic models of Moore FSM with the distribution of class codes.

Three approaches can be used for hardware reduction in P_{Ei} Moore FSM: (1) the replacement of logical conditions; (2) the maximal encoding of the collections of microoperations; (3) the encoding of the fields of compatible microoperations. Let us discuss these approaches.

5.4 Replacement of Logical Conditions and Distribution of Class Codes

In contrast to *MP* Moore FSM, there are two blocks of RLC in the MP_{Ei} FSMs. Let us discuss the general case for MP_{E1} Moore FSMs (Fig. 5.17). It includes blocks BRLC1 and BRLC2 executing the replacement of logical conditions.

The BRLC1 implements the system

$$P' = P'(T, X'), \tag{5.35}$$

whereas the BRLC2 the system

$$P'' = P''(\tau, X''). \tag{5.36}$$

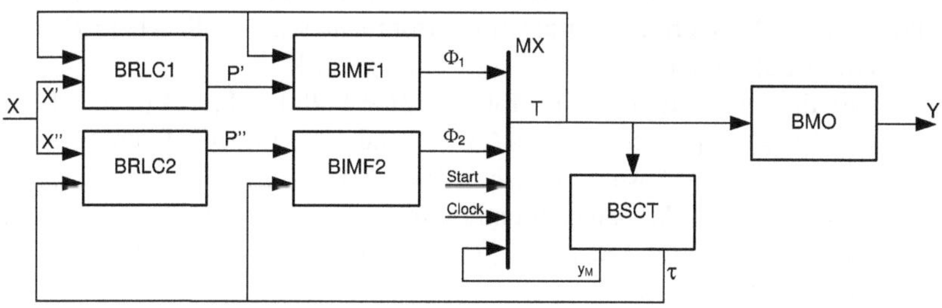

Fig. 5.17 General structural diagram of MP_{E1} Moore FSM

The BIMF1 implements the system

$$\Phi_1 = \Phi_1(T, P'), \tag{5.37}$$

whereas the BIMF2 the system

$$\Phi_2 = \Phi_2(\tau, P''). \tag{5.38}$$

There are the same functions executed by the BSCT and the LUTer3 (Fig. 5.7). The same is true for BMO and EMBer.

There are the following steps in design method for MP_{E1} Moore FSM:

1. Constructing the set of states for Moore FSM.
2. Constructing the set $\Pi_A = \{B_1, \ldots, B_I\}$.
3. Executing the optimal state assignment.
4. Constructing the sets Π_B and Π_C.
5. Executing the encoding of classes $B_i \in \Pi_B$.
6. Finding the sets P' and P''.
7. Executing the RLC for $x_e \in X'$.
8. Executing the RLC for $x_e \in X''$.
9. Constructing the transformed table for BIMF1.
10. Constructing the transformed table for BIMF2.
11. Constructing the table of BSCT.
12. Constructing the table of MX.
13. Constructing the table of BMO.
14. Implementing the FSM logic circuit.

Let us discuss an example of design for Moore FSM $MP_{E14}(\Gamma_6)$. The subscript 14 means that both circuits of BIMF1 and BIMF2 are implemented as EMBers. Let us point out that the BMO is implemented as EMBer, whereas the BSCT as LUTer. Let us use the same state codes as it is shown in Fig. 5.9. It means that the steps 1–5 are already executed for the discussed example.

Let us form the system of GFTs for $P_E(\Gamma_6)$. It is the system (5.22). The set X' includes logical conditions determining transitions from states $a_m \in B_i$ where $B_i \in \Pi_C$. Analysis of (5.22) gives the set $X' = \{x_1, x_3, x_4\}$. The set X'' includes logical conditions $x_e \in X$ determining transitions from states $a_m \in B_i$ where $B_i \in \Pi_B$. Analysis of (5.22) gives the set $X'' = \{x_2, x_3, x_4, x_5, x_6\}$.

There are the following sets $X(B_i)$ for $B_i \in \Pi_C$: $X(B_1) = \{x_1\}, X(B_3) = \{x_3, x_4\}$, $X(B_5) = \{\}, X(B_6) = \{x_1\}$. So, there is $G_1 = 2$ and $P' = \{p_1, p_2\}$. Let us replace the logical conditions $x_e \in X'$ as it is shown in Table 5.4.

The following sets $X(B_i)$ can be found for the classes $B_i \in \Pi_B$: $X(B_2) = \{x_2, x_3\}$, $X(B_4) = \{x_5, x_6\}$ and $X(B_7) = \{x_4, x_5, x_6\}$. It gives the set $P'' = \{p_3, p_4, p_5\}$. The RLC is shown in Table 5.5 for $x_e \in X''$.

Table 5.4 Table of RLC for $x_e \in X'$

p_g	B_i			
	B_1	B_3	B_5	B_6
p_1	x_1	x_3	–	x_1
p_2	–	x_4	–	–
$K(B_i)$	0000	001*	1*10	0110

Table 5.5 Table of RLC for $x_e \in X''$

p_g	B_i		
	B_2	B_4	B_7
p_3	x_2	x_5	x_5
p_4	x_3	x_6	x_6
p_5	–	–	x_4
$K(B_i)$	01	10	11

Table 5.4 is used for deriving the system (5.35). It is the following system:

$$\begin{aligned} p_1 &= \bar{T}_1\bar{T}_2\bar{T}_3\bar{T}_4x_1 \vee \bar{T}_1\bar{T}_2T_3x_3 \vee \bar{T}_1T_2T_3\bar{T}_4x_1; \\ p_2 &= x_4. \end{aligned} \quad (5.39)$$

Table 5.5 is used for deriving the system (5.36). It is the following system:

$$\begin{aligned} p_3 &= \bar{\tau}_1\tau_2x_2 \vee \tau_1x_5; \\ p_4 &= \bar{\tau}_1\tau_2x_3 \vee \tau_1x_6; \\ p_5 &= x_4. \end{aligned} \quad (5.40)$$

The system (5.39) determines the LUTer1 consisting from a single LUTerp_1. The system (5.40) determines the LUTer2 consisting from LUTerp_3 and LUTerp_4. There is no need in implementing the circuits for variables p_2 and p_5.

There are the same columns in tables for LUTerp_1, LUTerp_3 and LUTerp_4. But these columns differ in content. There are the following columns in these tables: $K(B_i), X(P_g), p_g, q$. For example, the LUTerp_3 is represented by Table 5.6.

This table is constructed on the base of the corresponding equation from (5.40). Acting in the same way, it is possible to construct tables for all functions $p_g \in P' \cup P''$.

Two steps are necessary to construct the transformed tables of BIMF1 and BIMF2. Firstly, it is necessary to construct the tables similar to Tables 5.1 and 5.2. Secondly, it is necessary to replace the column X_h by the column P_h. There is the transformed table of BIMF1 represented by Table 5.7.

The transformed table of BIMFi ($i = \overline{1,2}$) is used for constructing the tables for EMBer$i$ ($i = \overline{1,2}$). Each table includes the columns: $K(B_i)$, P (an address of a cell), Φ, q. It is necessary $H(B_i)$ cells to represent the transitions from the states $a_m \in B_i$, where:

$$H(B_i) = 2^{G_j} \quad (j = \overline{1,2}). \quad (5.41)$$

Table 5.6 Table of LUTerp_3 for Moore FSM $MP_{E14}(\Gamma_6)$

$K(B_i)$	$X(P_3)$	p_3	q	$K(B_i)$	$X(P_3)$	p_3	q
$\tau_1\tau_2$	x_2x_5			$\tau_1\tau_2$	x_2x_5		
00	00	0	1	10	00	0	9
00	01	0	2	10	01	1	10
00	10	0	3	10	11	0	11
00	11	0	4	10	11	1	12
01	00	0	5	11	00	0	13
01	01	0	6	11	01	1	14
01	10	1	7	11	10	0	15
01	11	1	8	11	11	1	16

Table 5.7 Transformed table of BIMF1 for Moore FSM $MP_{E14}(\Gamma_6)$

B_i	$K(B_i)$	a_s	$K(a_s)$	P_h	Φ_h	h
B_1	0000	a_2	0001	p_1	D_4	1
		a_3	1101	$\bar{p_1}$	$D_1D_2D_4$	2
B_3	001*	a_6	0111	p_1	$D_2D_3D_4$	3
		a_8	1110	$\bar{p_1}x_4$	$D_1D_2D_3$	4
		a_9	1010	$\bar{p_1}\bar{x_4}$	D_1D_3	5
B_5	1*10	a_{10}	0110	1	D_2D_3	6
B_6	0110	a_{12}	1001	p_1	D_1D_4	7
		a_2	0001	$\bar{p_1}$	D_4	8

Table 5.8 Part of the table of EMBer1 for Moore FSM $MP_{E14}(\Gamma_6)$

$K(B_i)$	P	Φ	q	h	B_i
$T_1T_2T_3T_4$	p_1x_4	$D_1D_2D_3D_4$			
0000	00	1101	1	2	B_1
0000	01	1101	2	2	
0000	10	0001	3	1	
0000	11	0001	4	1	
0010	00	1010	5	5	B_3
0010	01	1110	6	4	
0010	10	0111	7	3	
0010	11	0111	8	3	
0011	00	1010	9	5	
0011	01	1110	10	4	
0011	10	0111	11	3	
0011	11	0111	12	3	

A part of the table of EMBer1 is represented by Table 5.8.

We add the columns h, B_i to show the correspondence between Tables 5.7 and 5.8. There is no "*" in the code $K(B_i)$. Because of it, it is necessary 4 rows to represent the corresponding transitions. There is a single "*" in the $K(B_3)$. So, it is necessary 8 rows to represent the transitions for the codes 0010 and 0011 (Table 5.8).

All other tables can be constructed as it is for the FSM $P_{E1}(\Gamma_6)$. These steps are already executed.

5.5 Increasing the Number of Class Variables

The variables $T_r \in T$ are called state variables [4]. By an analogy with state variables, let us name the variables $\tau_r \in \tau$ class variables. Let us discuss the following approach for design of FPGA-based Moore FSMs.

Let us find the set $\Pi_A = \{B_1, \ldots, B_I\}$ where $B_i \in \Pi_A$ is a class of PES. Let us represent the set Π_A as

$$\Pi_A = \Pi_{A1} \cup \Pi_{A2} \cup \cdots \cup \Pi_{AK}. \tag{5.42}$$

The following relation takes place for sets Π_{Ak}:

$$\Pi_{Ai} \cap \Pi_{Aj} = \emptyset \quad (i \neq j, \quad i, j \in \{1, \ldots, K\}). \tag{5.43}$$

Let us encode the classes $B_i \in \Pi_{Ak}$ by codes $K(B_i)$ having R_{Ak} bits:

$$R_{Ak} = \lceil \log_2 I_{Ak} \rceil \quad (k = \overline{1, K}). \tag{5.44}$$

In (5.44), there is $I_{Ak} = |\Pi_{Ak}|$ $(k = \overline{1, K})$. Let us use the variables $\tau_r \in \tau^k$ to encode the classes $B_i \in \Pi_{Ak}$.

Let us use the variables $z_r \in Z$ to represent the relations $B_i \in \Pi_{Ak}$ $(k = \overline{1, K})$. It is necessary R_K variables to encode each of above mentioned relations:

$$R_K = \lceil \log_2 K \rceil. \tag{5.45}$$

Let it be calculated the value of Δ_t for some Moore FSM $P(\Gamma_j)$ and EMBs of a particular FPGA chip. To calculate Δ_t, the expression (5.5) is used. Let us discuss the case (5.8) when there is $\Delta_t = 0$. It leads to P_{KE1} Moore FSM (Fig. 5.18).

The set τ combines the sets $\tau^1, \ldots, \tau^K$. The MX is controlled by variables $z_r \in Z$. This approach can be viewed as a Moore FSM having K sources of class codes.

The MX executes the choice of Φ_k as a source for functions $T_r \in T$. So, it executes the function

$$\Phi = \bigvee_{k=1}^{K} Z_k \Phi_k. \tag{5.46}$$

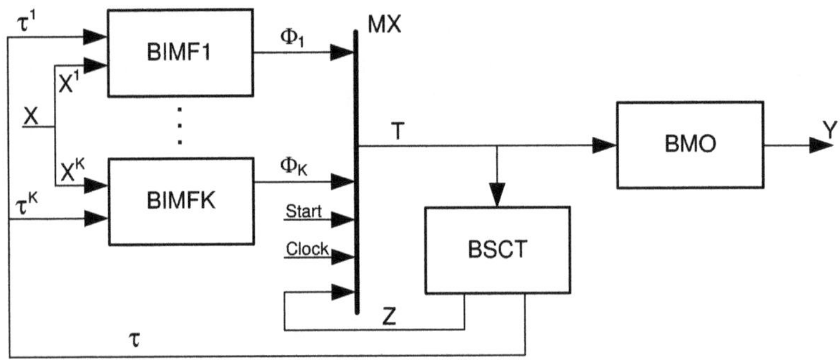

Fig. 5.18 Structural diagram of P_{KE1} Moore FSM

In (5.46), the symbol Z_k stands for the conjunction of variables $z_r \in Z$ corresponding to the relation $B_i \in \Pi_{Ak}$.

Let it be S inputs in the LUTs of an FPGA chip used for implementing the FSM's circuit. To minimize the hardware amount in MX, it is necessary that each of functions $D_r \in \Phi$ is implemented using a single LUT. It is possible if the following relation takes place:

$$K + R_K \leq S. \tag{5.47}$$

Using (5.47), one can find the maximum possible amount of blocks representing the BIMF.

Let the following relation take place:

$$\Delta_t \geq \sum_{k=1}^{K} R_{Ak} + R_K. \tag{5.48}$$

In this case, the BMO generates the functions Z and τ. It leads to P_{KE2} Moore FSM (Fig. 5.19).

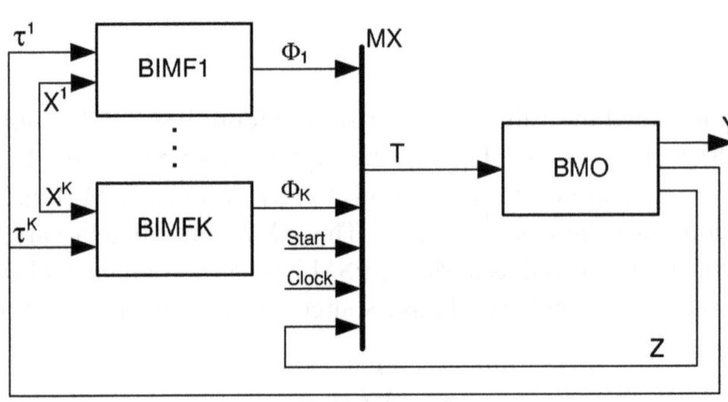

Fig. 5.19 Structural diagram of P_{KE2} Moore FSM

In the case of P_{KE2} Moore FSM, the block BSCT is absent. It leads to the decreasing for the number of LUTs in comparison with P_{KE1} Moore FSM.

At last, it is possible that BMO can implement only a part of functions. It is true if the following condition takes place:

$$0 < \Delta_t < \sum_{k=1}^{K} R_{Ak} + R_K. \tag{5.49}$$

It leads to P_{KE3} Moore FSM (Fig. 5.20).

In this case, the set τ is divided by sets τ_{BMO} and τ_{CT} ($\tau_{BMO} \cap \tau_{CT} = \emptyset$; $\tau_{BMO} \cup \tau_{CT} = \tau$). As in previous cases, the BIMFk implements the functions

$$\Phi_k = \Phi_k(\tau^k, X^k) \quad (k = \overline{1, K}). \tag{5.50}$$

Obviously, the P_{KE3} is the most general model of Moore FSM with K sources of class codes. There are the following steps in its design method:

1. Constructing the set of states A.
2. Constructing the set $\Pi_A = \{B_1, \ldots, B_I\}$.
3. Constructing the sets $\Pi_{Ak}(k = \overline{1, K})$.
4. Executing the state assignment.
5. Executing the encoding of classes $B_i \in \Pi_A$.
6. Constructing set Z and encoding of directions.
7. Constructing the tables of BIMFk ($k = \overline{1, K}$).
8. Constructing the table of BSCT.
9. Constructing the table of BMO.
10. Constructing the table of MX.
11. Implementing the FSM logic circuit.

Let us discuss an example of design for Moore FSM $P_{2E1}(\Gamma_6)$. It is possible if the EMB in use has the configuration 16×8 bits. In this case, there is $\Delta_t = 0$ (because

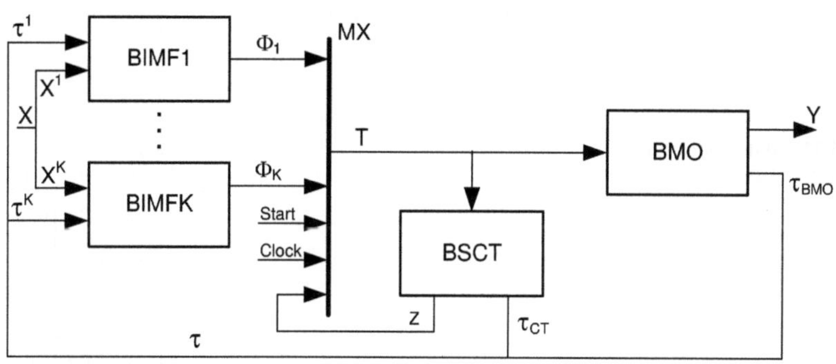

Fig. 5.20 Structural diagram of P_{KE3} Moore FSM

of $N = 8$). The GSA Γ_6 is shown in Fig. 5.8. There is the set $A = \{a_1, \ldots, a_{13}\}$. These states are distributed among $I = 7$ classes $B_i \in \Pi_A$. There is the set $\Pi_A = \{B_1, \ldots, B_7\}$, where $B_1 = \{a_1\}$, $B_2 = \{a_2, a_3\}$, $B_3 = \{a_4\}$, $B_4 = \{a_5, a_6, a_7\}$, $B_5 = \{a_8, a_9\}$, $B_6 = \{a_{10}\}$, $B_7 = \{a_{11}, a_{12}, a_{13}\}$.

As follows from the formula P_{2E1}, there is $K = 2$. So, it is necessary to distribute the classes $B_i \in \Pi_A$ between the sets Π_{A1} and Π_{A2}. Let us use two rules for the distribution. Firstly, let the amount of classes be approximately the same in the classes $\Pi_{Ak} (k = \overline{1, K})$. Secondly, let the following condition take place:

$$|A(\Pi_{Ai}) \cap A(\Pi_{Aj})| \rightarrow min(i, j \in \{1, \ldots, K\}). \quad (5.51)$$

In (5.51), the symbol $A(\Pi_{Ak})$ means the set of states of transition for the states from classes $B_i \in \Pi_{Ak}$.

Using these rules, the following sets can be constructed. The set $\Pi_{A1} = \{B_1, B_2, B_7\}$ corresponds to the set $A(\Pi_{A1}) = \{a_1, a_7, a_{13}\}$. The set $\Pi_{A2} = \{B_3, \ldots, B_6\}$ corresponds to the set $A(\Pi_{A2}) = \{a_1, a_8, \ldots, a_{13}\}$.

Let us encode states $a_m \in A$ as it is shown in Fig. 5.21. Analysis of this Karnaugh map shows that: (1) there is $T_1 = 0$ for states $a_m \in A(\Pi_{A1})$ and (2) there is $T_1 = 1$ for states $a_m \in A(\Pi_{A2})$ (the only difference is for a_{13}).

The are $I_{A1} = 3$ and $I_{A2} = 4$ in the discussed case. Using (5.44), the following values can be found: $R_{A1} = R_{A2} = 2$. It gives the set $\tau = \{\tau_1, \ldots, \tau_4\}$ with $\tau^1 = \{\tau_1, \tau_2\}$ and $\tau^2 = \{\tau_3, \tau_4\}$. Let us encode the classes $B_i \in \Pi_A$ as the following: $K(B_1) = K(B_3) = 00$, $K(B_2) = K(B_4) = 01$, $K(B_5) = K(B_7) = 10$ and $K(B_6) = 11$.

There are two sources of codes used by the MX. So, there is $K = 2$; it gives $R_K = 1$ and the set $Z = \{z_1\}$. Let $z_1 = 0$ mean that the input memory functions are generated by the BIMF1. Obviously, the BIMF2 is determined by $z_1 = 1$.

The table of BIMF*k* is constructed on the base of the system of GFT. It is the system (5.22) in the discussed case. Table 5.9 represents the table of BIMF1. Table 5.10 represents the BIMF2.

Table 5.9 is used to derive the system

$$\Phi_1 = \Phi_1(\tau^1, X^1). \quad (5.52)$$

Fig. 5.21 State codes for Moore FSM $P(\Gamma_6)$

T_1T_2 \ T_3T_4	00	01	11	10
00	a_1	a_8	a_{10}	a_9
01	a_{11}	a_{12}	a_6	a_{13}
11	a_5	*	*	a_7
10	a_2	a_3	*	a_4

Table 5.9 Table of BIMF1 for Moore FSM $P_{2E1}(\Gamma_6)$

B_i	$K(B_i)$	a_s	$K(a_s)$	X_h	Φ_h	h
B_1	00	a_2	1000	x_1	D_1	1
		a_3	1001	$\bar{x_1}$	D_1D_4	2
B_2	01	a_4	1010	x_2	D_1D_3	3
		a_5	1100	$\bar{x_2}x_3$	D_1D_2	4
		a_6	0111	$\bar{x_2}\bar{x_3}$	$D_2D_3D_4$	5
B_7	10	a_1	0000	x_4	–	6
		a_{13}	0110	$\bar{x_4}x_5$	D_2D_3	7
		a_7	1110	$\bar{x_4}\bar{x_5}x_6$	$D_1D_2D_3$	8
		a_5	1101	$\bar{x_4}\bar{x_5}\bar{x_6}$	$D_1D_2D_4$	9

Table 5.10 Table of BIMF2 for Moore FSM $P_{2E1}(\Gamma_6)$

B_i	$K(B_i)$	a_s	$K(a_s)$	X_h	Φ_h	h
B_3	00	a_6	0111	x_3	D_2D_3D4	
		a_8	0001	$\bar{x_3}x_4$	D_4	
		a_9	0010	$\bar{x_3}\bar{x_4}$	D_3	
B_4	01	a_8	0001	x_5	D_4	
		a_{11}	0100	$\bar{x_5}x_6$	D_2	
		a_{13}	0110	$\bar{x_5}\bar{x_6}$	D_2D_3	
B_5	10	a_{10}	0011	1	D_3D_4	
B_6	11	a_{12}	0101	x_1	D_2D_4	
		a_1	0000	$\bar{x_1}$	–	

Table 5.10 is used to derive the system

$$\Phi_2 = \Phi_2(\tau^2, X^2). \tag{5.53}$$

For example, the following equations can be derived from Table 5.9:

$$\begin{aligned}
D_1 &= \bar{\tau_1}\bar{\tau_2} \vee \bar{\tau_1}\tau_2(x_2 \vee x_3) \vee \tau_1\bar{\tau_2}\bar{x_4}\bar{x_5};\\
D_2 &= \bar{\tau_1}\tau_2\bar{x_2} \vee \tau_1\bar{\tau_2}\bar{x_4};\\
D_3 &= \bar{\tau_1}\tau_2(x_2 \vee \bar{x_3}) \vee \tau_1\bar{\tau_2}(\bar{x_4}x_5 \vee \bar{x_4}\bar{x_5}x_6);\\
D_4 &= \bar{\tau_1}\bar{\tau_2}\bar{x_1} \vee \bar{\tau_1}\tau_2\bar{x_2}\bar{x_3} \vee \tau_1\bar{\tau_2}\bar{x_4}\bar{x_5}\bar{x_6}.
\end{aligned} \tag{5.54}$$

Let us point out that some function $D_r = 1$ if the variable $T_r = 1$ in the codes of states $a_m \in A(\Pi_{Ak})$. If a variable $T_r = 0$ for all codes of states $a_m \in A(\Pi_{Ak})$, then the variable $D_r = 0$ for all rows of the corresponding table of BIMFk. There are no such variables in Table 5.9, but the variable $T_1 = 0$ for all rows of Table 5.10.

The following equations can be derived from Table 5.10:

$$\begin{aligned}
D_1 &= 0;\\
D_2 &= \bar{\tau}_3\bar{\tau}_4x_3 \vee \bar{\tau}_3\tau_4\bar{x}_5 \vee \tau_3\tau_4x_1;\\
D_3 &= \bar{\tau}_3\bar{\tau}_4(x_3 \vee \bar{x}_4) \vee \bar{\tau}_3\tau_4\bar{x}_5 \vee \tau_3\bar{\tau}_4;\\
D_4 &= \bar{\tau}_3\bar{\tau}_4(x_3 \vee x_4) \vee \bar{\tau}_3\tau_4x_5 \vee \tau_3\bar{\tau}_4 \vee \tau_3\tau_4\bar{x}_1.
\end{aligned} \tag{5.55}$$

If LUTs are used for implementing, for example, the BIMF1, then the Eq. (5.54) are transformed into four tables of LUTerD_1–LUTerD_4. If EMBs are used to implement the circuit of BIMF1, then Table 5.9 are transformed into the table of EMBer1. The same is true for the BIMF2.

In the common case, there are the following fields in a code $K(B_i)$: $K(MX)$ is a code of a direction for MX, $K(B_i)_1$ is a code of a class $B_i \in \Pi_{A1}, \ldots, K(B_i)_K$ is a code of a class $B_i \in \Pi_{AK}$. In the discussed case, there are 5 bits in the code $K(B_i)$:

$$K(B_i) = \langle z_1, \tau_1, \tau_2, \tau_3, \tau_4 \rangle. \tag{5.56}$$

So, there are five outputs in the block BSCT in the discussed case.

There are the following columns in the table of BSCT: $K(a_m), Z, \tau, m$. It is Table 5.11 in the discussed example.

Let us explain how this table is constructed. The code 0000 belongs to the state $a_1 \in B_1$. There is the following relation $B_1 \in \Pi_{A1}$. So, there is $z_1 = 0$ (input memory functions are taken from the BIMF1) and $\tau_1 = \tau_2 = 0$ (it corresponds to the $K(B_1) = 00$), whereas $\tau_3 = \tau_4 = *$ ("don't care"). The first row is completed. The second row contains the code $K(a_8)$ and $a_8 \in B_5$ where $B_5 \in \Pi_{A2}$. So, there are $z_1 = 1, \tau_1 = \tau_2 = *$ and $\tau_1 = 1, \tau_2 = 0$ in the second row of the table. And so on.

The table of BSCT is used for deriving the systems $Z = Z(T)$ and $\tau = \tau(T)$. They are the systems (4.8) and (4.4), correspondingly. If the BSCT is implemented as a LUTer, then the equations should be minimized. For example, the following equations can be derived from Table 5.11:

Table 5.11 Table of BSCT for Moore FSM $P_{2E1}(\Gamma_6)$

$K(a_m)$	Z	τ	m	$K(a_m)$	Z	τ	m
$T_1T_2T_3T_4$	z_1	$\tau_1\tau_2\tau_3\tau_4$		$T_1T_2T_3T_4$	z_1	$\tau_1\tau_2\tau_3\tau_4$	
0000	0	00**	1	1000	0	01**	2
0001	1	**10	8	1001	0	01**	3
0010	1	**10	9	1010	1	**00	4
0011	1	**11	10	1011	*	****	*
0100	0	10**	11	1100	1	**01	5
0101	0	10**	12	1101	*	****	*
0110	0	10**	13	1110	1	**01	7
0111	1	**01	6	1111	*	****	*

$$
\begin{aligned}
z_1 &= T_1T_2 \vee T_3T_4 \vee \bar{T}_1\bar{T}_2T_4 \vee \bar{T}_2T_3;\\
\tau_1 &= T_2;\\
\tau_2 &= T_1;\\
\tau_3 &= T_1T_2;\\
\tau_4 &= T_1T_2 \vee T_3T_4 = \tau_3 \vee T_3T_4.
\end{aligned}
\tag{5.57}
$$

We do not discuss the tables of MX and BMO for this example. These tables are constructed as we do it before.

Now let us discuss an example of design for $P_{3E3}(\Gamma_7)$. There is the GSA Γ_7 shown in Fig. 5.22.

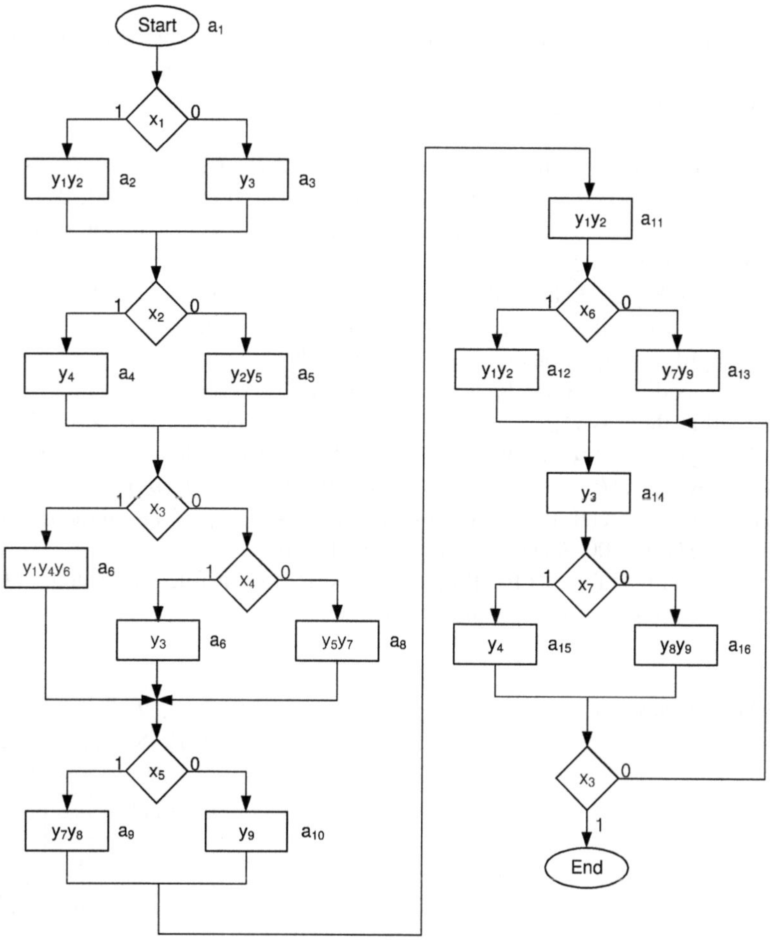

Fig. 5.22 Initial GSA Γ_7

There are the following sets derived from GSA Γ_7: $A = \{a_1, \ldots, a_{16}\}$, $X = \{x_1, \ldots, x_7\}$, $Y = \{y_1, \ldots, y_9\}$. So, there are the following sets and parameters: $M = 16$, $R_1 = 4$, $T = \{T_1, \ldots, T_4\}$, $\Phi = \{D_1, \ldots, D_4\}$, $L = 7$, $N = 9$. There are the following classes of PES: $B_1 = \{a_1\}$, $B_2 = \{a_2, a_3\}$, $B_3 = \{a_4, a_5\}$, $B_4 = \{a_6, a_7, a_8\}$, $B_5 = \{a_9, a_{10}\}$, $B_6 = \{a_{11}\}$, $B_7 = \{a_{12}, a_{13}\}$, $B_8 = \{a_{14}\}$, $B_9 = \{a_{15}, a_{16}\}$. So, there is $I = 9$ and $\Pi_A = \{B_1, \ldots, B_9\}$. It means that there is $R_A = 4$.

There is the following equality in the discussed case:

$$R_1 = R_A. \tag{5.58}$$

It means that the transformation of state codes does not reduce the number of literals in functions Φ.

Let the FPGA chip include LUTs having $S = 5$. Using (5.47), we can find that: $K \leq 3$, $R_K \leq 2$. Let us choose $K = 3$. It gives $R_K = 2$ and $Z = \{z_1, z_2\}$. Let us encode the sets $\Pi_{Ak} \subseteq \Pi_A$ as the following: $K(\Pi_{A1}) = 00$, $K(\Pi_{A2}) = 01$ and $K(\Pi_{A3}) = 10$.

Let us form the following sets Π_{Ak}: $\Pi_{A1} = \{B_1, B_2, B_3\}$, $\Pi_{A2} = \{B_4, B_5, B_6\}$, $\Pi_{A3} = \{B_7, B_8, B_9\}$. It gives the sets $A(\Pi_{A1}) = \{a_2, a_3, a_4, \ldots, a_8\}$, $A(\Pi_{A2}) = \{a_9, a_{10}, a_{11}, a_{12}, a_{13}\}$, $A(\Pi_{A3}) = \{a_{14}, a_{15}, a_{16}, a_1\}$. Let us encode the states $a_m \in A$ as it is shown in Fig. 5.23.

Analysis of Fig. 5.23 shows that there is $T_1 = 0$ for all states $a_m \in A(\Pi_{A1})$. Also, there is $T_1 = 1$ for all states $a_m \in A(\Pi_{A2})$ and for states $a_{14}, a_{15}, a_{16} \in A(\Pi_{A3})$. It allows simplifying equations for functions $D_r \in \Phi$.

Let us encode the classes $B_i \in \Pi_A$. There is $I_{Ak} = 3$ for $\Pi_{Ak} \subseteq \Pi_A (k = \overline{1,3})$. It means that there is $R_{Ak} = 2$. Now we have the following sets: $Z = \{z_1, z_2\}$, $\tau^1 = \{\tau_1, \tau_2\}$, $\tau^2 = \{\tau_3, \tau_4\}$ and $\tau^3 = \{\tau_5, \tau_6\}$. Let us encode the classes $B_i \in \Pi_A$ as the following: $K(B_1) = K(B_4) = K(B_7) = 00$; $K(B_2) = K(B_5) = K(B_8) = 01$; $K(B_3) = K(B_6) = K(B_9) = 10$. Let us encode the sets $\Pi_{Ak} \subseteq \Pi_A$ in the following manner: $K(\Pi_{A1}) = 00$, $K(\Pi_{A2}) = 01$, $K(\Pi_{A3}) = 10$.

To construct the tables of BIMFk, it is necessary to construct the system of GFT. It is the following system in the discussed case:

Fig. 5.23 State codes for $P_{3K3}(\Gamma_7)$

T_1T_2 \ T_3T_4	00	01	11	10
00	a_1	a_2	a_4	a_3
01	a_5	a_6	a_8	a_7
11	a_{13}	a_{14}	a_{15}	a_{16}
10	a_9	a_{10}	a_{12}	a_{11}

$$\begin{aligned} \Pi_{A1} &\begin{cases} B_1 \rightarrow x_1a_2 \vee \bar{x}_1a_3; \\ B_2 \rightarrow x_2a_4 \vee \bar{x}_2a_5; \\ B_3 \rightarrow x_3a_6 \vee \bar{x}_3x_4a_7 \vee \bar{x}_3\bar{x}_4a_8; \end{cases} \\ \Pi_{A2} &\begin{cases} B_4 \rightarrow x_5a_9 \vee \bar{x}_5a_{10}; \\ B_5 \rightarrow a_{11}; \\ B_6 \rightarrow x_6a_{12} \vee \bar{x}_6a_{13}; \end{cases} \\ \Pi_{A3} &\begin{cases} B_7 \rightarrow a_{14}; \\ B_8 \rightarrow x_7a_{15}; \\ B_9 \rightarrow x_3a_1 \vee \bar{x}_3a_{14} \end{cases} \end{aligned} \tag{5.59}$$

The table of BIMFk is constructed using the part $\Pi_{Ak}(k = \overline{1,3})$ from the system (5.59). These blocks are represented by Tables 5.12, 5.13 and 5.14.

Let us denote as D_r^k the function $D_r \in \Phi$ implemented by BIMFk $(k = \overline{1,K})$. In the discussed case, there is the following system of equations derived from

Table 5.12 Table of BIMF1 for Moore FSM $P_{3E3}(\Gamma_7)$

B_i	$K(B_i)$	a_s	$K(a_s)$	X_h	Φ_h	h
B_1	00	a_2	0001	x_1	D_4	1
		a_3	0010	$\bar{x}_1$	D_3	2
B_2	01	a_4	0011	x_2	D_3D_4	3
		a_5	0100	$\bar{x}_2$	D_2	4
B_3	10	a_6	0101	x_3	D_2D_4	5
		a_7	0110	$\bar{x}_3x_4$	D_2D_3	6
		a_8	0111	$\bar{x}_3\bar{x}_4$	$D_2D_3D_4$	7

Table 5.13 Table of BIMF2 for Moore FSM $P_{3E3}(\Gamma_7)$

B_i	$K(B_i)$	a_s	$K(a_s)$	X_h	Φ_h	h
B_4	00	a_9	1000	x_5	D_1	1
		a_{10}	1001	$\bar{x}_5$	D_1D_4	2
B_5	01	a_{11}	1010	1	D_1D_3	3
B_6	10	a_{12}	1011	x_6	$D_1D_3D_4$	4
		a_{13}	1100	$\bar{x}_6$	D_1D_2	5

Table 5.14 Table of BIMF3 for Moore FSM $P_{3E3}(\Gamma_7)$

B_i	$K(B_i)$	a_s	$K(a_s)$	X_h	Φ_h	h
B_7	00	a_{14}	1101	1	$D_1D_2D_4$	1
B_8	01	a_{15}	1111	x_7	$D_1D_2D_3D_4$	2
		a_{16}	1110	$\bar{x}_7$	$D_1D_2D_3$	3
B_9	10	a_1	0000	x_3	–	4
		a_{14}	1101	$\bar{x}_3$	$D_1D_2D_4$	5

Tables 5.12, 5.13 and 5.14:

$$\begin{aligned}
D_1^1 &= 0;\\
D_2^1 &= \tau_2\bar{x}_2 \vee \tau_1;\\
D_3^1 &= \bar{\tau}_1\bar{\tau}_2\bar{x}_1 \vee \tau_2 x_2 \vee \tau_1\bar{x}_3;\\
D_4^1 &= \bar{\tau}_1\bar{\tau}_2 x_1 \vee \tau_2 x_2 \vee \tau_1 x_3 \vee \tau_1\bar{x}_4;\\
D_1^2 &= 1;\\
D_2^2 &= \tau_3\bar{x}_6;\\
D_3^2 &= \tau_4 \vee \tau_3 x_6;\\
D_4^2 &= \bar{\tau}_3\bar{\tau}_4\bar{x}_5 \vee \tau_3 x_6;\\
D_1^3 &= \bar{\tau}_5 \vee \tau_5\bar{x}_3 = D_2^3;\\
D_3^3 &= \tau_6;\\
D_4^3 &= \bar{\tau}_5\bar{\tau}_6 \vee \tau_6 x_7 \vee \tau_5\bar{x}_3.
\end{aligned} \tag{5.60}$$

Analysis of system (5.60) shows that it can be implemented using 8 LUTs having $S \leq 5$. There are no LUTs necessary for implementing the functions D_2^1, D_1^2, D_2^3 and D_3^3.

The MX is represented using the following equations:

$$\begin{aligned}
D_1 &= \bar{z}_1\bar{z}_2 \cdot 0 \vee z_2 D_1^2 \vee z_1 D_1^3 = z_2 \vee z_1 D_1^3;\\
D_2 &= \bar{z}_1\bar{z}_2 D_2^1 \vee z_2 D_2^2 \vee z_1 D_2^3;\\
D_3 &= \bar{z}_1\bar{z}_2 D_3^1 \vee z_2 D_3^2 \vee z_1 D_3^3;\\
D_4 &= \bar{z}_1\bar{z}_2 D_4^1 \vee z_2 D_4^2 \vee z_1 D_4^3.
\end{aligned} \tag{5.61}$$

There are the following columns in the table of BSCT: a_m, $K(a_m)$, B_i, $K(B_i)$, Π_{Ak}, $K(\Pi_{Ak})$, m. The EMBer can implement Δ_t functions $f \in ZU\tau$. To find these functions, let us find the minimized functions $Z(T)$ and $\tau(T)$.

Let $A(B_i^k) \subseteq A$ be a set of states from the classes $B_i \in \Pi_{Ak}$. There are three such sets in the discussed case: $A(B_i^1) = \{a_1, \ldots, a_5\}$, $A(B_i^2) = \{a_6, \ldots, a_{11}\}$, $A(B_i^3) = \{a_{12}, \ldots, a_{16}\}$. Let us construct the Karnaugh maps for functions $\tau_1, \tau_2 \in \tau$ (Fig. 5.24).

Fig. 5.24 Karnaugh maps for functions $\tau_1(a)$ and $\tau_2(b)$

(a)

T_1T_2 \ T_3T_4	00	01	11	10
00	0	0	1	0
01	1	*	*	*
11	*	*	*	*
10	*	*	*	*

(b)

T_1T_2 \ T_3T_4	00	01	11	10
00	0	1	0	1
01	0	*	*	*
11	*	*	*	*
10	*	*	*	*

These maps are filled using the following equations:

$$\begin{aligned} B_1 &= A_1; \\ B_2 &= A_2 \vee A_3; \\ B_3 &= A_4 \vee A_5. \end{aligned} \quad (5.62)$$

It gives the following equations:

$$\begin{aligned} \tau_1 &= A_4 \vee A_5; \\ \tau_2 &= A_2 \vee A_3. \end{aligned} \quad (5.63)$$

The BIMF1 operates if there is $z_1 = z_2 = 0$. So, for $z_1 = z_2 = 0$ all a_m give "don't care" values if $a_m \in A(B_i^1)$. It explains the signs "*" in the Karnaugh maps from Fig. 5.24.

The following equations can be obtained from (5.63):

$$\begin{aligned} \tau_1 &= T_2 \vee T_3 T_4; \\ \tau_2 &= \bar{T}_3 T_4 \vee T_3 \bar{T}_4. \end{aligned} \quad (5.64)$$

Acting in the same way, the following equations can be obtained:

$$\begin{aligned} z_1 &= T_1 T_2 \vee T_1 T_3 T_4; \\ z_2 &= \bar{T}_1 T_2 T_4 \vee \bar{T}_1 T_2 T_3 \vee T_1 \bar{T}_2 \bar{T}_3 \vee T_1 \bar{T}_2 \bar{T}_4; \\ \tau_3 &= T_1 T_3; \\ \tau_4 &= T_1 \bar{T}_3; \\ \tau_5 &= T_2 T_3; \\ \tau_6 &= \bar{T}_3 T_4. \end{aligned} \quad (5.65)$$

Next, we should choose the equations requiring the most complicated circuits for implementation with EMBer. For example, if there is $S = 2$, then the equations for τ_1, z_1 and z_2 should be implemented by EMBer. If there is $S = 3$, then it is a must to implement the equation z_2 by EMBer.

In the discussed case, EMBs in use have $t_F = 12$. It means that $\Delta_t = 3$. Let us take the functions z_1, z_2 and τ_1 for implementing with EMBer. So, there are 5 of different tables for LUTs from the BSCT. These tables are constructed in the trivial way.

There are the following columns in the table of BMO: a_m, $K(a_m)$, $Y(a_m)$, $Z(a_m)$, $\tau(a_m)$, m. In the discussed case, it is Table 5.15.

We take the microoperations $y_n \in Y(a_m)$ from Fig. 5.22. The state codes are taken from Fig. 5.23. The column $Z(a_m)$ includes the variables $z_r \in Z$ equal to 1 for the states $a_m \in A(B_i^k)$. So, there is the value of $z_1 = 1$ for the states $a_m \in A(B_i^3)$, the value of $z_2 = 1$ is for the states $a_m \in A(B_i^2)$. It is found due to the codes $K(\Pi_{A1}) = 00$, $K(\Pi_{A2}) = 01$ and $K(\Pi_{A3}) = 10$. The column $\tau(a_m)$ includes the function τ_1 for the classes $B_i \in \Pi_{A1}$. In the discussed case it is the class $B_3 = \{a_4, a_5\}$.

Table 5.15 Table of BMO for Moore FSM $P_{3E3}(\Gamma_7)$

a_m	$K(a_m)$	$Y(a_m)$	$Z(a_m)$	$\tau(a_m)$	m
a_1	0000	–	–	–	1
a_2	0001	y_1y_2	–	–	2
a_3	0010	y_3	–	–	3
a_4	0011	y_4	–	τ_1	4
a_5	0100	y_2y_5	–	τ_1	5
a_6	0101	$y_1y_4y_6$	z_2	–	6
a_7	0110	y_3	z_2	–	7
a_8	0111	y_5y_7	z_2	–	8
a_9	1000	y_7y_8	z_2	–	9
a_{10}	1001	y_9	z_2	–	10
a_{11}	1010	y_1y_2	z_2	–	11
a_{12}	1011	y_1y_2	z_1	–	12
a_{13}	1100	y_7y_9	z_1	–	13
a_{14}	1101	y_3	z_1	–	14
a_{15}	1110	y_4	z_1	–	15
a_{16}	1111	y_8y_9	z_1	–	16

Let us point out that two options are possible for each block BIMFk ($k = \overline{1, K}$). The block's circuit can be implemented either as LUTer or as EMBer. It gives 2^K different models for each FSM $P_{KEi}(i = \overline{1, 3})$. We do not discuss these models in our book.

Let us point out that the models P_{E1}–P_{E6} discussed before have $K = 2$. It means that they should correspond to the variable $z_1 \in Z$. It gives the set $Z = \{z_1\}$. In the next Chapter, we use these denotations for above mentioned models.

To diminish the number of LUTs in the circuit of BIMFk ($k = \{1, 2, \ldots, K\}$), it is necessary to diminish the number of literals in the functions $\Phi_k \subseteq \Phi$. It can be achieved by the RLC. There are other methods of hardware reduction discussed in the Chap. 6 together with the RLC for P_{KEi} Moore FSMs.

References

1. Altera: http://www.altera.com. Accessed Jan 2015
2. Atmel: http://www.atmel.com. Accessed Jan 2015
3. P. Bacchetta, L. Daldos, D. Sciuto, C. Silvano, Low-power state assignment techniques for finite state machines, in *Proceedings of the 2000 IEEE International Symposium on Circuits and Systems (ISCAS'2000)*, vol. 2, Geneva. IEEE (2000), pp. 641–644
4. S. Baranov, *Logic Synthesis of Control Automata* (Kluwer Academic Publishers, Dordrecht, 1994)

5. A. Barkalov, Principles of logic optimization for Moore microprogram automaton. Cybern. Syst. Anal. **34**(1), 54–60 (1998)
6. A. Barkalov, L. Titarenko, S. Chmielewski, Reduction in the number of PAL macrocells int the circuit of a Moore FSM. Int. J. Appl. Math. Comput. Sci. **17**(4), 565–675 (2007)
7. A. Barkalov, L. Titarenko, S. Chmielewski, Decrease of hardware amount in logic circuit of Moore FSM. Przegląd Telekomunikacyjny i Wiadomości Telokomunikacyjne (6), 750–752 (2008)
8. A. Barkalov, L. Titarenko, S. Chmielewski, Optimization of Moore control unit with refined state encoding, in *Proceedings of the 15th International Conference MIXDES 2008*, Poznań, Poland. Department of Microelectronics and Computer Science, Technical University of Łódz (2008), pp. 417–420
9. A. Barkalov, L. Titarenko, S. Chmielewski, Optimization of Moore FSM on system-on-chip using PAL technology, in *Proceedings of the International Conference TCSET 2008*, Lviv-Slavsko, Ukraina. Ministry of Education and Science of Ukraine, Lviv Polytechnic National University, Lviv, Publishing House of Lviv Polytechnic (2008), pp. 314–317
10. A. Barkalov, L. Titarenko, S. Chmielewski, Hardware reduction in CPLD-based Moore FSM. J. Circuits, Syst., Comput. **23**(6), 1450086–1–1450086–21 (2014)
11. S. Chmielewski, Using structural peculiarities of Moore FSM for reduction of number of PALS. Ph.D. thesis, University of Zielona Góra (2014)
12. Cypress Semiconductor Corporation: http://www.cypress.com. Accessed Jan 2015
13. Cypress Semiconductor Corporation: Cypress programmable logic: delta 39K. Data sheet, http://cypress.com/pld/delta39k.html. Accessed Jan 2015
14. R. Czerwinski, D. Kania, *Finite State Machine Logic Synthesis for Complex Programmable Logic Devices*. Lecture Notes in Electrical Engineering, vol. 23 (Springer, Berlin, 2013)
15. J. Rho, F. Hatchel, R. Somenzi, R. Jacoby, Exact and heuristic algorithms for the minimization of incompletely specified state machines. IEEE Trans. Comput.-Aided Des. **13**(2), 167–177 (1994)
16. Xilinx: http://www.xilinx.com. Accessed Jan 2015

Chapter 6
Hardware Reduction in Multidirectional Moore FSMs

6.1 Hardware Reduction in Two-Directional Moore FSMs

We show in Chap. 5 that it could be K of sources of state codes in Moore FSMs. Let us name such FSMs K-directional Moore FSMs. Let us start from an approach of hardware reduction for $K = 2$.

The structural diagram of P_{2E1} Moore FSM is shown in Fig. 6.1. It can be applied if the condition (5.8) takes place.

If $K = 2$, then the set Π_A is represented as $\Pi_{A1} \cup \Pi_{A2}$. Using (5.44), we can find $R_K = 1$ and $Z = \{z_1\}$. Let us encode the sets Π_{A1}, Π_{A2} in the following way: $K(\Pi_{A1}) = 0$ and $K(\Pi_{A2}) = 1$. If $z_1 = 0$, then the functions Φ_1 are transited through the MX. If $z_1 = 1$, the MX transfers the functions Φ_2. So, the functions $\Phi_2(\Phi_1)$ are ignored if there is $z_1 = 0(z_1 = 1)$. This property can be used for the hardware reduction.

The proposed approach is based on the optimal state assignment. So, let us denote as P_{20E1} FSM the P_{2E1} FSM based on the proposed method of hardware reduction. Let us explain this approach using an example of design for Moore FSM $P_{20E1}(\Gamma_8)$. The GSA Γ_8 is shown in Fig. 6.2.

Analysis of GSA Γ_8 gives the following sets and their cardinality numbers: $X = \{x_1, ..., x_7\}, L = 7, Y = \{y_1, ..., y_8\}, N = 8, A = \{a_1, ..., a_{15}\}, M_1 = 15$. Using (1.14), the value $R_1 = 4$ is obtained. It gives the sets $T = \{T_1, ..., T_4\}$ and $\Phi = \{D_1, ..., D_4\}$.

Using the definition of PES [1], the the following set Π_A is constructed: $\Pi_A = \{B_1, ..., B_7\}$. It includes the classes $B_1 = \{a_1\}, B_2 = \{a_2, a_3, a_4\}, B_3 = \{a_5, a_6\}, B_4 = \{a_7, a_8\}, B_5 = \{a_9, a_{10}\}, B_6 = \{a_{11}, a_{12}, a_{13}\}$ and $B_7 = \{a_{14}, a_{15}\}$.

Let us represent the set Π_A as $\Pi_{A1} \cup \Pi_{A2}$ where $\Pi_{A1} = \{B_2, B_3, B_4, B_5\}$ and $\Pi_{A2} = \{B_1, B_6, B_7\}$. Let us encode the states $a_m \in A$ in a way shown in Fig. 6.3.

There is $K = 2$. It gives $R_K = 1$ and $Z = \{z_1\}$. Let us encode the directions in the following way: $K(\Pi_{A1}) = 0$ and $K(\Pi_{A2}) = 1$. If $z_1 = 0$, then the state assignments are treated as insignificant for the states $a_m \in A(B_i^2)$. It gives the Karnaugh map shown in Fig. 6.4.

A. Barkalov et al., *Logic Synthesis for FPGA-Based Finite State Machines*, Studies in Systems, Decision and Control 38, DOI 10.1007/978-3-319-24202-6_6

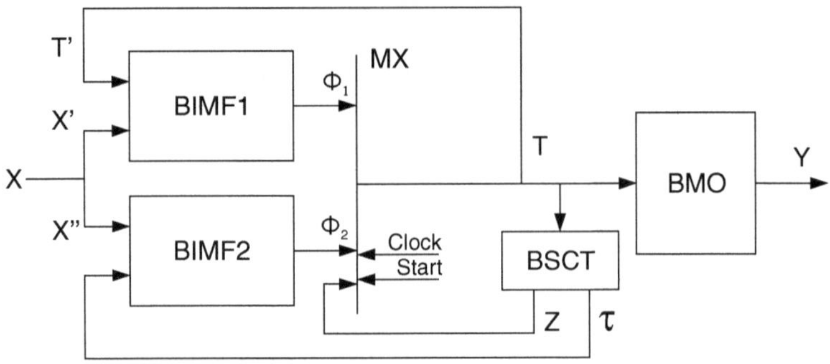

Fig. 6.1 Structural diagram of P_{2E1} Moore FSM

It is possible to derive the following class codes from Fig. 6.4: $K(B_2) = 00{*}{*}$, $K(B_3) = 01{*}{*}$, $K(B_4) = 11{*}{*}$ and $K(B_5) = 10{*}{*}$. It gives the set $T' = \{T_1, T_2\}$.

There are three classes in the set Π_{A2}. It gives $I_{A2} = 3$. Using (5.43), we can find $R_{A2} = 2$ and $\tau^2 = \{\tau_1, \tau_2\}$. Let us encode the classes $B_i \in \Pi_{A2}$ as the following: $K(B_1) = 00$, $K(B_6) = {*}1$ and $K(B_7) = 1{*}$. To get the codes of $B_6, B_7 \in \Pi_{A2}$, they use the "don't care" class assignment 11.

Let us construct the system of GFT for GSA Γ_8. It is the following one:

$$\begin{aligned}
B_1 &\rightarrow x_1a_2 \vee \bar{x}_1x_2a_3 \vee \bar{x}_1\bar{x}_2a_4;\\
B_2 &\rightarrow x_3a_5 \vee \bar{x}_3x_4a_6 \vee \bar{x}_3\bar{x}_4a_4;\\
B_3 &\rightarrow x_2a_7 \vee \bar{x}_2a_8;\\
B_4 &\rightarrow x_5a_7 \vee \bar{x}_5x_6a_5 \vee \bar{x}_5\bar{x}_6a_{10};\\
B_5 &\rightarrow x_3x_7a_{11} \vee x_3\bar{x}_7a_{12} \vee \bar{x}_3a_{13};\\
B_6 &\rightarrow x_1a_4 \vee \bar{x}_1a_{15};\\
B_7 &\rightarrow x_4a_{14} \vee \bar{x}_4a_1.
\end{aligned} \tag{6.1}$$

Tables of BIMFk ($k = \overline{1,2}$) are constructed as we did it before. They include the following columns: B_i, $K(B_i)$, a_s, $K(a_s)$, X_h, Φ_h, h. The tables use the corresponding formulae from the system of GFT. There is the table of BIMF1 represented by Table 6.1, whereas the table of BIMF2 by Table 6.2. We use the state codes (Fig. 6.3) in these tables.

The table of BIMF1 is used for deriving the system (5.18). Analysis of Table 6.1 shows that there are the sets $T' = \{T_1, T_2\}$ and $X' = \{x_2, ..., x_7\}$. The table of BIMF2 is used for deriving the system (5.19). Analysis of Table 6.2 shows that there is the set $X'' = \{x_1, x_2, x_4\}$. The system (6.2) is derived from Table 6.1. The system (6.3) is derived from Table 6.2. Both systems are minimized using the rules of Boolean algebra [2].

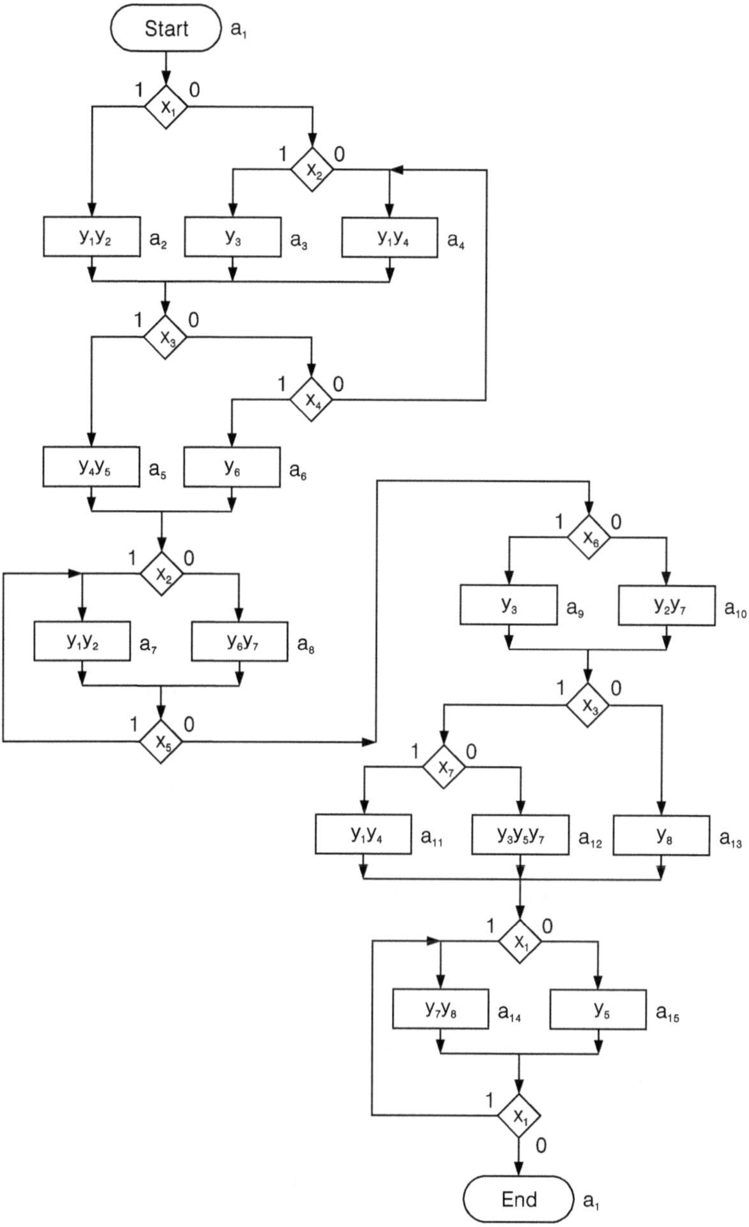

Fig. 6.2 Initial GSA Γ_8

Fig. 6.3 State codes for Moore FSM $P(\Gamma_8)$

T_1T_2 \ T_3T_4	00	01	11	10
00	a_1	a_2	a_4	a_3
01	a_5	a_6	a_{11}	a_{15}
11	a_7	a_8	a_{13}	*
10	a_9	a_{10}	a_{12}	a_{14}

Fig. 6.4 Karnaugh map for $A(B_i^1)$

T_1T_2 \ T_3T_4	00	01	11	10
00	*	a_2	a_4	a_3
01	a_5	a_6	*	*
11	a_7	a_8	*	*
10	a_9	a_{10}	*	*

Table 6.1 Table of BIMF1 for Moore FSM $P_{20E1}(\Gamma_8)$

B_i	$K(B_i)$	a_s	$K(a_s)$	X_h	Φ_h	h
B_2	00**	a_5	0100	x_3	D_2	1
		a_6	0101	$\bar{x_3}x_4$	D_2D_4	2
		a_4	0011	$\bar{x_3}\bar{x_4}$	D_3D_4	3
B_3	01**	a_7	1100	x_2	D_1D_2	4
		a_8	1101	$\bar{x_2}$	$D_1D_2D_4$	5
B_4	11**	a_7	1100	x_5	D_1D_2	6
		a_9	1000	$\bar{x_5}x_6$	D_1	7
		a_{10}	1001	$\bar{x_5}\bar{x_6}$	D_1D_4	8
B_5	10**	a_{11}	0111	x_3x_7	$D_2D_3D_4$	9
		a_{12}	1011	$x_3\bar{x_7}$	$D_1D_3D_4$	10
		a_{13}	1111	$\bar{x_3}$	$D_1D_2D_3D_4$	11

Table 6.2 Table of BIMF2 for Moore FSM $P_{20E1}(\Gamma_8)$

B_i	$K(B_i)$	a_s	$K(a_s)$	X_h	Φ_h	h
B_1	00	a_2	0001	x_1	D_4	1
		a_3	0010	$\bar{x_1}x_2$	D_3	2
		a_4	0011	$\bar{x_1}\bar{x_2}$	D_3D_4	3
B_3	*1	a_4	0011	x_1	D_3D_4	4
		a_{15}	0110	$\bar{x_1}$	D_2D_3	5
B_4	1*	a_{14}	1010	x_4	D_1D_3	6
		a_1	0000	$\bar{x_4}$	–	7

Table 6.3 Table of BSCT for Moore FSM $P_{20E1}(\Gamma_8)$

a_m	$K(a_m)$	$Z(a_m)$	$\tau(a_m)$	m	
a_1	0000	z_1	–	1	B_1
a_1	0000	z_1	τ_2	11	B_6
a_1	0000	z_1	τ_2	12	
a_1	0000	z_1	τ_2	13	
a_1	0000	z_1	τ_1	14	B_7
a_1	0000	z_1	τ_1	15	

$$\begin{aligned}
D_1^1 &= T_2 \vee T_1\bar{T}_2(\bar{x}_3 \vee \bar{x}_7);\\
D_2^1 &= \bar{T}_1\bar{T}_2(x_3 \vee x_4) \vee \bar{T}_1 T_2 \vee T_1 T_2 x_5 \vee T_1\bar{T}_2\bar{x}_3;\\
D_3^1 &= \bar{T}_1\bar{T}_2\bar{x}_3\bar{x}_4 \vee T_1\bar{T}_2;\\
D_4^1 &= \bar{T}_1\bar{T}_2\bar{x}_3 \vee \bar{T}_1 T_2\bar{x}_2 \vee T_1 T_2\bar{x}_5\bar{x}_6 \vee T_1\bar{T}_2.
\end{aligned} \tag{6.2}$$

$$\begin{aligned}
D_1^2 &= \tau_1 x_4;\\
D_2^2 &= \tau_2\bar{x}_1;\\
D_3^2 &= \bar{\tau}_1\bar{\tau}_2\bar{x}_1 \vee \tau_2 \vee \tau_1 x_4;\\
D_4^2 &= \bar{\tau}_1\bar{\tau}_2(x_1 \vee \bar{x}_2) \vee \tau_2 x_1.
\end{aligned} \tag{6.3}$$

Let it be the configuration 16×8 for EMBs used in this example. There is $N = 8$. Using (5.5), it is possible to find the value of $\Delta_t = 0$. So, the model of P_{20E1} can be used.

There are the following columns in the table of BSCT: $a_m, K(a_m), Z(a_m), \tau(a_m), m$. The table is constructed for the states $a_m \in A(B_i^2)$. It is Table 6.3 in the discussed case.

This table is used for deriving the systems $Z(T)$ and $\tau(T)$. After minimizing, the following equations are obtained:

$$\begin{aligned}
z_1 &= \bar{T}_1\bar{T}_2\bar{T}_3\bar{T}_4 \vee T_2 T_3 \vee T_1 T_3;\\
\tau_1 &= T_2 T_3\bar{T}_4 \vee T_1 T_3\bar{T}_4;\\
\tau_2 &= T_2 T_3 T_4 \vee T_1 T_3 T_4.
\end{aligned} \tag{6.4}$$

There are the following columns in the table of BMO: $a_m, K(a_m), Y(a_m), m$. It is constructed in the trivial way. We do not show it for the discussed example.

Let it be LUTs having $S = 6$ for the FPGA chip used in the discussed case. In this case, any equation from (6.2)–(6.4) is implemented as a single LUT. The same is true to the equations representing MX:

$$\begin{aligned}
D_1 &= \bar{z}_1 D_1^1 \vee z_1 D_1^2;\\
D_2 &= \bar{z}_1 D_2^1 \vee z_1 D_2^2;\\
D_3 &= \bar{z}_1 D_3^1 \vee z_1 D_3^2;\\
D_4 &= \bar{z}_1 D_4^1 \vee \bar{z}_1 D_4^2.
\end{aligned} \tag{6.5}$$

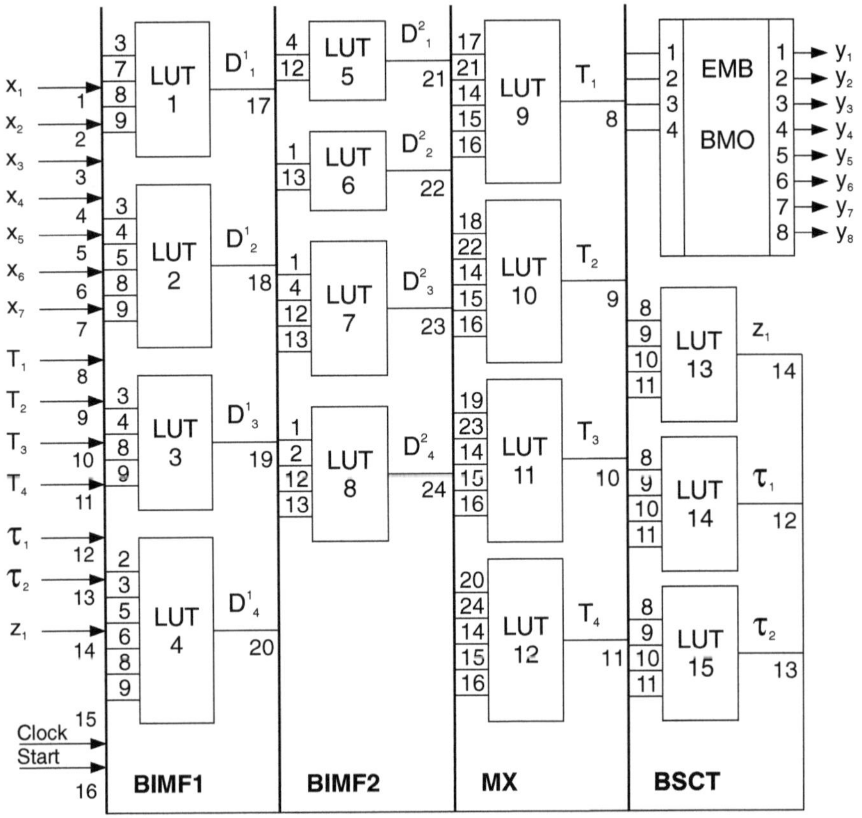

Fig. 6.5 Logic circuit of Moore FSM $P_{20E1}(\Gamma_8)$

There is the logic circuit of Moore FSM $P_{20E1}(\Gamma_8)$ shown in Fig. 6.5.

The circuit of BIMF1 is implemented by LUT1–LUT4. The circuit of BIMF2 is implemented by LUT5–LUT8. The circuit of MX is implemented by LUT9–LUT12. Let us point out that only 3 inputs are required in LUT9–LUT12. But we show that the pulses *Clock* and *Start* are used to control the D flip-flops inside these LUTs. So, the register RG is hidden inside the LUTs of MX. The circuit of BMO is implemented as a single EMB. The tables for each LUT can be constructed using the corresponding equations from the systems (6.2)–(6.5).

Obviously, the functions Z can be implemented by the BIMF2 as

$$Z = Z(\tau). \tag{6.6}$$

It leads to P_{20E2} Moore FSM (Fig. 6.6).

Let us point out that it is possible to optimize the circuit of BSCT for $P_{20E1}(\Gamma_8)$. Obviously, it is correct to treat the states $a_m \in A(B_i^1)$ as insignificant for $z_1 = 1$. It gives the following Karnaugh map (Fig. 6.7).

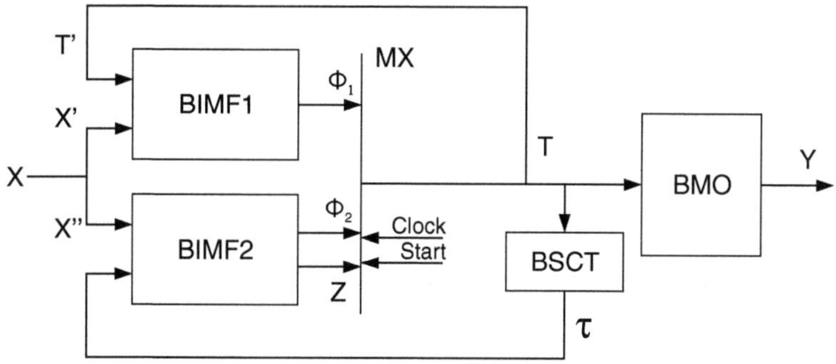

Fig. 6.6 Structural diagram of P_{20E2} Moore FSM

Fig. 6.7 Karnaugh map for $A(B_i^2)$

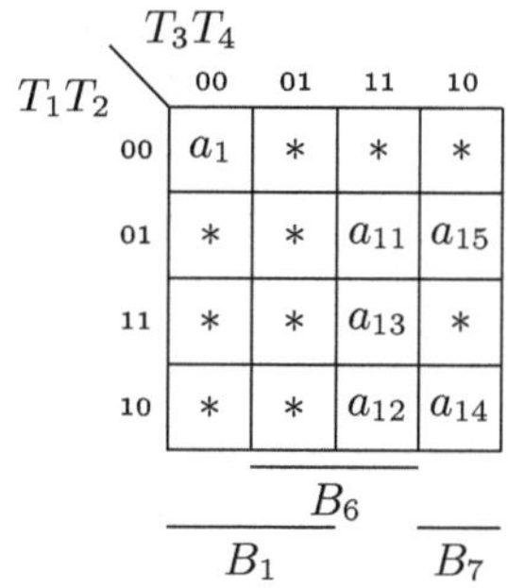

There are the following class codes found from Fig. 6.7: $K(B_1) = **0*$, $K(B_6) = ***1$, $K(B_7) = **10$. It means that the functions $D_r \in \Phi_2$ can be represented as

$$\Phi_2 = \Phi_2(X'', T''). \tag{6.7}$$

It leads to P^1_{20E1} Moore FSM (Fig. 6.8).

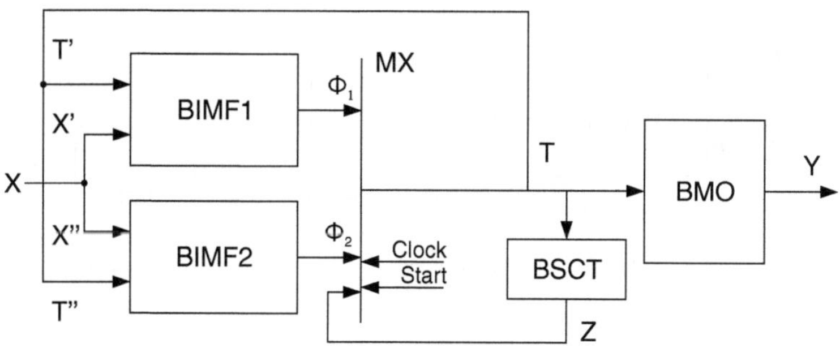

Fig. 6.8 Structural diagram of P^1_{20E1} Moore FSM

Table 6.4 Table of BIMF2 for Moore FSM $P^1_{20E1}(\Gamma_8)$

B_i	$K(B_i)$	a_s	$K(a_s)$	X_h	Φ_h	h
B_1	$**0*$	a_2	0001	x_1	D_4	1
		a_3	0010	$\bar{x}_1x_2$	D_3	2
		a_4	0011	$\bar{x}_1\bar{x}_2$	D_3D_4	3
B_6	$***1$	a_4	0011	x_1	D_3D_4	4
		a_{15}	0110	$\bar{x}_1$	D_2D_3	5
B_7	$**10$	a_{14}	1010	x_4	D_1D_3	6
		a_1	0000	$\bar{x}_4$	–	7

The BSCT implements only the equation $z_1(T)$ for P^1_{20E1} Moore FSM. There is the table of BIMF2 for the Moore FSM $P^1_{20E1}(\Gamma_8)$ represented by Table 6.4.

The system (6.7) is derived from the table of BIMF2. It is the following system in the discussed case:

$$\begin{aligned} D_1^2 &= T_3\bar{T}_4x_4; \\ D_2^2 &= T_4\bar{x}_1; \\ D_3^2 &= \bar{T}_3\bar{x}_1 \vee T_3 \vee T_3\bar{T}_4x_4; \\ D_4^2 &= \bar{T}_3(x_1 \vee \bar{x}_2) \vee T_4x_1. \end{aligned} \tag{6.8}$$

The BSCT should implement the first equation of (6.4). Obviously, there are no LUT14 and LUT15 in the circuit of $P^1_{20E1}(\Gamma_8)$. So, this approach allows saving two LUTs (and 8 interconnections) in comparison with $P_{20E1}(\Gamma_8)$.

So, there are three models of the two-directional Moore FSMs for the case $\Delta_t = 0$. They are the models P_{20E1}, P^1_{20E1} and P_{20E2}.

Let the following condition take place:

$$\Delta_t \geq R_{A2} + 1. \tag{6.9}$$

In this case, the model of P_{20E3} Moore FSM (Fig. 6.9) can be used.

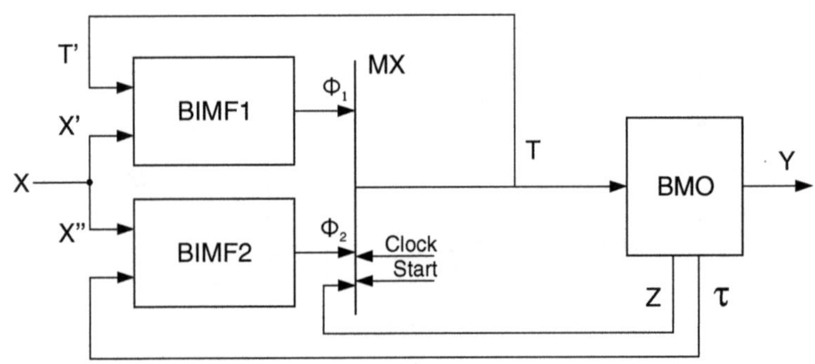

Fig. 6.9 Structural diagram of P_{20E3} Moore FSM

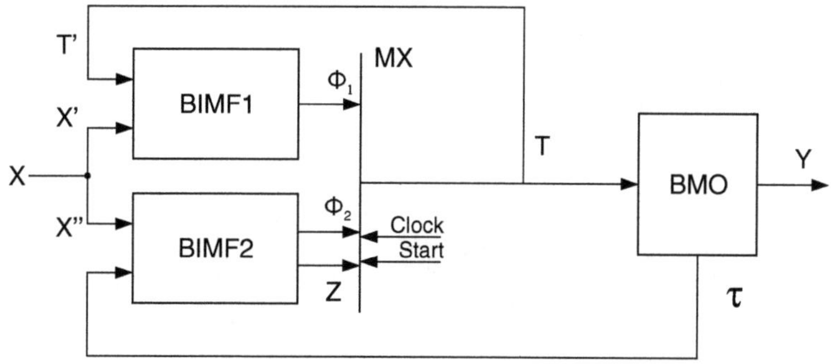

Fig. 6.10 Structural diagram of P_{20E4} Moore FSM

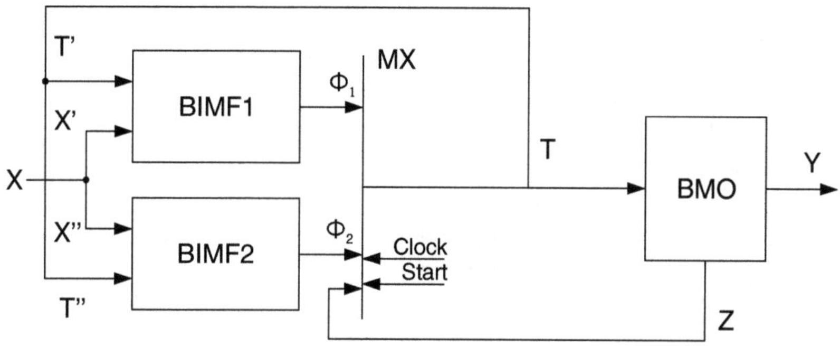

Fig. 6.11 Structural diagram of P^1_{20E3} Moore FSM

The function z_1 can be implemented by the BIMF2. It leads to P_{20E4} Moore FSM (Fig. 6.10).

This model should be applied if the following condition takes place:

$$\Delta_t = R_{A2}. \tag{6.10}$$

Obviously, it is possible to use the model of P^1_{20E3} (Fig. 6.11). It can be used if

$$\Delta_t = 1. \tag{6.11}$$

Let the following condition take place:

$$0 < \Delta_t < R_{A2} + 1. \tag{6.12}$$

In this case, the P_{20E5} Moore FSM (Fig. 6.12) can be used.

Two more models are possible. In case of P^1_{20E5} FSM, the BMO implements only the function z_1 (Fig. 6.13). In case of P_{20E6} FSM (Fig. 6.14), the functions Z are implemented by BIMF2.

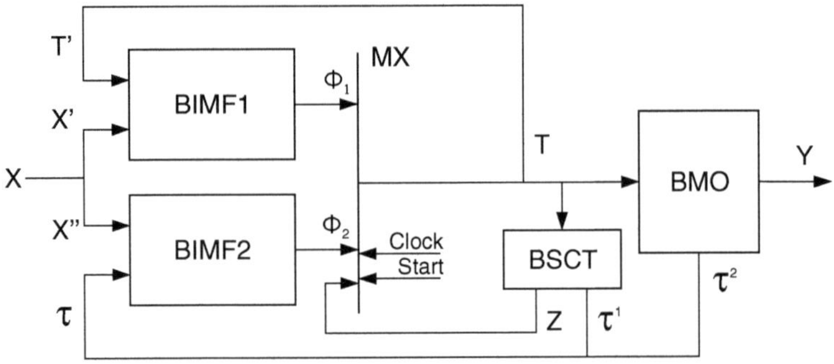

Fig. 6.12 Structural diagram of P_{20E5} Moore FSM

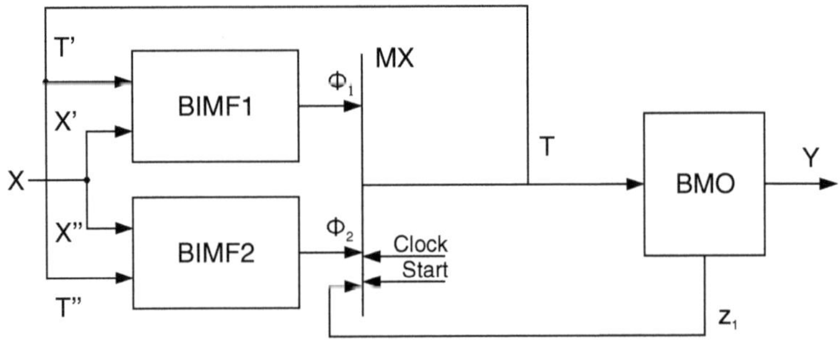

Fig. 6.13 Structural diagram of P^1_{20E5} Moore FSM

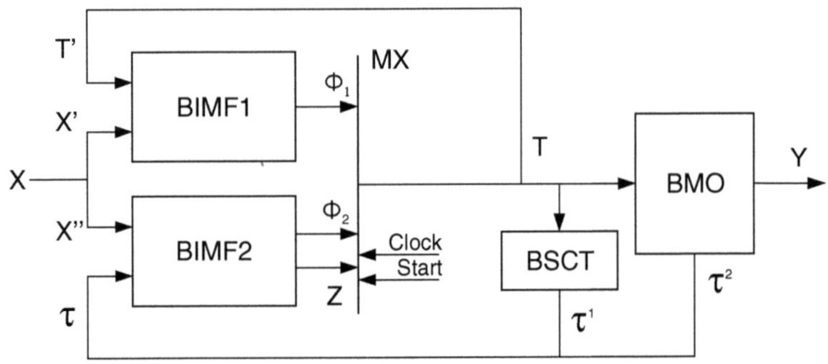

Fig. 6.14 Structural diagram of P_{20E6} Moore FSM

If functions $z_r \in Z$ are implemented as $Z(\tau)$, then it is necessary to use one code $K(B_i)$ for the relation $B_i \notin \Pi_{A2}$. In this case, the formula (5.43) is changed. Now it is the following formula:

$$R_{A2} = \lceil log_2(I_{A2} + 1) \rceil. \tag{6.13}$$

Table 6.5 Characteristics of two-directional Moore FSMs

№	FSM	BIMF1	BIMF2	BSCT	BMO
1	P_{20E1}	$\Phi_1 = \Phi_1(T', X')$	$\Phi_2 = \Phi_2(\tau, X'')$	$Z = Z(T)$ $\tau = \tau(T)$	$Y = Y(t)$
2	P_{20E2}	$\Phi_1 = \Phi_1(T', X')$	$\Phi_2 = \Phi_2(\tau, X'')$ $Z = Z(\tau)$	$\tau = \tau(T)$	$Y = Y(t)$
3	P^1_{20E1}	$\Phi_1 = \Phi_1(T', X')$	$\Phi_2 = \Phi_2(T'', X'')$	$Z = Z(T)$	$Y = Y(t)$
4	P_{20E3}	$\Phi_1 = \Phi_1(T', X')$	$\Phi_2 = \Phi_2(\tau, X'')$	–	$Y = Y(t)$ $Z = Z(T)$ $\tau = \tau(T)$
5	P_{20E4}	$\Phi_1 = \Phi_1(T', X')$	$\Phi_2 = \Phi_2(\tau, X'')$ $Z = Z(\tau)$	–	$Y = Y(t)$ $\tau = \tau(T)$
6	P^1_{20E3}	$\Phi_1 = \Phi_1(T', X')$	$\Phi_2 = \Phi_2(T'', X'')$	–	$Y = Y(t)$ $Z = Z(T)$
7	P_{20E5}	$\Phi_1 = \Phi_1(T', X')$	$\Phi_2 = \Phi_2(\tau, X'')$	$Z = Z(T)$ $\tau^1 = \tau^1(T)$	$Y = Y(t)$ $\tau^2 = \tau^2(T)$
8	P_{20E6}	$\Phi_1 = \Phi_1(T', X')$	$\Phi_2 = \Phi_2(\tau, X'')$ $Z = Z(\tau)$	$\tau^1 = \tau^1(T)$	$Y = Y(t)$ $\tau^2 = \tau^2(T)$
9	P^1_{20E5}	$\Phi_1 = \Phi_1(T', X')$	$\Phi_2 = \Phi_2(T'', X'')$	–	$Y = Y(t)$ $Z = Z(T)$

There are 9 different structures of two-directional Moore FSM (Table 6.5). Let us point out that each of these models has 4 modifications. A modification is determined by the elements used for implementing circuits of BIMF1 and BIMF2: $\langle LUTer, LUTer\rangle$, $\langle LUTer, EMBer\rangle$, $\langle EMBer, LUTer\rangle$ and $\langle EMBer, EMBer\rangle$.

These FSMs are designed in the ways we discussed in this section. Analysis of Table 6.5 shows that the BIMF1 always implements the functions (5.18). The functions of other blocks depend on the model. The type of the model to be used depends on the relations between such parameters as Δ_t and R_K.

6.2 Hardware Reduction in 3-Directional Moore FSMs

They can increase the number of directions till the relation (5.46) is true. If there is $S = 6$, then it is possible to have $K \in \{2, 3, 4\}$. We discussed the case $K = 2$ in the previous section. Now, let us discuss the case when $K = 3$.

Let us encode the states $a_m \in A(\Pi_{A1})$ in the optimal way. In this case, the BIMF1 implements the functions (5.18). It leads to P_{30E1} Moore FSM (Fig. 6.15). This model is used if the condition (5.8) takes place.

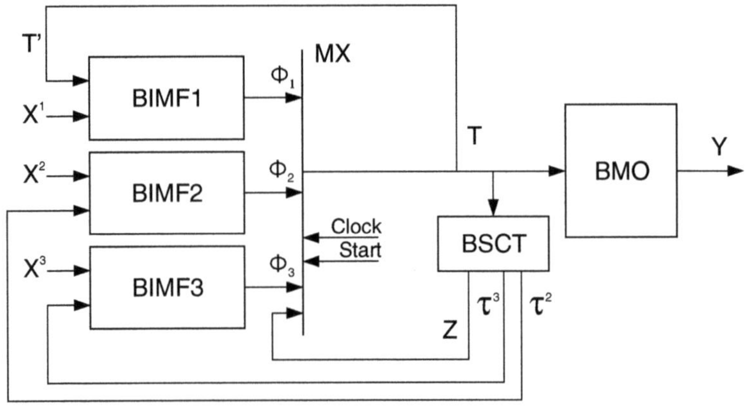

Fig. 6.15 Structural diagram of P_{30E1} Moore FSM

There are the following steps in the design method of P_{30E1} Moore FSM:

1. Constructing the set of states A.
2. Constructing the set $\Pi_A = \{B_1, ..., B_I\}$.
3. Representing the set Π_A as $\Pi_{A1} \cup \Pi_{A2} \cup \Pi_{A3}$.
4. Executing the optimal state assignment.
5. Executing the encoding of the sets $\Pi_{Ak} \subset \Pi_A$.
6. Executing the class assignment for $B_i \in \Pi_{A2} \cup \Pi_{A3}$.
7. Constructing the tables of BIMFk ($k = \overline{1,3}$).
8. Constructing the table of MX.
9. Constructing the table of BMO.
10. Constructing the table of BSCT.
11. Implementing the FSM logic circuit.

Let us discuss an example of design for $P_{30E1}(\Gamma_7)$. The GSA Γ_7 is shown in Fig. 5.22. It includes 16 states divided by $I_A = 9$ classes of PES: $B_1 = \{a_1\}$, $B_2 = \{a_2, a_3\}$, $B_3 = \{a_4, a_5\}$, $B_4 = \{a_6, a_7, a_8\}$, $B_5 = \{a_9, a_{10}\}$, $B_6 = \{a_{11}\}$, $B_7 = \{a_{12}, a_{13}\}$, $B_8 = \{a_{14}\}$ and $B_9 = \{a_{15}, a_{16}\}$. Let us execute the step 3 of the proposed procedure.

Let us use the following rules for creating the set Π_1: 1) it includes 4 of classes $B_i \in \Pi_A$; 2) each class $B_i \in \Pi_{A1}$ should include not more than 4 states. Other sets $\Pi_{Ak}(k = \overline{2,3})$ should include approximately the same amount of classes. Using these rules, the following sets can be found: $\Pi_{A1} = \{B_2, B_3, B_4, B_5\}$, $\Pi_{A2} = \{B_1, B_6, B_7\}$ and $\Pi_{A3} = \{B_8, B_9\}$. Let us encode the states $a_m \in A$ as it is shown in Fig. 6.16a.

There is the part of a map corresponding to the set $\Pi_{A1} \subset \Pi_A$ (Fig. 6.16b). There are the signs "*" for the states $a_m \notin A(B_i^1)$. The following codes can be found from the Karnaugh map (Fig. 6.16b): $K(B_2) = {*}{*}00$, $K(B_3) = {*}{*}01$, $K(B_4) = {*}{*}10$ and $K(B_5) = {*}{*}11$. So, there is $T' = \{T_3, T_4\}$.

Fig. 6.16 State codes of FSM $P_{30E1}(\Gamma_7)$ (**a**) and the part of codes for Π_{A1} (**b**)

(a)

T_1T_2 \ T_3T_4	00	01	11	10
00	a_1	a_{11}	a_{12}	a_{14}
01	a_2	a_4	a_9	a_6
11	a_3	a_5	a_{10}	a_7
10	a_{15}	a_{16}	a_{13}	a_8

(b)

T_1T_2 \ T_3T_4	00	01	11	10
00	*	*	*	*
01	a_2	a_4	a_9	a_6
11	a_3	a_5	a_{10}	a_7
10	*	*	*	a_8

Table 6.6 Class codes for Moore FSM $P_{30E1}(\Gamma_7)$

Π_{A2}	$K(B_i)$	Π_{A3}	$K(B_i)$
	$\tau_1\tau_2$		τ_3
B_1	00	B_8	0
B_6	*1	B_9	1
B_7	1*		

Using (5.44), the value $R_K = 2$ can be found. Let us encode the sets Π_{Ak} : $K(\Pi_{A1}) = 00, K(\Pi_{A2}) = *1$ and $K(\Pi_{A3}) = 1*$. The Karnaugh map (Fig. 6.16b) corresponds to the code $K(\Pi_{A1}) = 00$.

There are $I_{k2} = 3$ and $I_{k3} = 2$. It gives the sets $\tau^2 = \{\tau_1, \tau_2\}$ and $\tau^3 = \{\tau_3\}$ due to $R_{k1} = 2$ and $R_{k2} = 1$. Let us encode the classes $B_i \notin \Pi_{A1}$ in the following way (Table 6.6).

There are the same columns in tables of BIMFk ($k = \overline{1,3}$): B_i, $K(B_i)$, a_s, $K(a_s)$, X_h, Φ_h, h. These tables are constructed on the base of a system of GFT. It is the system (5.57) in the discussed case.

The BIMF1 is represented by Table 6.7. This table is used to derive the functions (5.18). The BIMF2 is represented by Table 6.8. This table is used for deriving the system

$$\Phi_2 = \Phi_2(\tau^2, X^2). \tag{6.14}$$

Table 6.7 Table of BIMF1 for Moore FSM $P_{30E1}(\Gamma_7)$

B_i	$K(B_i)$	a_s	$K(a_s)$	X_h	Φ_h	h
B_2	**00	a_4	0101	x_2	D_2D_4	1
		a_5	1101	$\bar{x}_2$	$D_1D_2D_4$	2
B_3	**01	a_6	0110	x_3	D_2D_3	3
		a_7	1110	$\bar{x}_3x_4$	$D_1D_2D_3$	4
		a_8	1010	$\bar{x}_3\bar{x}_4$	D_1D_3	5
B_4	**10	a_9	0111	x_5	$D_2D_3D_4$	6
		a_{10}	1111	$\bar{x}_5$	$D_1D_2D_3D_4$	7
B_5	**11	a_{11}	0001	1	D_4	8

Table 6.8 Table of BIMF2 for Moore FSM $P_{30E1}(\Gamma_7)$

B_i	$K(B_i)$	a_s	$K(a_s)$	X_h	Φ_h	h
B_1	00	a_2	0101	x_1	D_2	1
		a_3	1101	$\bar{x}_1$	D_1D_2	2
B_6	$*1$	a_{12}	0111	x_6	D_3D_4	3
		a_{13}	1111	$\bar{x}_6$	$D_1D_3D_4$	4
B_7	$1*$	a_{14}	0001	1	D_3	5

Table 6.9 Table of BIMF3 for Moore FSM $P_{30E1}(\Gamma_7)$

B_i	$K(B_i)$	a_s	$K(a_s)$	X_h	Φ_h	h
B_8	0	a_{15}	1000	x_7	D_1	1
		a_{16}	1001	$\bar{x}_7$	D_1D_4	2
B_9	1	a_1	0000	x_3	–	3
		a_{14}	0010	$\bar{x}_3$	D_3	4

The BIMF3 is represented by Table 6.9. This table is used for deriving the system

$$\Phi_3 = \Phi_3(\tau^3, X^3). \tag{6.15}$$

There are the state codes from Fig. 6.16 used in those three tables.

There is the system Φ_1 represented as (6.16), the system Φ_2 as (6.17) and the system Φ_3 as (6.18). There are the corresponding superscripts for functions $D_r^k \in \Phi(k = \overline{1,3};\ r = \overline{1,R_1})$ The equations are minimized for all those systems.

$$\begin{aligned} D_1^1 &= \bar{T}_3\bar{T}_4\bar{x}_2 \vee \bar{T}_3T_4\bar{x}_3 \vee T_3\bar{T}_4\bar{x}_5;\\ D_2^1 &= \bar{T}_4 \vee \bar{T}_3T_4(x_3 \vee x_4);\\ D_3^1 &= \bar{T}_3T_4 \vee T_3\bar{T}_4;\\ D_4^1 &= \bar{T}_4 \vee T_3T_4. \end{aligned} \tag{6.16}$$

$$\begin{aligned} D_1^2 &= \bar{\tau}_1\bar{\tau}_2\bar{x}_1 \vee \tau_2\bar{x}_6;\\ D_2^2 &= \bar{\tau}_1\bar{\tau}_2;\\ D_3^2 &= \tau_2 \vee \tau_1;\\ D_4^2 &= \tau_2. \end{aligned} \tag{6.17}$$

$$\begin{aligned} D_1^3 &= \bar{\tau}_3;\\ D_2^3 &= 0;\\ D_3^3 &= \tau_3\bar{x}_3;\\ D_4^3 &= \bar{\tau}_3\bar{x}_7. \end{aligned} \tag{6.18}$$

If LUTs are used for implementing the circuit of BIMFk, then the corresponding equations of the system Φ_k are transformed into the tables of LUTs. If EMBs are

Table 6.10 Table of MX for Moore FSM $P_{30E1}(\Gamma_7)$

k	$K(\Pi_{Ak})$	D_1	D_2	D_3	D_4
1	00	D_1^1	D_2^1	D_3^1	D_4^1
2	*1	D_1^2	D_2^2	D_3^2	D_4^2
3	1*	D_1^3	D_2^3	D_3^3	D_4^3

Table 6.11 Table of BSCT for Moore FSM $P_{30E1}(\Gamma_7)$

a_m	$K(a_m)$	B_i	$K(B_i)$	Π_{Ak}	$K(\Pi_{Ak})$	τ_m	Z_m	m
a_1	0000	B_1	00	Π_{A2}	01	–	z_2	1
a_{11}	0001	B_6	01	Π_{A2}	01	τ_2	z_2	2
a_{12}	0011	B_7	10	Π_{A2}	01	τ_1	z_2	3
a_{13}	1011	B_7	10	Π_{A2}	01	τ_1	z_2	4
a_{14}	0010	B_8	0	Π_{A3}	10	–	z_1	5
a_{15}	1000	B_9	1	Π_{A3}	10	τ_3	z_1	6
a_{16}	1001	B_9	1	Π_{A3}	10	τ_3	z_1	7

used, then the corresponding table of BIMFk is transformed into the table of EMBerk ($k = \overline{1,3}$).

There is the table of MX represented by Table 6.10. It is used to derive the equations for LUTs implementing the circuit of MX.

For example, the following equation can be derived from Table 6.10:

$$D_1 = \bar{z}_1\bar{z}_2 D_1^1 \vee z_2 D_1^2 \vee z_1 D_1^3. \tag{6.19}$$

Similar equations can be obtained for each of functions generated by the MX.

The table of BMO is constructed in the way we have discussed before. There are the following columns in the table of BSCT: $a_m, K(a_m), B_i, K(B_i), \Pi_{Ak}, K(\Pi_{Ak}), \tau_m, Z_m, m$. The meaning of these columns follows from Table 6.11.

This table is used to derive the systems $\tau(T)$ and $Z(T)$. For example, the following equations can be derived from Table 6.11: $\tau_3 = T_1\bar{T}_2\bar{T}_3$; $z_1 = \bar{T}_1\bar{T}_2T_3\bar{T}_4 \vee T_1\bar{T}_2\bar{T}_3$. Next, the equations can be transformed into tables of corresponding LUTs.

Let us point out, that there are no states $a_m \in A(B_i^1)$ in Table 6.11. It follows from the fact that there is $K(\Pi_{A1}) = 00$.

Now let us discuss the structural diagram of P_{30E1}^1 Moore FSM (Fig. 6.17). As follows from Fig. 6.16, the functions $\tau_r \in \tau$ are necessary only for the BIMF3. The functions $D_r \in \Phi_2$ are represented by the system:

$$\Phi_2 = \Phi_2(T'', X^2). \tag{6.20}$$

Let us construct the Karnaugh map (Fig. 6.18) for the states $a_m \in A(B_i^2)$. It is constructed on the base of the Karnaugh map (Fig. 6.16) for the FSM $P_{30E1}(\Gamma_7)$.

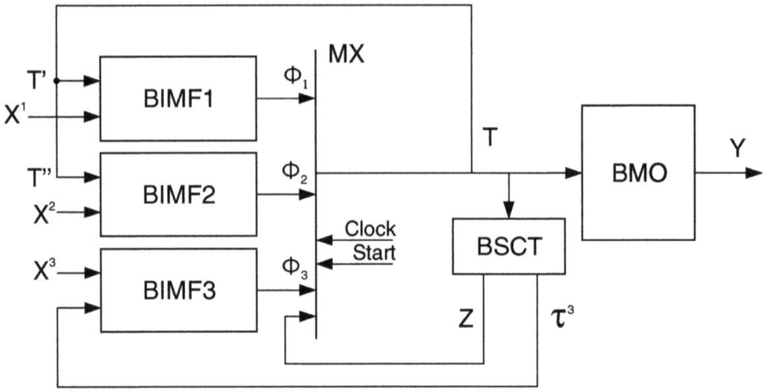

Fig. 6.17 Structural diagram of P^1_{30E1} Moore FSM

Fig. 6.18 State codes for Π_{A2}

T_1T_2 \ T_3T_4	00	01	11	10
00	a_1	a_{11}	a_{12}	$*$
01	$*$	$*$	$*$	$*$
11	$*$	$*$	$*$	$*$
10	$*$	$*$	a_{13}	$*$

The analysis of this Karnaugh map gives the following codes: $K(B_1) = ***0$, $K(B_6) = **01$ and $K(B_7) = **1*$. So, there is $T'' = \{T_3, T_4\}$. Each class $B_i \in \Pi_{A2}$ is represented by a single interval of 4-dimensional Boolean space. So, it is possible to use the model of P^1_{30E1} Moore FSM in the case of GSA Γ_7.

There is no change for tables of BIMF1, BIMF3, MX and BMO in the case of P^1_{30E1} FSM in comparison with the equivalent P_{30E1} FSM. In the discussed case, the BIMF2 is represented by Table 6.12, whereas the BSCT by Table 6.13.

The table of BIMF2 is used to derive the system (6.20). It is the following system in the discussed case:

Table 6.12 Table of BIMF2 for Moore FSM $P^1_{30E1}(\Gamma_7)$

B_i	$K(B_i)$	a_s	$K(a_s)$	X_h	Φ_h	h
B_1	$***0$	a_2	0100	x_1	D_1	1
		a_3	1100	$\bar{x_1}$	D_1D_4	2
B_6	$**01$	a_{12}	0011	x_6	–	3
		a_{13}	1011	$\bar{x_6}$	D_3	4
B_7	$**1*$	a_{14}	0010	1	D_3	4

Table 6.13 Table of BSCT for Moore FSM $P^1_{30E1}(\Gamma_7)$

a_m	$K(a_m)$	B_i	$K(B_i)$	Π_{Ak}	$K(\Pi_{Ak})$	τ_m	Z_m	m
a_1	0000	B_1	***0	Π_{A2}	01	–	z_2	1
a_{11}	0001	B_6	**01	Π_{A2}	01	–	z_2	2
a_{12}	0011	B_7	**1*	Π_{A2}	01	–	z_2	3
a_{13}	1011	B_7	**1*	Π_{A2}	01	–	z_2	4
a_{14}	0010	B_8	0	Π_{A3}	10	–	z_1	5
a_{15}	1000	B_9	1	Π_{A3}	10	τ_3	z_1	6
a_{16}	1001	B_9	1	Π_{A3}	10	τ_3	z_1	7

$$
\begin{aligned}
D_1^2 &= \bar{T}_4\bar{x}_1 \vee \bar{T}_3 T_4 \bar{x}_6;\\
D_2^2 &= \bar{T}_4;\\
D_3^2 &= T_3 \vee T_4;\\
D_4^2 &= \bar{T}_3 T_4.
\end{aligned}
\tag{6.21}
$$

As follows from Table 6.13, there is the set $\tau^3 = \{\tau_1\}$. The following equations can be derived from Table 6.13:

$$
\begin{aligned}
\tau_1 &= T_1\bar{T}_2\bar{T}_3;\\
z_1 &= \bar{T}_1\bar{T}_2 T_3\bar{T}_4 \vee T_1\bar{T}_2\bar{T}_3;\\
z_2 &= \bar{T}_1\bar{T}_2\bar{T}_3 \vee \bar{T}_2 T_3 T_4.
\end{aligned}
\tag{6.22}
$$

Next, let us discuss the structural diagram of P^2_{30E1} Moore FSM (Fig. 6.19). In this case, the BSCT generates only the functions $Z = Z(T)$. There is no need in functions $\tau_r \in \tau$.

In this case, the BIMF3 implements the system

$$\Phi_3 = \Phi_3(T''', X^3). \tag{6.23}$$

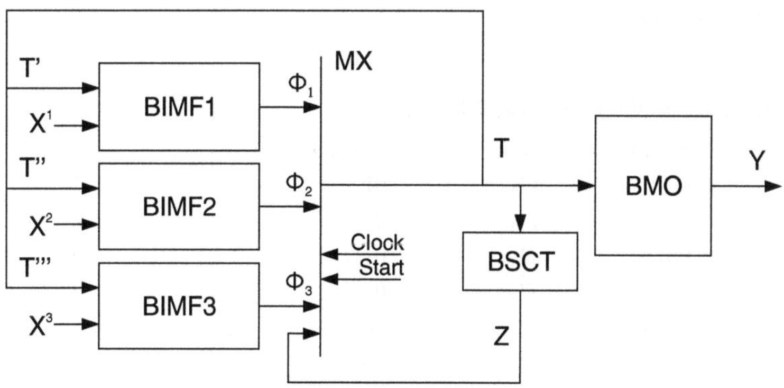

Fig. 6.19 Structural diagram of P^2_{30E1} Moore FSM

Fig. 6.20 State codes for Π_{A3}

T_1T_2 \ T_3T_4	00	01	11	10
00	*	*	*	a_{14}
01	*	*	*	*
11	*	*	*	*
10	a_{15}	a_{16}	*	*

Table 6.14 Table of BIMF3 for Moore FSM $P^2_{30E1}(\Gamma_7)$

B_i	$K(B_i)$	a_s	$K(a_s)$	X_h	Φ_h	h
B_8	$**1*$	a_{15}	1000	x_7	D_1	1
		a_{16}	1001	$\bar{x_7}$	D_1D_4	2
B_9	$**0*$	a_1	0000	x_3	–	3
		a_{13}	0010	$\bar{x_3}$	D_3	4

The BIMF1 implements the functions (5.18), the BIMF2 the functions (6.7).

Let us analyse the Karnaugh map (Fig. 6.20). It corresponds to the set Π_{A3} of the Moore FSM $P_{30E1}(\Gamma_7)$.

The following codes can be found from this Karnaugh map: $K(B_8) = **1*$ and $K(B_9) = **0*$. Now,there is the following table of BIMF3 (Table 6.14).

There is the following system (6.23) derived from Table 6.14:

$$\begin{aligned} D_1^3 &= T_3; \\ D_2^3 &= 0; \\ D_3^3 &= \bar{T}_3\bar{x_3}; \\ D_4^3 &= \bar{T}_3 x_7. \end{aligned} \tag{6.24}$$

The BIMF1 is represented by the system (6.16), the BIMF2 by the system (6.21).

There is $\tau = \emptyset$ for P^2_{30E1} Moore FSMs. So, the BSCT is represented by Table 6.15 in the discussed case.

Table 6.15 Table of BSCT for Moore FSM $P^2_{30E1}(\Gamma_7)$

a_m	$K(a_m)$	Π_{Ak}	$K(\Pi_{Ak})$	Z_m	m
a_1	0000	Π_{A2}	01	z_2	1
a_{11}	0001	Π_{A2}	01	z_2	2
a_{12}	0011	Π_{A2}	01	z_2	3
a_{13}	1011	Π_{A2}	01	z_2	4
a_{14}	0010	Π_{A3}	10	z_1	5
a_{15}	1000	Π_{A3}	10	z_1	6
a_{16}	1001	Π_{A3}	10	z_1	7

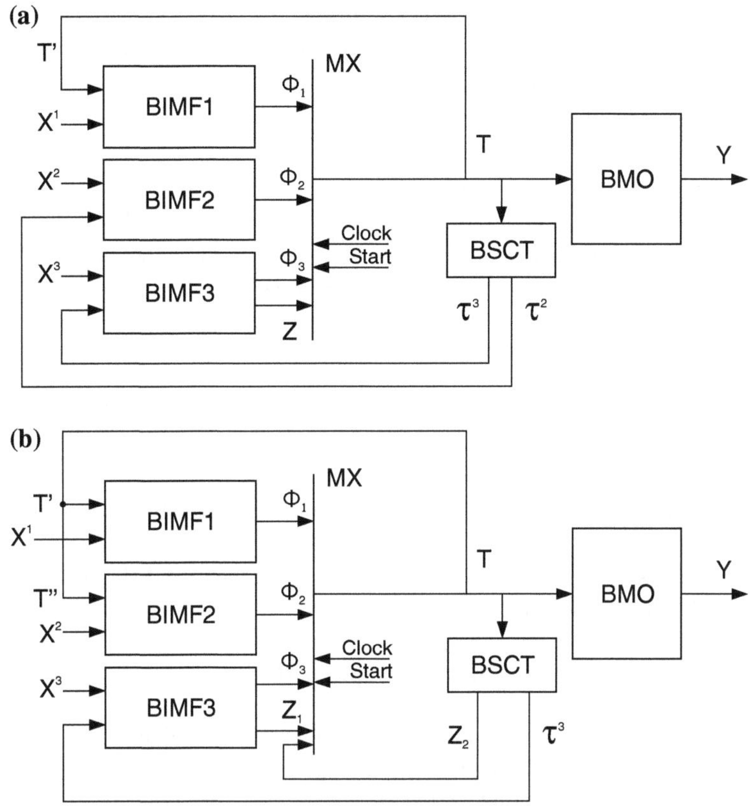

Fig. 6.21 Structural diagram of P_{20E1} (**a**) and P_{30E3} (**b**) Moore FSM

There are the same equations for $z_1, z_2 \in Z$ as they are in system (6.22).

Analysis of Table 6.14 shows that it is possible to distribute the functions $z_1, z_2 \in Z$ between the BIMF2 and BIMF3. It can be done for P_{30E1} Moore FSM. It is possible to implement the functions z_1 by the BIMF3 in the case of P^1_{30E1} Moore FSM. To do it, the code should be added to indicate that $B_i \notin \Pi_{A2}$ and $B_i \notin \Pi_{A3}$. In the case of P^2_{30E1} FSM, the functions Z can be implemented only by BSCT. So, two additional models are possible for the 3-directional Moore FSM with $\Delta_t = 0$. They are shown in Fig. 6.21.

So, there are five different models of Moore FSM with $K = 3$ for $\Delta_t = 0$. Their characteristics are shown in Table 6.16.

Let us point out that the BIMF1 always executes the functions (5.18), whereas the BMO the functions (1.5). Due to it, we do not show these blocks in Table 6.16.

Now let us discuss the case when

$$\Delta_t \neq 0. \tag{6.25}$$

Table 6.16 Characteristics of 3-directional Moore FSMs ($\Delta_t = 0$)

№	FSM	BIMF2	BIMF3	BSCT
1	P_{30E1}	$\Phi_2 = \Phi_2(T^2, X^2)$	$\Phi_3 = \Phi_3(\tau^3, X^3)$	$Z = Z(T)$
				$\tau^2 = \tau^2(T)$
				$\tau^3 = \tau^3(T)$
2	P^1_{30E2}	$\Phi_2 = \Phi_2(T'', X^2)$	$\Phi_3 = \Phi_3(\tau^3, X^3)$	$Z = Z(T)$
				$\tau^3 = \tau^3(T)$
3	P^2_{30E1}	$\Phi_2 = \Phi_2(T'', X^2)$	$\Phi_3 = \Phi_3(T''', X^3)$	$Z = Z(T)$
4	P_{30E2}	$\Phi_2 = \Phi_2(\tau^2, X^2)$	$\Phi_3 = \Phi_3(\tau^3, X^3)$	$\tau^2 = \tau^2(T)$
		$z_2 = z_2(\tau^2)$	$z_1 = z_1(\tau^3)$	$\tau^3 = \tau^3(T)$
5	P_{30E3}	$\Phi_2 = \Phi_2(T'', X^2)$	$\Phi_3 = \Phi_3(\tau^3, X^3)$	$\tau_3 = \tau^3(T)$
			$z_1 = z_1(\tau^3)$	$z_2 = z_2(T)$

In this case, the BMO can execute a part of functions executed by the BSCT (or even all of them).

Let the states $a_m \in A$ be encoded in the way when the model P_{30E1} can be used. Let the following condition take place:

$$\Delta_t = R_{A2} + R_{A3} + R_K. \tag{6.26}$$

In this case, the BMO can implement the functions $z_r \in Z$ and $\tau_r \in \tau^2 \cup \tau^3$. It leads to P_{30E4} Moore FSM shown in Fig. 6.22.

Let the following condition take place:

$$\Delta_t = R_{A2} + R_{A3}. \tag{6.27}$$

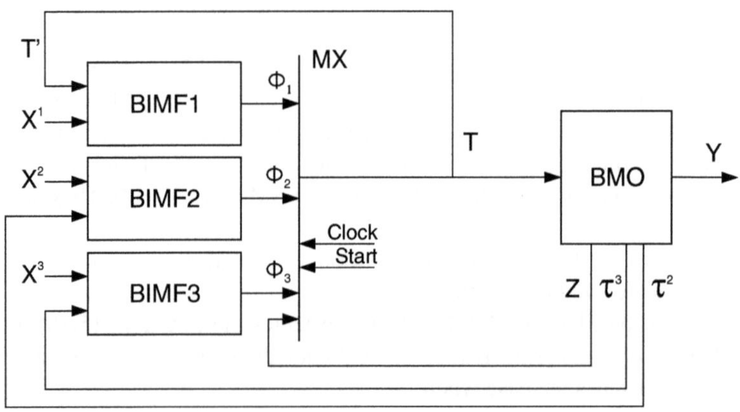

Fig. 6.22 Structural diagram of P_{30E4} Moore FSM

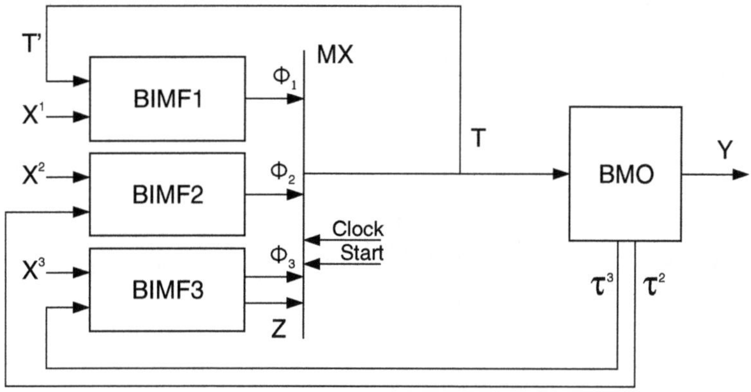

Fig. 6.23 Structural diagram of P_{30E5} Moore FSM

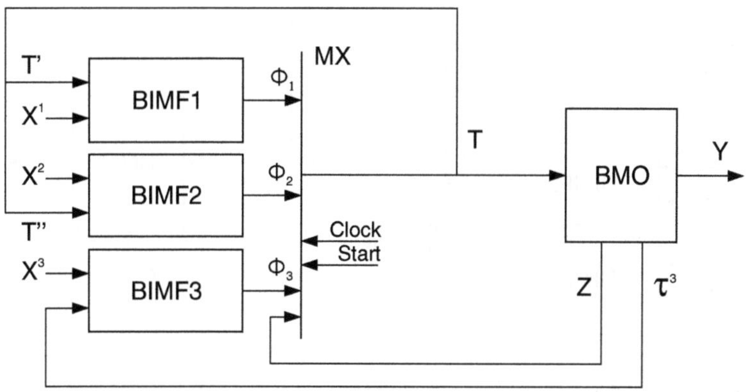

Fig. 6.24 Structural diagram of P^1_{30E4} Moore FSM

In this case, the functions $z_r \in Z$ can be implemented by the BIMF2 and BIMF3. It leads to P_{30E5} Moore FSM (Fig. 6.23).

Let it be possible to use the model P^1_{30E1} Moore FSM. Let the following condition take place:

$$\Delta_t = R_{A3} + R_K. \tag{6.28}$$

In this case, the BMO implements the functions $z_r \in Z$ and $\tau_r \in \tau^3$. It leads to P^1_{30E4} Moore FSM (Fig. 6.24).

Let the following condition take place:

$$\Delta_t = R_{A3} + 1. \tag{6.29}$$

In this case, the function $z_1 \in Z$ is implemented by the BIMF3. The BMO implements the functions $z_2 \in Z$ and τ^3. It results in P_{30E6} Moore FSM (Fig. 6.25).

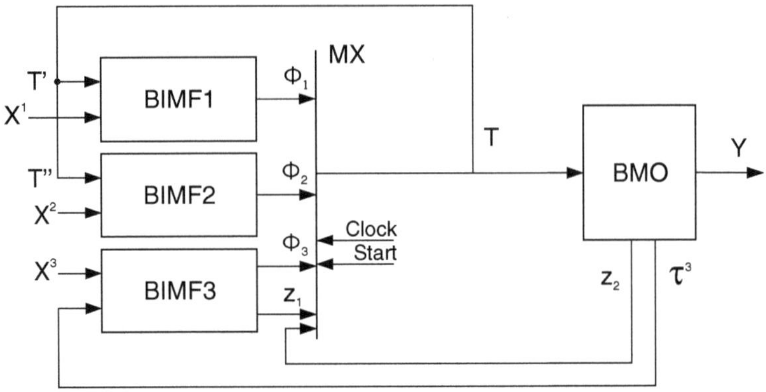

Fig. 6.25 Structural diagram of P_{30E6} Moore FSM

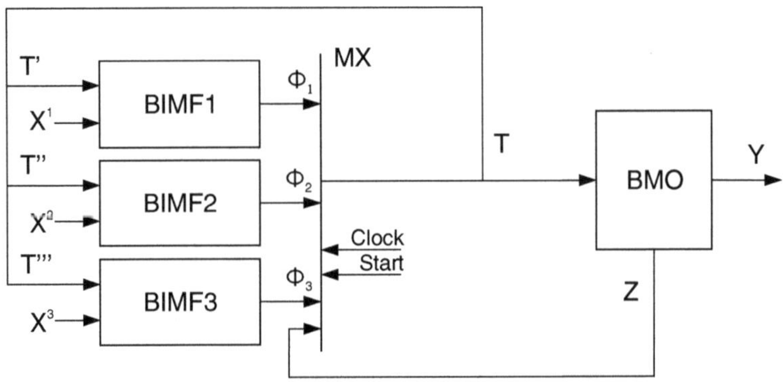

Fig. 6.26 Structural diagram of P^2_{30E4} Moore FSM

Let it be possible to use the model of P^2_{30E1} Moore FSM. Let the following condition take place:

$$\Delta_t = R_K. \qquad (6.30)$$

In this case, the BMO implements the functions $z_r \in Z$. It leads to P^2_{30E4} Moore FSM (Fig. 6.26).

So, there are five models of 3-directional Moore FSMs which do not use the BSCT. Its functions are executed by the BMO. Table 6.17 shows those models. As for Table 6.16, there are the same functions of BIMF1 as they are in Table 6.17. Also, there is no column "BSCT" because this block is not used for those five models. The column "Cond." contains the number of a formula showing the condition for using a given model.

Comparison of Tables 6.16 and 6.17 shows that there are the same functions of blocks for P_{30Ei} and $P_{30E(i+3)}$, where $i \in \{1, 2, 3\}$. So, the same methods are used for implementing the corresponding pairs of models $\langle P_{30E1}, P_{30E4}\rangle$, $\langle P^1_{30E1}, P^1_{30E4}\rangle$, ...,

Table 6.17 Characteristics of 3-directional Moore FSMs without BSCT

№	FSM	BIMF2	BIMF3	BSCT	Cond.
1	P_{30E4}	$\Phi_2 = \Phi_2(T^2, X^2)$	$\Phi_3 = \Phi_3(\tau^3, X^3)$	$Y = Y(T)$	6.26
				$Z = Z(T)$	
				$\tau^2 = \tau^2(T)$	
				$\tau^3 = \tau^3(T)$	
2	P^1_{30E4}	$\Phi_2 = \Phi_2(T'', X^2)$	$\Phi_3 = \Phi_3(\tau^3, X^3)$	$Y = Y(T)$	6.27
				$Z = Z(T)$	
				$\tau^3 = \tau^3(T)$	
3	P^2_{30E4}	$\Phi_2 = \Phi_2(T'', X^2)$	$\Phi_3 = \Phi_3(T''', X^3)$	$Y = Y(T)$	6.28
				$Z = Z(T)$	
4	P_{30E5}	$\Phi_2 = \Phi_2(\tau^2, X^2)$	$\Phi_3 = \Phi_3(\tau^3, X^3)$	$Y = Y(T)$	6.29
		$z_2 = z_2(\tau^2)$	$z_1 = z_1(\tau^3)$	$\tau^2 = \tau^2(T)$	
				$\tau^3 = \tau^3(T)$	
5	P_{30E6}	$\Phi_2 = \Phi_2(T'', X^2)$	$\Phi_3 = \Phi_3(\tau^3, X^3)$	$Y = Y(T)$	6.30
			$z_1 = z_1(\tau^3)$	$\tau_3 = \tau^3(T)$	
				$z_2 = z_2(T)$	

$\langle P_{30E3}, P_{30E6}\rangle$. There is the only difference: the functions of BSCT are executed by BMO. So, there are different tables of BMO for the corresponding pairs of models. In the case of models from Table 6.16, the table of BMO contains only the columns $K(a_m)$, $Y(a_m)$, m. In the case of models from Table 6.17, the columns $Z(a_m)$, $\tau^2(a_m)$ or $\tau^3(a_m)$ can be added.

Now, let us discuss the case when the BMO can implement only some part of BSCT's functions. It is possible if the condition (6.25) is true.

Let it be possible to use the model P_{30E1}. Let the following condition take place:

$$R_{A2} + R_{A3} = \Delta_t. \tag{6.31}$$

It leads to P_{30E7} Moore FSM (Fig. 6.27).

In the case of P_{30E7} FSM, the BSCT implements the functions $z_r \in Z$, whereas the functions $\tau^2, \tau^3 \in \tau$ are generated by the BMO.

Let it be possible to use the model of P^1_{30E1} Moore FSM. Let the following condition take place:

$$R_{A3} = \Delta_t. \tag{6.32}$$

It leads to the P^1_{30E7} Moore FSM (Fig. 6.28).

Let it be possible to use the model of P^2_{30E1} Moore FSM. Let the following condition take place:

$$\Delta_t = 1. \tag{6.33}$$

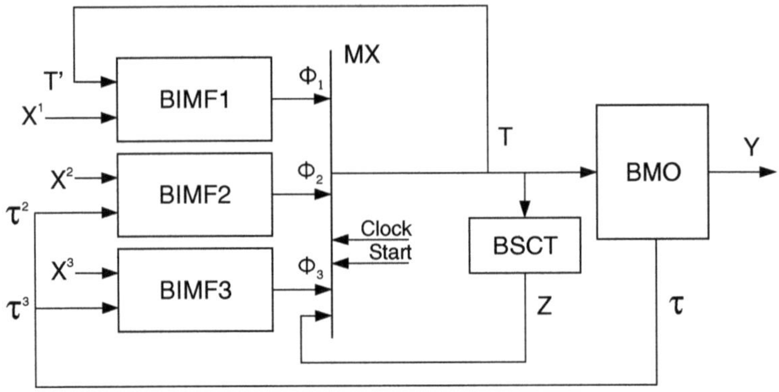

Fig. 6.27 Structural diagram of P_{30E7} Moore FSM

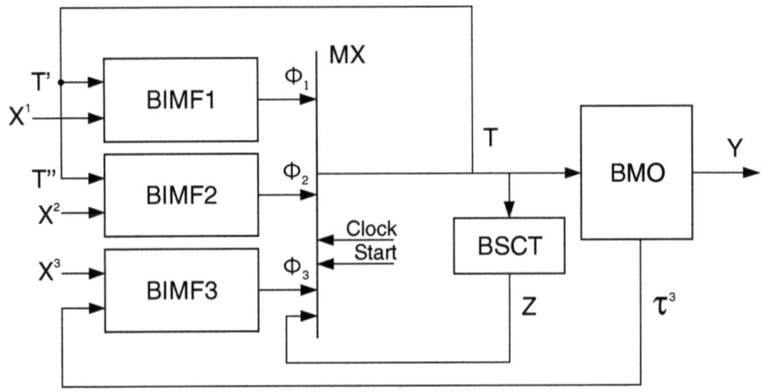

Fig. 6.28 Structural diagram of P^1_{30E7} Moore FSM

In this case, the functions $z_r \in Z$ are distributed between the BMO and BSCT. It leads to P^2_{30E7} Moore FSM (Fig. 6.29).

It is possible to use the model of P_{30E8} Moore FSM (Fig. 6.30).

It is possible to use the model of P_{30E9} Moore FSM (Fig. 6.31).

Let us point out that these models correspond to the following conditions, respectively:

$$\Delta_t = R_{K2}. \tag{6.34}$$

$$\Delta_t = R_{K3}. \tag{6.35}$$

These models correspond to some extreme conditions. It is quite possible that functions τ are distributed between the BMO and BSCT for P_{30E7}. The same is true for functions Z. So, there are many modifications of the basic models shown in Figs. 6.27, 6.28, 6.29, 6.30, and 6.31. The number of these models is determined by the values of R_{A2}, R_{A3} and R_K. So, their number can be different for each GSA Γ.

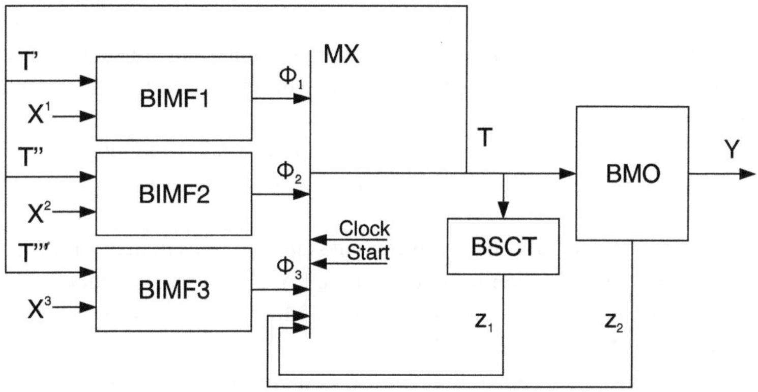

Fig. 6.29 Structural diagram of P^2_{30E7} Moore FSM

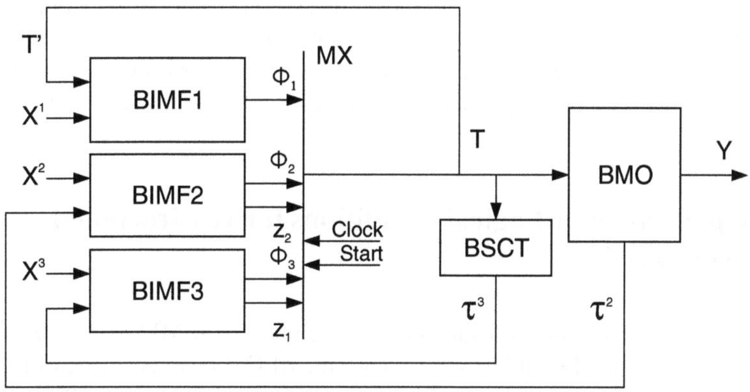

Fig. 6.30 Structural diagram of P_{30E8} Moore FSM

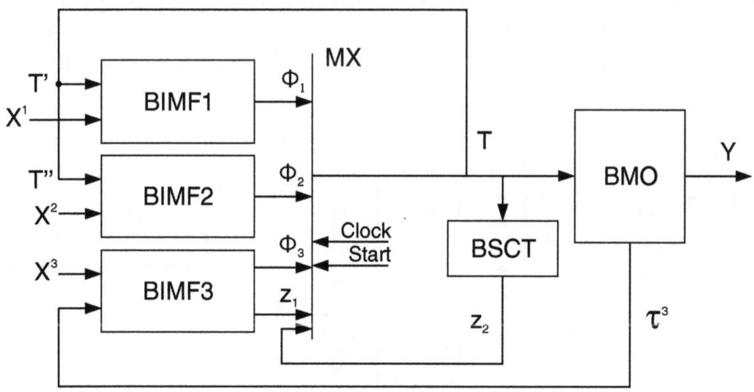

Fig. 6.31 Structural diagram of P_{30E9} Moore FSM

Let us point out that it is possible to use either LUTs or EMBs for implementing the circuits of BIMF1–BIMF3. It gives 8 different implementations for each of discussed models. For example, Table 6.16 determines 40 different circuits. The same is true for Table 6.17.

There are the following positive features of the proposed models. There is less amount of literals in the partial SOPs for functions D_r^k then for the function $D_r(k = \overline{1,k}; r = \overline{1,R_1})$.

Each function D_r^k depends on less amount of state variables than it is for P_0 Moore FSMs. Each function D_r^k depends on less amount of class variables than it is for P_E Moore FSMs. So, it allows to get the circuit of Moore FSM having less hardware than it is for equivalent P_0 or P_E FSMs.

But there is a rapid growth for the number of possible solutions with the increasing for the number of directions, K. It results in complex algorithms of: 1) state assignment and 2) distribution of classes $B_i \in \Pi_A$ among the sets $\Pi_{Ak}(k = \overline{1,K})$. Both problems are NP-complex [3]. It means that it is necessary to work out heuristic algorithms for solution of above mentioned problems.

We do not discuss the Moore FSMs with $K > 3$. The corresponding models are developed using the same approach as the one used for $K = 2$ and $K = 3$.

6.3 Replacement of Logical Conditions for K-Directional Moore FSMs

It is possible to reduce the hardware amount in circuits of BIMFk $(k = \overline{1,K})$ due to using the method of RLC. It has sense to use the RLC if there is saving of LUTs in the particular block in comparison with its implementing without the RLC.

Let us, for example, discuss the structural diagram of $M_3P_{30E1}^1$ Moore FSM (Fig. 6.32). There are three blocks BRLCk in this FSM. It is shown by the subscript 3 in M_3. The block BRLCk implements the system of additional variables P^k used for the replacement of logical conditions $x_e \in X$. So, the set P is represented as

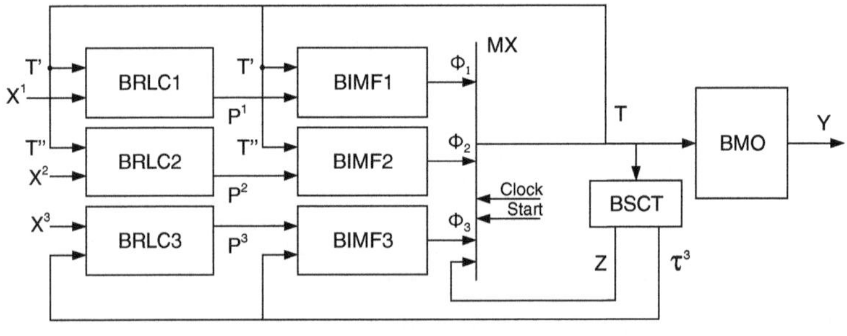

Fig. 6.32 Structural diagram of $M_3P_{30E1}^1$ Moore FSM

$$P = \bigcup_{k=1}^{K} P^k. \tag{6.36}$$

In $M_3P^1_{30E1}$ Moore FSM, the blocks BRLCk and BIMFk execute the following functions, respectively:

$$P^1 = P^1(T', X^1); \tag{6.37}$$

$$P^2 = P^2(T'', X^2); \tag{6.38}$$

$$P^3 = P^3(\tau^3, X^3); \tag{6.39}$$

$$\Phi_1 = \Phi_1(T', P^1); \tag{6.40}$$

$$\Phi_2 = \Phi_2(T'', P^2); \tag{6.41}$$

$$\Phi_3 = \Phi_3(\tau^3, P^3). \tag{6.42}$$

The blocks BSCT, MX and BMO execute the same functions as in P^1_{30E1} Moore FSM.

There are the following steps in the design method for $M_3P^1_{30E1}$ Moore FSMs:

1. Constructing the set of states $A = \{a_1, ..., a_{M1}\}$.
2. Constructing the set $\Pi_A = \{B_1, ..., B_I\}$.
3. Distribution the classes $B_i \in \Pi_A$ among the sets Π_{Ak}.
4. Executing the optimal state assignment.
5. Executing the encoding of the sets $\Pi_{Ak} \in \Pi_A$.
6. Executing the class assignment for $B_i \in \Pi_{Ak}$.
7. Finding the sets P^k for classes $B_i \in \Pi_{Ak}$.
8. Constructing the tables of BRLCk.
9. Constructing the tables of BIMFk.
10. Constructing the table of BMO.
11. Constructing the table of MX.
12. Constructing the table of BSCT.
13. Implementing the FSM logic circuit.

Let us discuss an example of design for $M_3P^1_{30E1}(\Gamma_7)$. The GSA Γ_7 is shown in Fig. 5.22. It includes $M_1 = 16$ states distributed among $I_A = 9$ classes of PES. There are the following classes: $B_1 = \{a_1\}, B_2 = \{a_2, a_3\}, B_3 = \{a_4, a_5\}, B_4 = \{a_6, a_7, a_8\}, B_5 = \{a_9, a_{10}\}, B_6 = \{a_{11}\}, B_7 = \{a_{12}, a_{13}\}, B_8 = \{1_{14}\}$ and $B_9 = \{a_{15}, a_{16}\}$. The steps 1–6 are already executed in the previous sections. There are the optimal state codes shown in Fig. 6.16. There are the following sets $\Pi_{Ak} \subset \Pi$: $\Pi_{A1} = \{B_2, B_3, B_4, B_5\}, \Pi_{A2} = \{B_1, B_6, B_7\}$ and $\Pi_{A3} = \{B_8, B_9\}$. They have the codes $K(\Pi_{A1}) = 00, K(\Pi_{A2}) = 01$ and $K(\Pi_{A3}) = 10$. The classes $B_i \in \Pi_{A3}$ are encoded as the following ones: $K(B_8) = 0$ and $K(B_9) = 1$. There are the sets $Z = \{z_1, z_2\}$ and $\tau^3 = \{\tau_1\}$.

Table 6.18 Table of BRLC1 for Moore FSM $M_3P^1_{30E1}(\Gamma_7)$

P_g	B_i			
	B_2	B_3	B_4	B_5
p_1	x_2	x_3	–	–
p_2	–	x_4	x_5	–
$K(B_i)$	$**00$	$**01$	$**10$	$**11$

Table 6.19 Table of BRLC2 for Moore FSM $M_3P^1_{30E1}(\Gamma_7)$

P_g	B_i		
	B_1	B_6	B_7
p_3	x_1	x_6	–
$K(B_i)$	$***0$	$**01$	$**1*$

Table 6.20 Table of BRLC3 for Moore FSM $M_3P^1_{30E1}(\Gamma_7)$

P_g	B_i	
	B_8	B_9
p_4	x_7	x_3
$K(B_l)$	0	1

Let us analyse the system of GFT (5.57). It allows finding the sets P^k : $P^1 = \{p_1, p_2\}$ used for replacing $x_e \in X' = \{x_2, ..., x_5\}$; $P^2 = \{p_3\}$ used for replacing the logical conditions $x_e \in X^2 = \{x_1, x_6\}$; $P^3 = \{p_4\}$ used for replacing the logical conditions from the set $X^3 = \{x_3, x_7\}$. Let us construct the tables of BRLCk (Tables 6.18, 6.19, and 6.20).

We take the class codes from Table 6.7 to use them into Table 6.18. The codes from Table 6.12 are used in Table 6.19. The class codes $B_i \in \Pi_{A3}$ are taken from Table 6.8 and used in Table 6.20.

These tables are used to derive the equations for systems (6.37)–(6.39). There are the following equations:

$$\begin{aligned}
p_1 &= \bar{T}_4 x_2 \vee T_4 x_3;\\
p_2 &= \bar{T}_3 x_4 \vee T_3 x_5;\\
p_3 &= \bar{T}_4 x_1 \vee \bar{T}_3 T_4 x_6;\\
p_4 &= \bar{\tau}_1 x_7 \vee \tau_1 x_3.
\end{aligned} \tag{6.43}$$

The equations (6.43) are used to implement the circuits of BRLC1–BRLC3 in the discussed case.

The tables of BIMFk are constructed on the base of Tables 6.7, 6.12 and 6.9 for $k = 1, 2, 3$ correspondingly. They are the Tables 6.21, 6.22, and 6.23.

The following systems of equations can be derived from Tables 6.21, 6.22, and 6.23:

Table 6.21 Table of BIMF1 for Moore FSM $M_3P^1_{30E1}(\Gamma_7)$

a_m	$K(a_m)$	a_s	$K(a_s)$	P_h	Φ_h	h
B_2	$**00$	a_4	0101	p_1	D_2D_4	1
		a_5	1101	$\bar{p_1}$	$D_1D_2D_4$	2
B_3	$**01$	a_6	0110	p_1	D_2D_3	3
		a_7	1101	$\bar{p_1}p_2$	$D_1D_2D_3$	4
		a_8	1010	$\bar{p_1}\bar{p_2}$	D_1D_3	5
B_4	$**10$	a_9	0111	p_2	$D_2D_3D_4$	6
		a_{10}	1111	$\bar{p_2}$	$D_1D_2D_3D_4$	7
B_5	$**11$	a_{11}	0001	1	1	8

Table 6.22 Table of BIMF2 for Moore FSM $M_3P^1_{30E1}(\Gamma_7)$

a_m	$K(a_m)$	a_s	$K(a_s)$	P_h	Φ_h	h
B_1	$***0$	a_2	0100	p_3	D_2	1
		a_3	1100	$\bar{p_3}$	D_1D_2	2
B_6	$**01$	a_{12}	0011	p_3	D_3D_4	3
		a_{13}	1011	$\bar{p_3}$	$D_1D_3D_4$	4
B_7	$**1*$	a_{14}	0010	1	D_3	5

Table 6.23 Table of BIMF3 for Moore FSM $M_3P^1_{30E1}(\Gamma_7)$

a_m	$K(a_m)$	a_s	$K(a_s)$	P_h	Φ_h	h
B_8	0	a_{15}	1000	p_4	D_1	1
		a_{15}	1001	$\bar{p_4}$	D_1D_4	2
B_9	1	a_1	0000	p_4	–	3
		a_{14}	0010	$\bar{p_4}$	D_3	4

$$\begin{aligned} D_1^1 &= \bar{T_3} \vee T_3T_4\bar{p_2}; \\ D_2^1 &= \bar{T_3} \vee T_3\bar{T_4}; \\ D_3^1 &= \bar{T_3}T_4 \vee T_3\bar{T_4}; \\ D_4^1 &= \bar{T_4}. \end{aligned} \tag{6.44}$$

$$\begin{aligned} D_1^2 &= \bar{T_4}\bar{P_3} \vee \bar{T_3}T_4 \vee t_3; \\ D_2^2 &= \bar{T_4}; \\ D_3^2 &= \bar{T_3}T_4 \vee T_3; \\ D_4^2 &= \bar{T_3}T_4. \end{aligned} \tag{6.45}$$

$$\begin{aligned} D_1^2 &= \bar{\tau_1}; \\ D_2^2 &= 0; \\ D_3^2 &= \tau_1 p_4; \\ D_4^2 &= \bar{\tau_1}\bar{p_4}. \end{aligned} \tag{6.46}$$

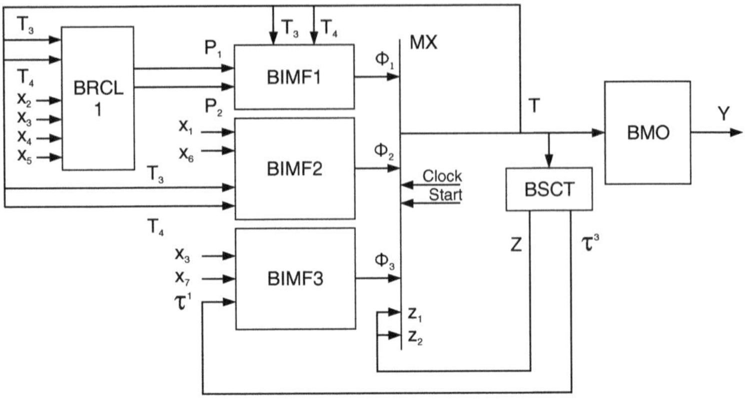

Fig. 6.33 Structural diagram of $M_1P^1_{30E1}(\Gamma_7)$ Moore FSM

Table 6.24 Table of BMO for Moore FSM $M_3P^1_{30E1}(\Gamma_7)$

a_m	$K(a_m)$	$Y(a_m)$
a_1	0000	–
a_2	0100	y_1y_2
a_3	1100	y_3
a_4	0101	y_4
a_5	1101	y_2y_5
a_6	0110	$y_1y_4y_6$
a_7	1110	y_3
a_8	1010	y_5y_7
a_9	0111	y_7y_8
a_{10}	1111	y_9
a_{11}	0001	y_1y_2
a_{12}	0011	y_1y_2
a_{13}	1011	y_7y_9
a_{14}	0010	y_3
a_{15}	1000	y_4
a_{16}	1001	y_8y_9

Comparison of systems (6.16) and (6.44) shows that the equations of (6.4) are much simpler. So, there is sense to use the BRLC1. Comparison of (6.21) and (6.45) shows that there is no saving. So, there is no sense in using the BRLC2. The same is true for the BRLC3. So, it is better to use the model $M_1P^1_{30E1}(\Gamma_7)$ shown in Fig. 6.33.

The MX is represented by Table 6.10. The BSCT is represented by Table 6.13. The table of BMO is constructed in the usual way. It is Table 6.24 in the discussed case.

Using this approach, it is possible to develop the design method for any model of Moore FSM with RLC. Let us point out that it is possible to add up to K blocks of BRLC in any K-directional model. It leads to increasing for the number of possible solutions and makes more difficult the problem of the choice of the best model for a given GSA Γ and a given chip of FPGA.

References

1. A. Barkalov, Principles of logic optimization for Moore microprogram automaton. Cybernetics and System Analysis **34**(1), 54–60 (1998)
2. E. McCluskey, *Logic Design Principles* (Prentice Hall, Englewood Cliffs, 1986)
3. G. De Micheli, *Synthesis and Optimization of Digital Circuits* (McGraw–Hill, 1994)

Chapter 7
Design of EMB-Based Mealy FSMs

7.1 Trivial Implementing Mealy FSMs

Let us remind some basic information about FPGAs. A as it is shown before, the majority of FPGAs include three main blocks: look-up table (LUT) elements connected with programmable flip-flops, embedded memory blocks (EMB), and a matrix of programmable interconnections [1, 2]. One LUT together with a flip-flop forms a logic element (LE), two LEs form a slice, two slices form a configurable logic block (CLB). The fast interconnections are used inside a CLB [3], but it is a very rear situation when only one CLB is enough for implementing an FSM logic circuit. The flip-flop of LE can be bypassed, so the output of LUT can be either registered or combinational. As a rule, the number of LUT's inputs is rather small ($S \leq 6$) [3, 4]. If the number of arguments of a Boolean function exceeds the number of LUT's inputs, then more than one LUT is necessary to implement the corresponding combinational circuit. In this case, the methods of functional decomposition are used [5–7]. It leads to increasing for the number of layers of logic in a resulting circuit and to complication for interconnections. In turn, it results in increasing for the propagation time and power consumption [8, 9]. To improve the parameters of an FSM circuit, the EMBs should be used for implementing some parts of a circuit [10–14].

As it is mentioned before, the EMBs of up-to-day FPGAs have a property of configurability. It means that such parameters as the number of cells and their outputs can be changed [1, 2]. Typical configurations of modern EMBs are the following: $16\text{K} \times 1, 8\text{K} \times 2, 4\text{K} \times 4, 2\text{K} \times 8, 1\text{K} \times 18, 512 \times 36$ (bits) [1, 2].

So, the modern EMBs are very flexible and can be tuned to meet a particular design project. Let an EMB contain V cells and t_F outputs. Let V_0 be a number of cells if there is $t_F = 1$. The number V can be determined as (2.1).

Let us discuss a case when a single EMB is enough for implementing an FSM's logic circuit. Let the following condition take place:

$$2^{L+R_0} \cdot (R_0 + N) \leq V_0. \tag{7.1}$$

A. Barkalov et al., *Logic Synthesis for FPGA-Based Finite State Machines*,
Studies in Systems, Decision and Control 38, DOI 10.1007/978-3-319-24202-6_7

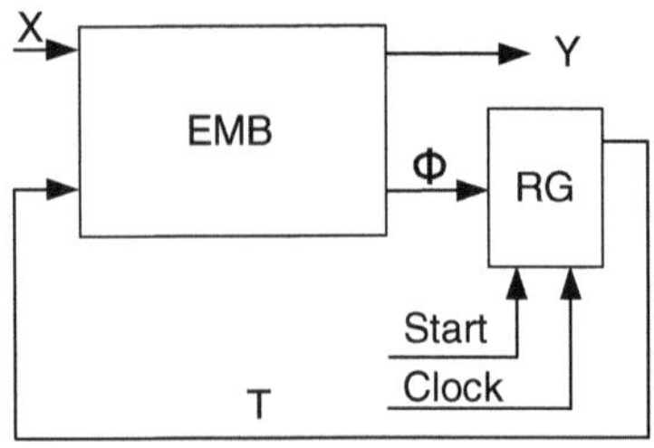

Fig. 7.1 Structural diagram of EMB-based *P* Mealy FSM

In this case a Mealy FSM can be implemented in a trivial way [15] using one EMB and *R* of flip-flops forming the register (Fig. 7.1). Obviously, it is EMB-based *P* Mealy FSM.

In P Mealy FSM, the EMB implements the functions (1.3)–(1.4). If an EMB is a synchronous one, then there is no need in the RG.

Let us discuss an example of design for EMB-based Mealy FSM $P(\Gamma_9)$. The GSA Γ_9 is shown in Fig. 7.2. The following sets and their parameters can be derived from Fig. 7.2: $A = \{a_1, a_2\}$, $M = 2$, $X = \{x_1, x_2\}$, $L = 2$, $Y = \{y_1, \dots, y_4\}$, $N = 4$, $R_o = 1$, $T = \{T_1\}$ and $\Phi = \{D_1\}$.
It is enough a single EMB having $S_A = 3$ and $t_F \geq 5$ to implement the circuit of $P(\Gamma_9)$. There are the same steps in the design of EMB-based *P* FSM as it is in [16]. But the step of the transformation of initial ST is added [17].

Let us encode the states of $P(\Gamma_9)$ in the following way $K(a_1) = 0$, $K(a_2) = 1$. Now it is possible to construct the ST (Table 7.1). It includes $H_0 = 5$ rows.

Let us transform Table 7.1 and construct the table of EMB (Table 7.2). The number of cells of EMB is determined by (3.17).

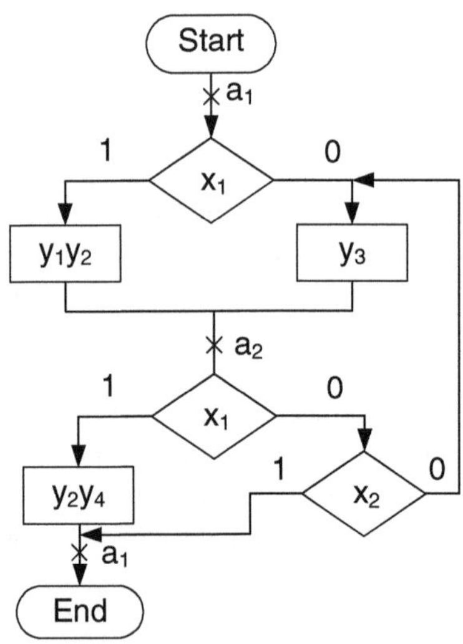

Fig. 7.2 Graph-scheme of algorithm Γ_9

Table 7.1 Structure table of Mealy FSM $P(\Gamma_9)$

a_m	$K(a_m)$	a_s	$K(a_s)$	X_h	Y_h	Φ_h	h
a_1	0	a_2	1	x_1	y_1y_2	D_1	1
		a_2	1	$\bar{x}_1$	y_3	D_1	2
a_2	1	a_1	0	x_1	y_2y_4	–	3
		a_1	0	$\bar{x}_1x_2$	–	–	4
		a_2	1	$\bar{x}_1\bar{x}_2$	y_3	D_1	5

Table 7.2 Table of EMB for Mealy FSM $P(\Gamma_9)$

$K(a_m)$	X	Y	Φ	v
T_1	x_1x_2	$y_1y_2y_3y_4$	D_1	
0	00	0010	1	1
0	01	0010	1	2
0	10	1100	1	3
0	11	1100	1	4
1	00	0010	1	5
1	01	0000	0	6
1	10	0101	0	7
1	11	0101	0	8

We have discussed similar tables in previous chapters. Let us point out that the number of transitions for each state $a_m \in A$ is determined as (3.18).

There is a functional circuit of $P(\Gamma_9)$ shown in Fig. 7.3. The LUT1 is used as D flip-flop. So, it is controlled by the pulses *Start* and *Clock*.

This solution leads to circuits with the highest performance. But the area of its practical application is limited [17]. It can be applied for rather simple GSAs.

We use standard benchmarks [18] to investigate different design methods discussed in this chapter This library includes 53 benchmarks taken from the practice of FSM design. There are characteristics of benchmarks shown in Table 7.3. It includes characteristics of Mealy and Moore FSMs.

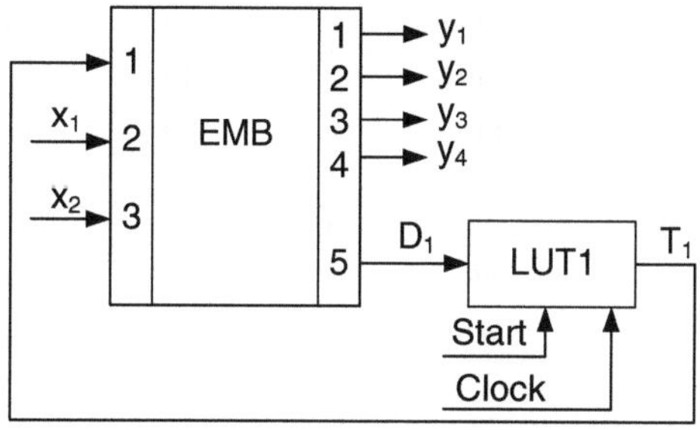

Fig. 7.3 Logic circuit of Mealy FSM $P(\Gamma_9)$

Table 7.3 Characteristics of benchmarks

No	Name	L	N	M_0	H_0	M_1	H_1
1	bbara	4	2	10	60	12	72
2	bbsse	7	7	16	56	26	104
3	bbtas	2	2	6	24	9	36
4	bbcount	3	4	7	28	10	40
5	cse	7	7	16	91	32	183
6	dk14	3	5	7	56	26	208
7	dk15	3	5	4	32	17	136
8	dk16	2	3	27	108	75	300
9	dk17	2	3	8	32	16	64
10	dk27	1	2	7	14	10	20
11	dk512	1	3	15	30	24	48
12	donfile	2	1	24	96	24	98
13	ex1	9	19	20	138	80	634
14	ex2	2	2	19	72	25	92
15	ex3	2	2	10	36	14	44
16	ex4	6	9	14	21	18	28
17	ex5	2	2	9	32	16	52
18	ex6	5	8	8	34	14	61
19	ex7	2	2	10	36	17	56
20	keyb	7	2	19	170	22	193
21	kirkman	12	6	16	328	138	3951
22	lion	2	1	4	11	5	14
23	lion9	2	1	9	25	11	31
24	mark1	5	16	15	36	22	50
25	mc	3	5	4	10	8	20
26	modulo12	1	1	12	24	12	24
27	opus	5	6	10	30	10	30
28	planet	7	19	48	115	103	248
29	planet1	7	19	48	115	103	248
30	pma	8	8	24	73	49	132
31	s1	8	6	20	107	20	107
32	s1488	8	19	48	251	168	912
33	s1494	8	19	48	250	168	1030
34	s1a	8	6	20	107	20	107
35	s208	11	2	18	153	37	309
36	s27	4	1	6	34	6	34
37	s298	3	6	218	1096	332	1669
38	s386	7	7	13	64	23	127

(continued)

Table 7.3 (continued)

No	Name	L	N	M_0	H_0	M_1	H_1
39	s420	19	2	18	137	37	282
40	s510	19	7	47	77	73	113
41	s8	4	1	5	20	5	20
42	s820	18	19	25	232	70	613
43	s832	18	19	25	245	70	707
44	sand	11	9	32	184	88	654
45	scf	27	56	121	286	141	330
46	shifreg	1	1	8	16	16	32
47	sse	7	7	16	56	26	104
48	styr	9	10	30	166	57	366
49	tav	4	4	4	49	27	322
50	tbk	6	3	32	1568	60	2940
51	tma	7	6	20	44	20	44
52	train11	2	1	11	25	14	34
53	train4	2	1	4	14	6	21

There are the most important characteristics of FSMs shown in Table 7.3. It includes the numbers of logical conditions (L), microoperations (N), states of Mealy FSMs (M_0), transitions of Mealy FSMs (H_0), states of Moore FSMs (M_1) and transitions of Moore FSMs (H_1).

To investigate different design methods, it is necessary to choose some chips for implementing the circuits. Let us use the XC5VLX30 chip of Virtex-5 by Xilinx as a base for investigations [19]. This chip includes 32 blocks of RAM (BRAM) having the following configurations: 36 K $\times$ 1, 18 K $\times$ 2, 9 K $\times$ 4, 4 K $\times$ 9, 2 K $\times$ 18, 1 K $\times$ 36, 512 $\times$ 72 (bits).

There are some results of investigations for some of benchmarks. All tables are organized in the same way. The first column of these tables contains the name of a benchmark, the next three columns contain numbers of slices, slice flip-flops and LUTs, respectively. The last column includes the number of EMBs required for implementing the corresponding logic circuit.

There are the characteristics of P FSMs shown in Table 7.4. We marked as "+" the case when a circuit can be implemented as s single EMB. There are 39 of benchmarks marked by "+" in Table 7.4. It means that up to 73.5 % of all FSMs (benchmarks) can be implemented as a single EMB using XC5VLX30 chip.

Our analysis of Tables 7.3 and 7.4 shows that the sign "+" marks rather simple FSMs. We cannot get the results for such complex benchmarks as $s420$, $s510$, $s820$ and so on. So, the area of trivial approach is rather limited. So, it is necessary to use different methods of structural decomposition to reduce the hardware amount in FSM circuits based on EMBs [17].

Table 7.4 Results of investigations for *P* Mealy FSMs

Name	Slice	Slice FF	LUT	BRAM	
bbara	1	4	0	1	+
bbsse	1	4	0	1	+
bbtas	1	3	0	1	+
bbcount	1	3	0	1	+
cse	16	17	40	0	+
dk14	1	3	0	1	+
dk15	1	2	0	1	+
dk16	2	5	0	1	+
dk17	1	3	0	1	+
dk27	1	3	0	1	+
dk512	1	4	0	1	+
donfile	2	5	0	1	+
ex1	2	5	0	12	–
ex2	5	11	12	0	+
ex3	3	10	6	0	+
ex4	1	4	0	1	+
ex5	3	9	5	0	+
ex6	1	3	0	1	+
ex7	3	9	5	0	+
keyb	27	15	74	0	+
kirkman	21	16	54	0	+
lion	1	2	0	1	+
lion9	3	9	5	0	+
mark1	11	19	20	0	+
mc	1	2	0	1	+
modulo12	1	4	0	1	+
opus	9	13	21	0	+
planet	2	6	0	7	–
planet1	2	6	0	7	–
pma	36	21	89	0	+
s1	2	5	0	3	–
s1488	2	6	0	13	–
s1494	2	6	0	13	–
s1a	2	5	0	3	–
s208	6	12	13	0	+
s27	1	3	0	1	+
s298	2	8	0	1	+

(continued)

Table 7.4 (continued)

Name	Slice	Slice FF	LUT	BRAM	
s386	1	4	0	1	+
s8	5	7	8	0	+
sand	2	5	0	28	–
shifreg	1	3	0	1	+
sse	1	4	0	1	+
styr	2	5	0	8	–
tav	1	2	0	1	+
tbk	2	5	0	1	+
tma	20	17	52	0	+
train11	4	9	5	0	+
train4	1	2	0	1	+

To investigate the methods discussed in this chapter, we developed special program tools. We call it K2F.

The software K2F was created using PHP language. The main task of this software is to translate the KISS2 file into a structure of FSM described using VHDL. On execution, the K2F program generates a file describing (with VHDL language) all the structures of FSMs presented in this chapter, which realize the control algorithm represented by the input KISS2 file.

The next step can be the synthesis and simulation of the results with the use of Active-HDL environment and the implementation using e.g. Xilinx ISE package. A typical design path with the use of K2F is shown on Fig. 7.4.

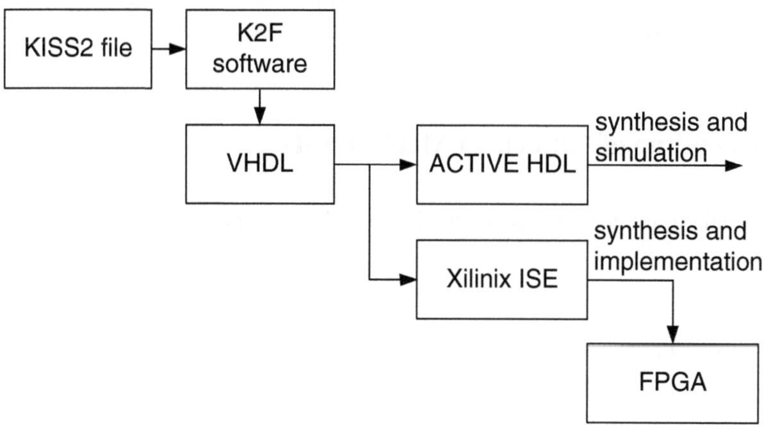

Fig. 7.4 Typical design path with the use of K2F

Fig. 7.5 Example of headers KISS2

```
.i <number of inputs>
.o <number of outputs>
.p <number of products>
.s <number of states used>
.r <reset state>
<input> <current-state > <next-state> <output>
                    .
                    .
                    .
<input> <current-state > <next-state> <output>
.e
```

The KISS2 format is a very popular text format for describing the behaviour of a control unit [18]. A KISS2 file is divided into two parts: header and a state transition table. The header contains generic parameters of control unit, i.e. the number of inputs, the number of outputs, the number of states and the number of terms. There is an example of headers in Fig. 7.5.

The field ⟨inputs⟩ is a sequence of ⟨number of inputs⟩ chars from the set $\{0, 1, -\}$. It represents a condition that must be met to make the transition from ⟨current state⟩ to ⟨next state⟩. The field ⟨output⟩ is a sequence of ⟨number of outputs⟩ chars from the set $\{0, 1, -\}$. It represents the state of outputs of the control unit after the transition to a new state. The symbol "–" represents don't care values. The optional attribute *.e* indicates the end of the state transition table.

We use the Xilinx ISE 14.1 package for synthesis and implementation of benchmarking FSMs. We use the platform XC5VLX30FF324 by Virtex-5 of Xilinx. It includes 32 blocks of RAM (BRAM) having the following configurations: 36 K $\times$ 1, 18 K $\times$ 2, 9 K $\times$ 4, 4 K $\times$ 9, 2 K $\times$ 18, 1 K $\times$ 36 and 512 $\times$ 72 (bits).

7.2 Encoding of Objects in Mealy FSMs

The trivial circuit can be implemented if the following condition takes place:

$$2^{L+R_0} \leq V_0. \tag{7.2}$$

In this case, the number of EMBs is determined by the following expression:

$$n_{EMB} = \left\lceil \frac{N + R_0}{t_F} \right\rceil. \tag{7.3}$$

In (7.3), the symbol t_F means the value of EMB's outputs in the case when the condition (7.2) takes place.

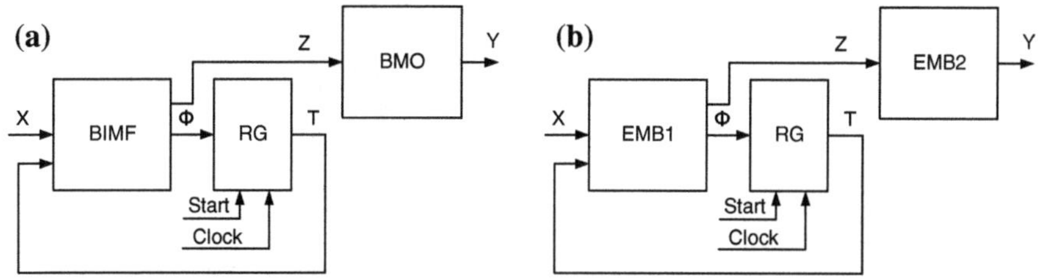

Fig. 7.6 Structural diagram of *PY* Mealy FSM for common case (**a**) and EMB-based (**b**)

If (7.2) is violated, then the number of EMBs increases drastically in the circuit of FSM [17]. Let us discuss the case when the condition (7.2) takes place. In this case, the value of n_{EMB} can be decreased due to encoding of: (1) collections of microoperations and (2) rows of ST. Let us name CMOs and rows as objects of FSM.

Let us start from the EMB-based *PY* Mealy FSM. The CMOs $Y_t \subseteq Y$ are encoded by binary codes $C(Y_t)$ having R_Z bits. The expression (1.32) is used for determining the value of R_Z. The variables $z_r \in Z$ are used for encoding of CMOs, where $|Z| = R_Z$. It leads to *PY* Mealy FSM (Fig. 7.6).

In the common case, the BIMF implements the systems (1.4) and (2.30). Functions Z enter the BMO (Fig. 7.6a). The BMO implements the functions (1.40). In the case of EMB-based *PY* FSM (Fig. 7.6b), the EMB1 corresponds to the BIMF and EMB2 to BMO.

There are the following steps in the design method of *PY* Mealy FSM:

1. Constructing the set of states A.
2. State assignment.
3. Constructing the structure table of Mealy FSM.
4. Encoding of collections of microoperations.
5. Constructing the transformed structure table.
6. Constructing the table of BIMF.
7. Constructing the table of BMO.
8. Implementing the FSM logic circuit with EMBs.

This method can be applied if the following conditions take places:

$$2^{L+R_0} \cdot (R_0 + R_Y) \leq V_0; \tag{7.4}$$

$$N \cdot 2^{R_Y} \leq V_0. \tag{7.5}$$

Let us discuss an example of design for Mealy FSM $PY(\Gamma_{10})$. The GSA Γ_{10} is shown in Fig. 7.7.

There are the following sets and their parameters derived from GSA Γ_{10}: $A = \{a_1, \ldots, a_5\}$, $M_0 = 5$, $X = \{x_1, x_2, x_3\}$, $L = 3$, $Y = \{y_1, \ldots, y_7\}$, $N = 7$, $R_0 = 3$, $T = \{T_1, T_2, T_3\}$ and $\Phi = \{D_1, D_2, D_3\}$. Let us encode the states $a_m \in A$ in the trivial way: $K(a_1) = 000, \ldots, K(a_5) = 100$.

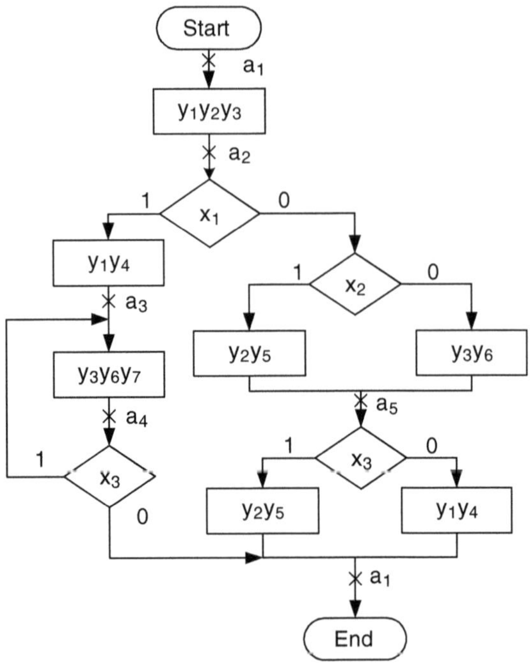

Fig. 7.7 Initial GSA Γ_{10}

Let us use the FPGA chip having EMBs with $V_0 = 384$. Let the following configurations exist for these EMBs: 64×6 and 32×1, bits. Using (7.1), it can be found that it is necessary 640 bits of memory to implement the circuit of $P(\Gamma_{10})$. So, this model cannot be used. Let us try using the model $PY(\Gamma_{10})$.

There are $T_0 = 6$ of CMOs in operator vertices of GSA Γ_{10}. They are the following: $Y_1 = \emptyset$, $Y_2 = \{y_1, y_2, y_3\}$, $Y_3 = \{y_2, y_4\}$, $Y_4 = \{y_3, y_6, y_7\}$, $Y_5 = \{y_2, y_5\}$, $Y_6 = \{y_3, y_6\}$. So, there are $R_Z = 3$ and $Z = \{z_1, z_2, z_3\}$.

Let us check the conditions (7.4)–(7.5). The expression (7.4) leads to the equality $384 = 384$. The expression (7.5) results in inequality $56 \leq 384$. So, there is possibility in using the model $PY(\Gamma_{10})$.

There are $H_0(\Gamma_{10}) = 9$ rows in the structure table of $P(\Gamma_{10})$. It is represented by Table 7.5.

Because EMBs are used for implementing all combinational parts of a circuit, the collections of microoperations can be encoded in the arbitrary manner. Let us encode them in the following manner: $C(Y_0) = 000$, $C(Y_1) = 001, \ldots, C(Y_5) = 101$.

To construct the transformed structure table, the column Y_h of initial table should be replaced by columns Y_t and $C(Y_t)$. Obviously, the number of rows in the transformed structure table is the same as for the initial structure table of Mealy FSM. The transformed structure table of Mealy FSM $PY(\Gamma_{10})$ is represented by Table 7.6

The transformed ST is used for constructing the table of EMB1. It includes the following columns: $K(a_m), X, Z, \Phi, q$. The columns $K(a_m), X$ form an address of a cell. The content of this cell is determined by the columns Z and Φ. There is $L = 3$,

Table 7.5 Structure table of Mealy FSM $P(\Gamma_{10})$

a_m	$K(a_m)$	a_s	$K(a_s)$	X_h	Y_h	Φ_h	h
a_1	000	a_2	001	1	$y_1y_2y_3$	D_3	1
a_2	001	a_3	010	x_1	y_1y_4	D_2	2
		a_5	100	$\bar{x}_1x_2$	y_2y_5	D_1	3
		a_5	100	$\bar{x}_1\bar{x}_2$	y_3y_6	D_1	4
a_3	010	a_4	011	1	$y_3y_6y_7$	D_2D_3	5
a_4	011	a_4	011	x_3	$y_3y_6y_7$	D_2D_3	6
		a_1	000	$\bar{x}_3$	–	–	7
a_5	100	a_1	000	x_3	y_2y_5	–	8
		a_1	000	$\bar{x}_3$	y_1y_4	–	9

Table 7.6 Transformed structure table of Mealy FSM $PY(\Gamma_{10})$

a_m	$K(a_m)$	a_s	$K(a_s)$	X_h	Y_t	$C(Y_t)$	Φ_h	h
a_1	000	a_2	001	1	Y_1	001	D_3	1
a_2	001	a_3	010	x_1	Y_2	010	D_2	2
		a_5	100	$\bar{x}_1x_2$	Y_3	011	D_1	3
		a_5	100	$\bar{x}_1\bar{x}_2$	Y_4	100	D_1	4
a_3	010	a_4	011	1	Y_5	101	D_2D_3	5
a_4	011	a_4	011	x_3	Y_5	101	D_2D_3	6
		a_1	000	$\bar{x}_3$	Y_0	000	–	7
a_5	100	a_1	000	x_3	Y_3	011	–	8
		a_1	000	$\bar{x}_3$	Y_2	010	–	9

Table 7.7 The part of table of EMB1 (for state a_2)

$K(a_m)$	X	Z	Φ	q	h
$T_1T_2T_3$	$x_1x_2x_3$	$z_1z_2z_3$	$D_1D_2D_3$		
001	000	100	100	9	4
001	001	100	100	10	4
001	010	011	100	11	3
001	011	011	100	12	3
001	100	010	010	13	2
001	010	010	010	14	2
001	110	010	010	15	2
001	111	010	010	16	2

so $H(a_m) = 8$. It means that: (1) there are 64 rows in the table of EMB1 and (2) only 40 rows carry useful information. The part of this table is represented by Table 7.7.

There are transitions from the state $a_2 \in A$ shown in Table 7.7. We add the column h to show the connection between Tables 7.6 and 7.7.

Table 7.8 Table of EMB2 for Mealy FSM $PY(\Gamma_{10})$

$C(Y_t)$	Y_t	q
$z_1z_2z_3$	$y_1y_2y_3y_4y_5y_6y_7$	
000	0000000	1
001	1110000	2
010	1001000	3
011	0100100	4
100	0010010	5
101	0010011	6
110	0000000	7
111	0000000	8

There are the following columns in the table of BMO: $C(Y_t)$, Y_t, q. It corresponds to the table of EMB2. This table has $2^{R_Z} = 8$ rows (Table 7.8). There are no CMOs encoded using the codes 110 and 111. So, these rows are filled by zeros in Table 7.8.

The Table 7.7 (together with other similar tables) is used for programming the EMB1. The Table 7.8 is used for programming the EMB2. There is the logic circuit of Mealy FSM $PY(\Gamma_{10})$ shown in Fig. 7.8. Let us point out that the EMB1 is organized as 64 × 6, whereas the EMB2 as 32 × 8 memory block.

Tables 7.9 presents the results of investigations for some of benchmarks. As we can see, sometimes the CAD tools do not allow using EMBs in the circuit. We cannot overcome this tendency of Xilinx.

It is possible to use the method of encoding of the fields of compatible MOs for optimising the EMB-based Mealy FSMs. Of course, it is necessary to use LUTs for implementing the final decoders [17]. In this case, the CMOs are used as objects, too.

There is the structural diagram of EMB-based *PD* Mealy FSM (Fig. 7.9). It is similar to *PY* FSM Fig. 7.6b. But there is the LUTer instead of EMB2. There are additional steps in the design method of *PD* FSM. They are connected with: (1)

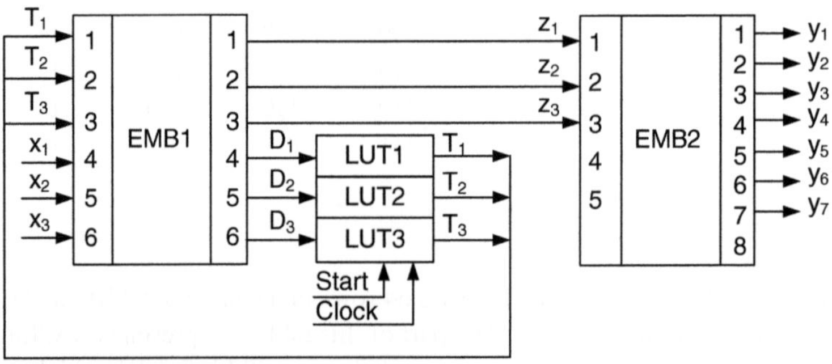

Fig. 7.8 Logic circuit of Mealy FSM $PY(\Gamma_{10})$

Table 7.9 Results of investigations for *PY* Mealy FSMs

Name	Slices	FF	LUTs	BRAMs
bbara	1	4	0	2
bbsse	1	4	0	2
bbtas	1	3	0	1
bbcount	1	3	0	1
cse	1	4	0	2
dk14	1	3	0	2
dk15	1	2	0	1
dk16	5	8	3	1
dk17	3	6	3	1
dk27	1	3	0	1
dk512	1	4	0	1
donfile	3	6	1	1
ex1	2	5	0	7
ex2	13	13	27	0
ex3	6	13	9	0
ex4	1	4	0	2
ex5	5	11	7	0
ex6	1	3	0	1
ex7	2	5	1	1
keyb	3	7	2	1
kirkman	29	16	77	1
lion	2	3	1	1
lion9	5	10	5	0
mark1	6	12	13	1
mc	1	2	0	2
modulo12	0	0	0	0
opus	8	11	17	1
planet	2	6	0	5
planet1	2	6	0	5
pma	30	15	85	1
s1	2	5	0	3
s1488	2	6	0	7
s1494	2	6	0	7
s1a	0	0	0	0
s208	2	5	0	15
s27	2	4	1	1
s298	7	14	5	1
s386	1	4	0	2

(continued)

Table 7.9 (continued)

Name	Slices	FF	LUTs	BRAMs
s8	0	0	0	0
sand	2	5	0	23
shifreg	2	4	0	1
sse	1	4	0	2
styr	2	5	0	6
tav	1	2	0	1
tbk	4	8	3	1
tma	23	15	59	1
train11	5	11	7	0
train4	2	3	1	1

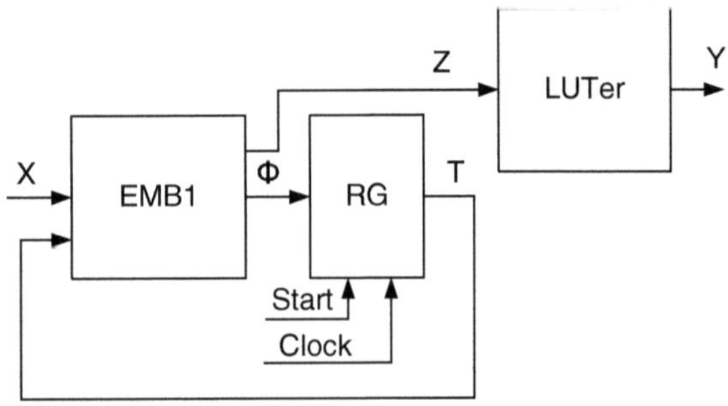

Fig. 7.9 Structural diagram of EMB-based *PD* Mealy FSM

constructing the sets of compatible MOs and (2) executing the encoding of the fields of compatible MOs [17].

This model can be applied if the following condition takes place:

$$2^{K+R_0} \cdot (R_0 + R_D) \le V_0. \tag{7.6}$$

In (7.6), the symbol R_D stands for the general amount of bits required for the encoding of all fields. It is determined as (3.41). Let us discuss an example of design for Mealy FSM $PD(\Gamma_{10})$. Some steps of design are already executed in the previous example. Now, let us find the partition $\Pi_Y = \{Y^1, \ldots, Y^K\}$ for the discussed example. Remind that Y^k is a k-th class of compatible MOs.

There are the following classes Y^k in the discussed case: $Y^1 = \{y_1, y_5, y_6\}$, $Y^2 = \{y_2, y_4, y_7\}$ and $Y^3 = \{y_3\}$. Therefore, $K = 3$, $N_1 = N_2 = 3$ and $N_3 = 1$. Using (1.35) and (3.41) the following values can be found: $R_1 = R_2 = 2$, $R_3 = 1$ and $R_D = 5$. It means that $Z^1 = \{z_1, z_2\}$, $Z^2 = \{z_3, z_4\}$ and $Z^3 = \{z_5\}$.

Let the FPGA chip in use have $V_0 = 512$ (bits) and let the configuration 64×8 exist. It was found that $V_0 \ge 640$ is required for implementing the logic circuit of $P(\Gamma_{10})$. So, it is impossible to use this model. The expression (7.6) produces the equality $512 = 512$. Therefore, it is possible to use the model $PD(\Gamma_{10})$.

Let us encode the microoperations $y_n \in Y^k$ as it is shown in Table 7.10.

There are the same columns in the transformed table of *PD* as it is for *PY* FSM. It is Table 7.11 in the discussed case.

Let us explain how the column $K(Y_t)$ is filled. The code $K(Y_t)$ of a collection of microoperations $Y_t \subseteq Y$ can be represented as a concatenation of codes $K(y_n)^k$ $(k = \overline{1, K})$, where $y_n \in Y_t$:

$$Y_t = K(y_n)^1 * K(y_n)^2 * \cdots * K(y_n)^K. \tag{7.7}$$

In (7.7), the sign * denotes the concatenation.

For example, there is $Y_1 = \{y_1, y_2, y_3\}$. From Table 7.10 we can derive that $K(y_1) = 01$, $K(y_2) = 01$ and $K(y_3) = 1$. Therefore, the first row of Table 7.11 should include the code 01011 in the column $C(Y_t)$. The collection $Y_4 = \{y_3, y_6\}$ and $y_3 \in Y^3$, $y_6 \in Y^1$. So, this collection does not include microoperations $y_n \in Y^2$. It means that the code $K(\emptyset)^2 = 00$ should be used. It gives the code $C(Y_4) = 11001$. All other codes $C(Y_t)$ are formed in this very manner.

Table 7.11 is used for constructing the table of EMB1. It includes 64 rows in the discussed case. Some part of this table is represented by Table 7.12. It contains the transitions for the state $a_2 \in A$.

Table 7.10 Codes of microoperations for FSM $PD(\Gamma_{10})$

$y_n \in Y^1$	$K(y_n)^1$	$y_n \in Y^2$	$K(y_n)^2$	$y_n \in Y^3$	$K(y_n)^3$
	z_1z_2		z_3z_4		z_5
$\emptyset$	0 0	$\emptyset$	0 0	$\emptyset$	0
y_1	0 1	y_2	0 1	y_3	1
y_5	1 0	y_4	1 0		
y_6	1 1	y_7	1 1		

Table 7.11 Transformed ST of Mealy FSM $PD(\Gamma_{10})$

a_m	$Ka_m)$	a_s	$K(a_s)$	X_h	Y_t	$C(Y_t)$	Φ_h	h
a_1	000	a_2	001	1	Y_1	01011	D_3	1
a_2	001	a_3	010	x_1	Y_2	01100	D_2	2
		a_5	100	$\bar{x}_1x_2$	Y_3	10010	D_1	3
		a_5	100	$\bar{x}_1\bar{x}_2$	Y_4	11001	D_1	4
a_3	010	a_4	011	1	Y_5	11111	D_2D_3	5
a_4	011	a_4	011	x_3	Y_5	11111	D_2D_3	6
		a_1	000	$\bar{x}_3$	Y_0	00000	–	7
a_5	100	a_1	000	x_3	Y_3	10010	–	8
		a_1	000	$\bar{x}_3$	Y_2	01100	–	9

Table 7.12 Table of EMB1 for state $a_2 \in A$ of Mealy FSM $PD(\Gamma_{10})$

$K(a_m)$	X	Z	Φ	q	h
$T_1T_2T_3$	$x_1x_2x_3$	$z_1z_2z_3z_4z_5$	$D_1D_2D_3$		
001	000	11001	100	9	4
001	001	11001	100	10	4
001	010	10010	100	11	3
001	011	10010	100	12	3
001	100	01100	010	13	2
001	101	01100	010	14	2
001	110	01100	010	15	2
001	111	01100	010	16	2

In the case of PD FSM, there is no need in the table of BMO. The system $Y(Z)$ can be derived from the table with codes of microoperations. In the discussed example, the following system can be derived from Table 7.10: $y_1 = \bar{z}_1 z_2$, $y_2 = \bar{z}_3 z_4$, $y_3 = z_5$, $y_4 = z_1 \bar{z}_2$, $y_6 = z_1 z_2$, $y_7 = z_3 z_4$.

The logic circuit of FSM $PD(\Gamma_{10})$ is shown in Fig. 7.10.

The circuit of BIMF is implemented using a single EMB having 64×8 bits. Three logic elements are used for implementing the circuit of RG. At last, six LUTs are used to implement the circuit of BMO. Obviously, the equation $y_3 = z_5$ is implemented without LUTs.

Now let us discuss Mealy FSM based on the encoding of rows of ST. There is the structural diagram of EMB-based PH Mealy FSM shown in Fig. 7.11.

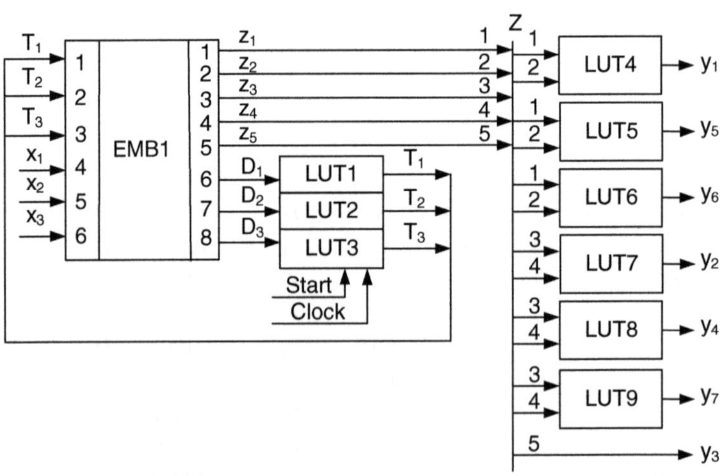

Fig. 7.10 Logic circuit of Mealy FSM $PD(\Gamma_{10})$

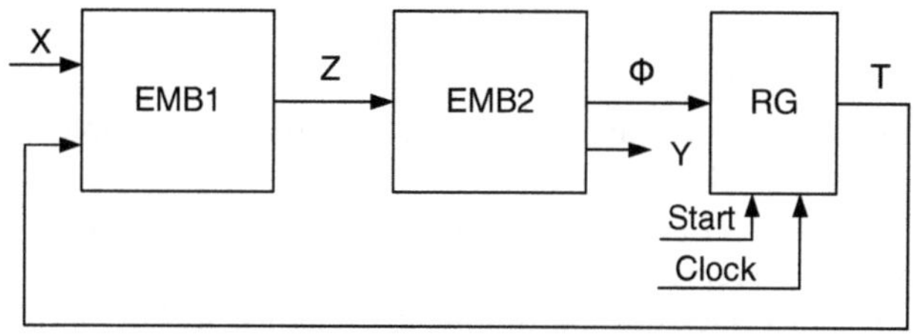

Fig. 7.11 Structural diagram of EMB-based *PH* Mealy FSM

The EMB1 represents the block of encoding of rows (BER) implementing the system (2.30). The EMB2 represents the BIMF implementing the system (1.40). Let the following conditions take places:

$$2^{L+R_0} \cdot R_H \leq V_0; \tag{7.8}$$

$$(N + R_0) \cdot 2^{R_H} \leq V_0. \tag{7.9}$$

Remind, the value of R_H is determined as (1.38). It shows the number of bits in the codes $K(F_h)$, where F_h is a term of ST (1.8).

In conditions (7.8)–(7.9) take places, then there are the following steps in the design method of *PH* Mealy FSM:

1. Constructing the set of states A.
2. State assignment.
3. Constructing the structure table of Mealy FSM.
4. Encoding of the rows of structure table.
5. Constructing the transformed structure table.
6. Constructing the table of BER.
7. Constructing the table of BIMF.
8. Implementing the logic circuit of FSM using EMBs.

Let us discus an example of design for Mealy FSM $PH(\Gamma_{10})$. It is possible to find the following sets for GSA Γ_{10}: $A = \{a_1, \ldots, a_5\}$, $M = 5$, $X = \{x_1, x_2, x_3\}$, $L = 3$, $Y = \{y_1, \ldots, y_7\}$, $N = 7$, $R = 3$, $T = \{T_1, T_2, T_3\}$ and $\Phi = \{D_1, D_2, D_3\}$. Let us encode the states $a_m \in A$ in the trivial way: $K(a_1) = 000, \ldots, K(a_5) = 100$.

Let the FPGA chip in use have $V_0 = 256$ with configurations 256×1, 128×2, 64×4, 32×8 and 16×16 (bits). Because 640 bits are necessary for implementing the logic circuit of $P(\Gamma_{10})$, the model of P FSM cannot be used.

There is $H_0 = 9$ (see Table 7.5), so we can find $R_H = 4$ and $Z = \{z_1, \ldots, z_4\}$. There are the following conditions for $PH(\Gamma_{10})$: 64×4=256 and $10 \times 16 < 256$. Therefore, it has sense to use the model of Mealy FSM $PH(\Gamma_{10})$. Let us encode the rows $F_h \in F$ in the trivial way: $K(F_1) = 0000, \ldots, K(F_9) = 1000$.

To construct the transformed structure table, it is enough to replace the columns Y_h and Φ_h of the initial structure table by the column $K(F_h)$. Obviously, this column contains a code for corresponding row. The transformed structure table for $PH(\Gamma_{10})$ is represented by Table 7.13.

The transformed structure table is used for constructing the table of BER. In the case of *PH* FSM, this table includes the following columns: $K(a_m)$, X, $K(F_h)$, q. For the Mealy FSM $PH(\Gamma_{10})$, this table has 64 rows. There is the part of this table shown in Table 7.14. It represents the transitions from the state $a_2 \in A$.

The table of BIMF includes the following columns: $K(F_h)$, Φ, Y, h. The first column contains the address of the cell of EMB2. The contents of cells are determined by columns Φ and Y. This table is filled in the trivial way. The contents of columns Φ and Y are taken from the structure table. In the case of FSM $PH(\Gamma_{10})$, it is Table 7.15.

The logic circuit of FSM $PH(\Gamma_{10})$ is shown in Fig. 7.12. There are three of LUTs used for implementing the circuit of RG. The content of EMB1 (EMB2) is taken from Table 7.14 (Table 7.15).

We investigated the possibility of using this model for the benchmarks [18]. The results of investigations is shown in Table 7.16. Comparison of Tables 7.9 and 7.16 shows that there is a gain in the hardware amount in the case of *PH* FSMs. This gain increases with the growth of the complexness of a benchmarks. For example,

Table 7.13 Transformed structure table of Mealy FSM $PH(\Gamma_{10})$

a_m	$K(a_m)$	a_s	$K(a_s)$	X_h	$K(F_h)$	h
a_1	000	a_2	001	1	0000	1
a_2	001	a_3	010	x_1	0001	2
		a_5	100	$\bar{x}_1 x_2$	0010	3
		a_5	100	$\bar{x}_1 \bar{x}_2$	0011	4
a_3	010	a_4	011	1	0100	5
a_4	011	a_4	011	x_3	0101	6
		a_1	000	$\bar{x}_3$	0110	7
a_5	100	a_1	000	x_3	0111	8
		a_1	000	$\bar{x}_3$	1000	9

Table 7.14 The part of table of BER for Mealy FSM $PH(\Gamma_{10})$

$K(a_m)$	X	Z	q	h
$T_1T_2T_3$	$x_1x_2x_3$	$z_1z_2z_3z_4$		
001	000	0011	9	4
001	001	0011	10	4
001	010	0010	11	3
001	011	0010	12	3
001	100	0001	13	2
001	101	0001	14	2
001	110	0001	15	2
001	111	0001	16	2

Table 7.15 Table of EMB2 for Mealy FSM $PY(\Gamma_{10})$

$K(F_h)$	Φ	Y_t	q
$z_1z_2z_3z_4$	$D_1D_2D_3$	$y_1y_2y_3y_4y_5y_6y_7$	
0000	001	1110000	1
0001	010	1001000	2
0010	100	0100100	3
0011	100	0010010	4
0100	011	0010011	5
0101	011	0010011	6
0110	000	0000000	7
0111	000	0100100	8
1000	000	1001000	9

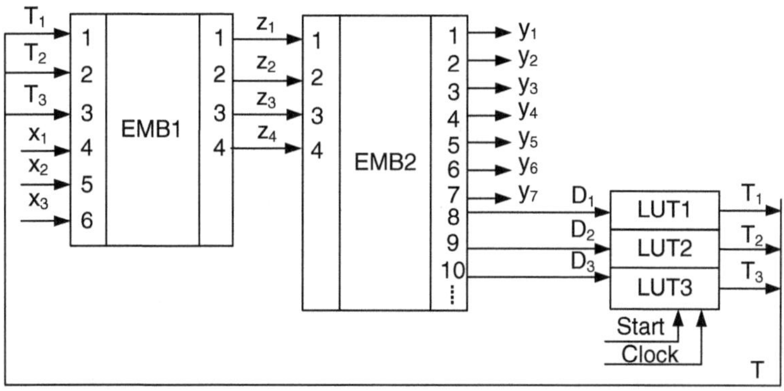

Fig. 7.12 Logic circuit of Mealy FSM $PH(\Gamma_{10})$

it is necessary 23 of BRAMs for *sand* (Table 7.9), whereas this number is decreased up to 17 or *PH* FSM (Table 7.16).

7.3 Replacement of Logical Conditions for Mealy FSMs

If condition (7.2) is violated, then the hardware reduction can be achieved due to applying the method of RLC. Let us find the value of G and form the set $P = \{p_1, \ldots, p_G\}$ for some Mealy FSM. Let the following condition take place:

$$2^{G+R_0} \cdot (N + R_0) \leq V_0. \tag{7.10}$$

In this case, it is possible to use the model of EMB-based *MP* Mealy FSM (Fig. 7.13). There is only a single EMB in this circuit.

Table 7.16 Results of investigations for *PH* Mealy FSMs

Name	Slices	FF	LUTs	BRAMs
bbara	1	4	0	1
bbsse	1	4	0	1
bbtas	1	3	0	1
bbcount	1	3	0	1
cse	12	15	19	1
dk14	1	3	0	1
dk15	1	2	0	1
dk16	2	5	0	1
dk17	1	3	0	1
dk27	1	3	0	1
dk512	1	4	0	1
donfile	2	5	0	1
ex1	2	5	0	5
ex2	6	11	11	1
ex3	4	10	6	1
ex4	1	4	0	1
ex5	1	4	0	1
ex6	1	3	0	1
ex7	3	9	5	1
keyb	20	12	44	1
kirkman	33	14	88	1
lion	1	2	0	1
lion9	3	9	5	1
mark1	1	4	0	2
mc	1	2	0	1
modulo12	1	4	0	1
opus	5	9	10	1
planet	2	6	0	3
planet1	2	6	0	3
pma	35	15	89	1
s1	2	5	0	3
s1488	2	6	0	5
s1494	2	6	0	5
s1a	2	5	0	3
s208	14	12	28	16
s27	1	3	0	1
s298	2	8	0	2
s386	1	4	0	1

(continued)

Table 7.16 (continued)

Name	Slices	FF	LUTs	BRAMs
s8	4	8	5	1
sand	2	5	0	17
shifreg	1	3	0	1
sse	1	4	0	1
styr	2	5	0	5
tav	1	2	0	1
tbk	2	5	0	2
tma	18	11	38	1
train11	4	9	5	1
train4	1	2	0	1

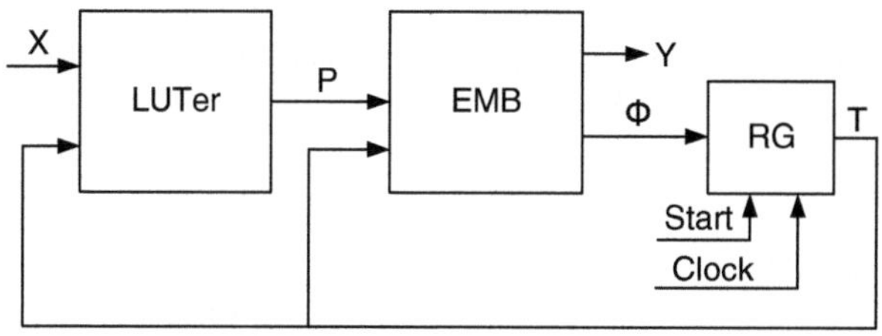

Fig. 7.13 Structural diagram of MP Mealy FSM

In the *MP* Mealy FSM, a block LUTer represents the block of replacement of logical conditions (BRLC). It is implemented using LUTs of an FPGA chip. The LUTer implements the system (1.47) which can be represented as the following:

$$\begin{aligned} p_1 &= p_1(T, X^1); \\ &\vdots \\ p_G &= p_G(T, X^G). \end{aligned} \tag{7.11}$$

In (7.11), the set X^g includes logical conditions $x_l \in X$ replaced by the variable $p_g \in P$. It is quite possible that the following relation is true:

$$X^i \cap X^j \neq \emptyset \quad (i \neq j;\ i, j \in \{1, \ldots, G\}). \tag{7.12}$$

A block EMB represents the block of input memory functions (BIMF). It implements the systems (1.48)–(1.49).

The design method of MP Mealy FSM includes the following steps:

1. Constructing the set of states A.
2. State assignment.
3. Constructing the structure table of P Mealy FSM.
4. Replacement of logical conditions.
5. Constructing the system (7.11).
6. Constructing the transformed structure table.
7. Constructing the table of BIMF.
8. Implementing the FSM logic circuit with EMB and LUTs of a particular FPGA chip.

Let us discuss an example of design for Mealy FSM $MP(\Gamma_{11})$. There is the GSA Γ_{11} shown in Fig. 7.14.

There are the following sets and their characteristics for Mealy FSM $MP(\Gamma_{11})$: $A = \{a_1, \ldots, a_4\}$, $M_0 = 4$, $X = \{x_1, \ldots, x_6\}$, $L = 6$, $Y = \{y_1, \ldots, y_6\}$, $N = 6$, $R_0 = 2$, $T = \{T_1, T_2\}$ and $\Phi = \{D_1, D_2\}$. Let us encode the states $a_m \in A$ in the trivial way: $K(a_1) = 00, \ldots, K(a_4) = 11$.

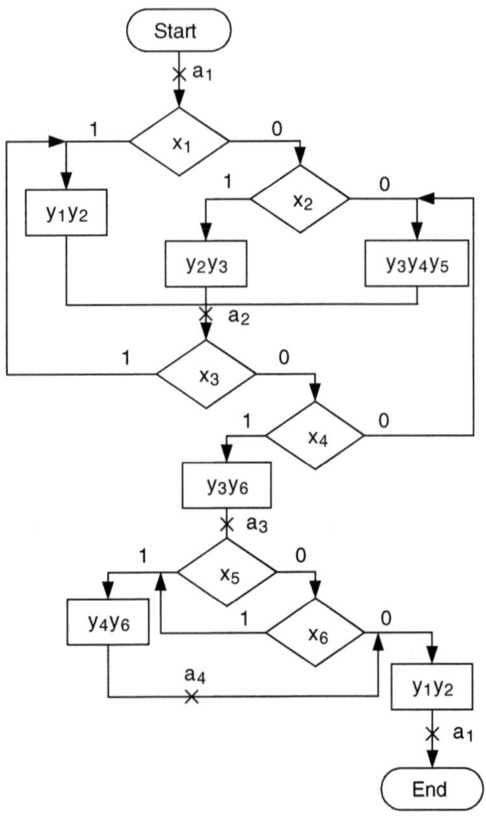

Fig. 7.14 Initial graph-scheme of algorithm Γ_{11}

Table 7.17 Structure table of Mealy FSM $P(\Gamma_{11})$

a_m	$K(a_m)$	a_s	$K(a_s)$	X_h	Y_h	Φ_h	h
a_1	00	a_2	01	x_1	y_1y_2	D_2	1
		a_2	01	$\bar{x}_1x_2$	y_2y_3	D_2	2
		a_2	01	$\bar{x}_1\bar{x}_2$	$y_3y_4y_5$	D_2	3
a_2	01	a_2	01	x_3	y_1y_2	D_2	4
		a_3	10	$\bar{x}_3x_4$	y_3y_6	D_1	5
		a_2	01	$\bar{x}_3\bar{x}_4$	$y_3y_4y_5$	D_2	6
a_3	10	a_4	11	x_5	y_4y_6	D_1D_2	7
		a_4	11	$\bar{x}_5x_6$	y_4y_6	D_1D_2	8
		a_1	00	$\bar{x}_5\bar{x}_6$	y_1y_2	–	9
a_4	11	a_1	00	1	y_1y_2	–	10

Let the FPGA chip in use have $V_0 = 128$ and let the following configurations of EMB exist: 128×1, 64×2, 32×4, 16×8 (bits). It is necessary 2048 bits of memory to implement the circuit of $P(\Gamma_{11})$. So, this model cannot be used.

There are 10 rows in ST of $P(\Gamma_{11})$ (Table 7.17). There are the following sets $X(a_m)$ derived from Table 7.17: $X(a_1) = \{x_1, x_2\}$, $X(a_2) = \{x_3, x_4\}$, $X(a_3) = \{x_5, x_6\}$, $X(a_4) = \emptyset$. Obviously, there is $G = 2$. It gives the set $P = \{p_1, p_2\}$. There are $G + R_0 = 4$ and $R_0 + N = 8$. So, the condition (7.10) is true. It means that the model of $MP(\Gamma_{11})$ can be used.

Let us replace the logical conditions as it is shown in the Table 7.18. There are no the same logical conditions in the sets $X(a_m) \subseteq X$ for the discussed case. Because of it, the distribution of the logical conditions among the variables $p_g \in P$ is executed in the trivial way. If there is $X(a_i) \cap X(a_j) \neq \emptyset$, then the distribution should be executed in such a way that the intersection for any pair of sets $X^g (g = \overline{1, G})$ has the minimum capacity [16].

The following system of equations can be derived from Table 7.18:

$$\begin{aligned} p_1 &= A_1x_1 \vee A_2x_3 \vee A_3x_5; \\ p_2 &= A_1x_1 \vee A_2x_4 \vee A_3x_6. \end{aligned} \tag{7.13}$$

If the variables $A_m (m = \overline{1, M})$ are replaced by corresponding state codes, the system (7.13) is transformed into the following one:

$$\begin{aligned} p_1 &= \bar{T}_1\bar{T}_2x_1 \vee \bar{T}_1T_2x_3 \vee T_1\bar{T}_2x_5; \\ p_2 &= \bar{T}_1\bar{T}_2x_1 \vee \bar{T}_1T_2x_4 \vee T_1\bar{T}_2x_6. \end{aligned} \tag{7.14}$$

Table 7.18 Table of replacement of logical conditions for Mealy FSM $MP(\Gamma_{11})$

a_m	a_1	a_2	a_3	a_4
p_1	x_1	x_3	x_5	–
p_2	x_2	x_4	x_6	–

Table 7.19 Transformed ST of Mealy FSM $MP(\Gamma_{11})$

a_m	$K(a_m)$	a_s	$K(a_s)$	P_h	Y_h	Φ_h	h
a_1	00	a_2	01	p_1	y_1y_2	D_2	1
		a_2	01	$\bar{p_1}p_2$	y_2y_3	D_2	2
		a_2	01	$\bar{p_1}\bar{p_2}$	$y_3y_4y_5$	D_2	3
a_2	01	a_2	01	p_1	y_1y_2	D_2	4
		a_3	10	$\bar{p_1}p_2$	y_3y_6	D_1	5
		a_2	01	$\bar{p_1}\bar{p_2}$	$y_3y_4y_5$	D_2	6
a_3	10	a_4	11	p_1	y_4y_6	D_1D_2	7
		a_4	11	$\bar{p_1}p_2$	y_4y_6	D_1D_2	8
		a_1	00	$\bar{p_1}\bar{p_2}$	y_1y_2	–	9
a_4	11	a_1	00	1	y_1y_2	–	10

This system corresponds to two of LUTs having $S = 5$. The tables of LUTs are constructed in the way discussed before.

There are practically the same columns in the ST and transformed ST [17]. But there is the column P_h in the transformed ST replacing the column X_h from the initial ST. In the discussed case, the transformation of tables leads to Table 7.19.

This table is a base for constructing the table of BIMF containing the following columns: $K(a_m)$, P, Y, Φ, q. The columns $K(a_m)$ and P form the addresses of cells. The table of BIMF includes $V(\Gamma_1) = 16$ rows for the discussed case. The number of rows $H(P)$ for representing transitions from any state is determined as:

$$H(P) = 2^G. \tag{7.15}$$

In the discussed case, there is $H(P) = 4$. The table of BIMF is represented by Table 7.20. The logic circuit of FSM $MP(\Gamma_{11})$ is shown in Fig. 7.15. We take LUTs having $S = 5$. In allows implementing each equation from (7.14) using only a single LUT. It is possible in the common case, if the following condition takes place:

$$R + |X^g| \leq S \quad (g = \overline{1, G}). \tag{7.16}$$

So, the are 4 of LUTs in the circuit of $MP(\Gamma_{11})$. The LUT3 and LUT4 are used to implement the circuit of RG.

This method can be applied for all benchmarks [18]. The results of investigations are shown in Table 7.21.

It is possible to use the method of RLC together with other methods of structural decomposition [17]. For example, the MPY Mealy FSM is based on combining the methods of RLC and maximal encoding of CMOs. Its structural diagram in shown in Fig. 7.16.

It is possible to combine the RLC and encoding of fields of compatible MOs. It leads to MPD Mealy FSM (Fig. 7.17).

Table 7.20 Table of BIMF of Mealy FSM $MP(\Gamma_{11})$

$K(a_m)$	P	Y	Φ	v	h
T_1T_2	p_1p_2	$y_1y_2y_3y_4y_5y_6$	D_1D_2		
00	00	001110	01	1	3
00	01	011000	01	2	2
00	10	110000	01	3	1
00	11	110000	01	4	1
01	00	001110	01	5	6
01	01	001001	10	6	5
01	10	110000	01	7	4
01	11	110000	01	8	4
10	00	110000	00	9	9
10	01	000101	11	10	8
10	10	000101	11	11	7
10	11	000101	11	12	7
11	00	110000	00	13	10
11	01	110000	00	14	10
11	10	110000	00	15	10
11	11	110000	00	16	10

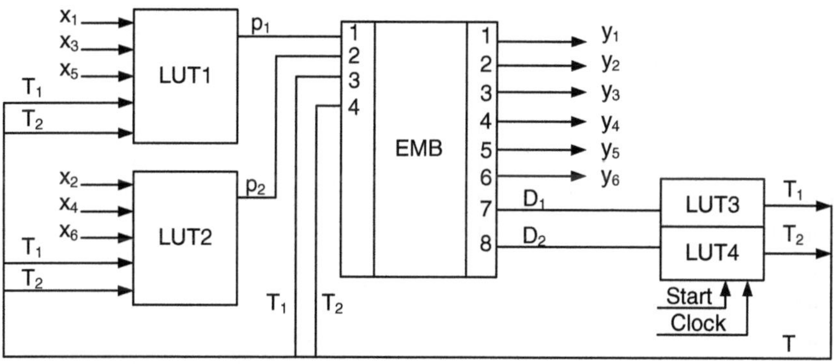

Fig. 7.15 Logic circuit of Mealy FSM $MP(\Gamma_{11})$

It is possible to combine the RLC and the encoding of the rows of ST. It leads to MPH Mealy FSM (Fig. 7.18).

We do not discuss these models in our book. The reader can combine corresponding methods to work out the design methods for FSM shown in Figs. 7.16, 7.17 and 7.18. We investigated these methods for the benchmarks [18]. Some results are shown in Tables 7.22 and 7.23. As follows from these tables, the proposed models can be used in practical cases.

Table 7.21 Results of investigations for *MP* Mealy FSMs

Name	Slice	Slice FF	LUT	BRAM
bbara	4	4	4	1
bbsse	6	4	8	1
bbtas	3	3	2	1
bbcount	3	3	3	1
cse	27	18	63	0
dk14	3	3	3	1
dk15	1	2	0	1
dk16	4	5	2	1
dk17	1	3	0	1
dk27	2	3	1	1
dk512	2	4	1	1
donfile	3	5	2	1
ex1	12	5	18	2
ex2	5	11	12	0
ex3	3	10	6	0
ex4	4	4	4	1
ex5	3	9	5	0
ex6	3	3	2	1
ex7	4	9	5	0
keyb	43	12	99	0
kirkman	79	17	188	0
lion	1	2	0	1
lion9	3	9	5	0
mark1	4	4	4	1
mc	3	2	2	1
modulo12	2	4	1	1
opus	3	4	6	1
planet	13	6	29	2
planet1	13	6	29	2
pma	9	5	15	1
s1	13	6	24	3
s1488	16	6	27	3
s1494	12	8	23	3
s1a	13	6	24	3
s208	12	12	19	0
s27	4	3	4	1
s298	5	8	4	1
s386	7	4	9	1

(continued)

Table 7.21 (continued)

Name	Slice	Slice FF	LUT	BRAM
s8	5	7	11	0
sand	13	5	30	2
shifreg	1	3	0	1
sse	9	4	9	1
styr	13	5	20	2
tav	1	2	0	1
tbk	2	5	0	1
tma	24	20	54	0
train11	4	9	5	0
train4	1	2	0	1

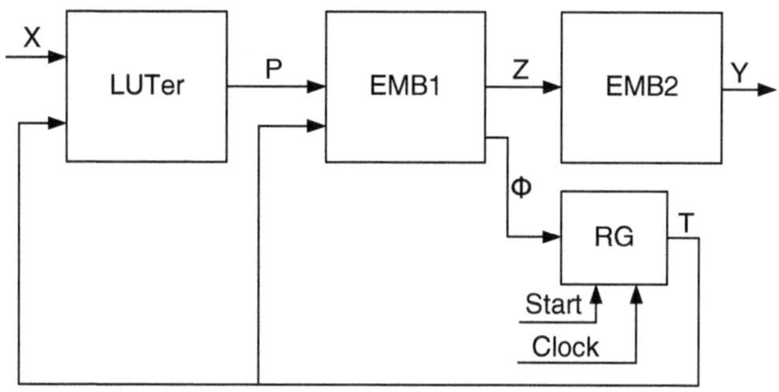

Fig. 7.16 Structural diagram of EMB-based *MPY* Mealy FSM

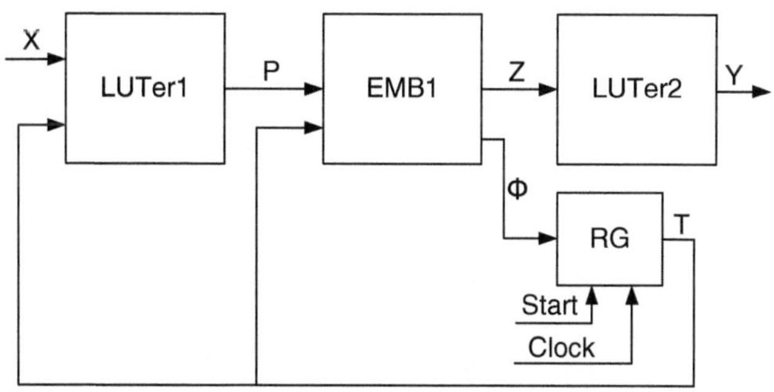

Fig. 7.17 Structural diagram of *MPD* Mealy FSM

To improve the characteristics of FSM with BRLC, it has sense to diminish the number of LUTs in this block. It can result in decreasing for both the propagation time and consumed energy of the final circuit.

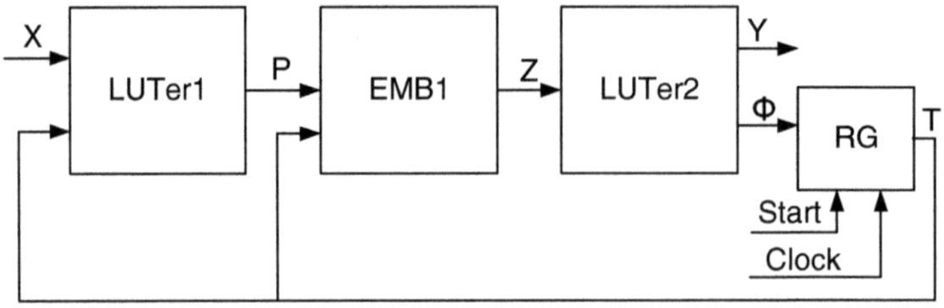

Fig. 7.18 Structural diagram of *MPH* Mealy FSM

Table 7.22 Results of investigations for *MPY* Mealy FSMs

Name	Slice	Slice FF	LUT	BRAM
bbara	3	4	4	1
bbsse	7	4	8	2
bbtas	3	3	2	1
bbcount	3	3	3	1
cse	6	4	8	2
dk14	3	3	3	1
dk15	1	2	0	1
dk16	6	8	5	1
dk17	3	6	3	1
dk27	2	3	1	2
dk512	2	4	1	1
donfile	5	6	3	1
ex1	10	5	18	2
ex2	7	13	15	0
ex3	6	3	9	0
ex4	3	4	4	2
ex5	4	11	7	0
ex6	2	3	2	1
ex7	3	5	3	1
keyb	11	7	25	1
kirkman	40	17	99	1
lion	2	3	1	1
lion9	4	10	5	0
mark1	4	4	4	1
mc	3	2	2	1
modulo12	0	0	0	0
opus	3	4	6	1

(continued)

Table 7.22 (continued)

Name	Slice	Slice FF	LUT	BRAM
planet	11	6	29	2
planet1	11	6	29	2
pma	9	5	15	1
s1	13	6	24	3
s1488	12	6	27	3
s1494	11	7	23	3
s1a	0	0	0	0
s208	6	5	9	2
s27	4	4	5	1
s298	9	14	9	1
s386	7	4	9	2
s8	0	0	0	0
sand	13	5	30	2
shifreg	2	4	0	1
sse	7	4	9	1
styr	11	5	20	2
tav	1	2	0	1
tbk	4	8	3	1
tma	20	15	49	1
train11	5	11	7	0
train4	2	3	1	0

Table 7.23 Results of investigations for *MPH* Mealy FSMs

Name	Slice	Slice FF	LUT	BRAM
bbara	3	4	4	1
bbsse	8	4	8	1
bbtas	2	3	2	1
bbcount	2	3	3	1
cse	13	15	27	1
dk14	2	3	3	1
dk15	1	2	0	1
dk16	3	5	2	1
dk17	1	3	0	1
dk27	2	3	1	1
dk512	2	4	1	1
donfile	3	5	2	1

(continued)

Table 7.23 (continued)

Name	Slice	Slice FF	LUT	BRAM
ex1	11	5	18	2
ex2	7	11	13	1
ex3	5	10	8	1
ex4	4	4	4	1
ex5	2	4	2	1
ex6	3	3	2	1
ex7	5	9	7	1
keyb	19	12	44	1
kirkman	136	15	409	1
lion	1	2	0	1
lion9	3	9	5	1
mark1	3	4	4	2
mc	3	2	2	1
modulo12	2	4	1	1
opus	4	4	6	1
planet	10	6	29	2
planet1	10	6	29	2
pma	11	5	15	1
s1	4	5	24	3
s1488	11	6	27	2
s1494	10	6	23	2
s1a	14	5	24	3
s208	17	12	37	1
s27	5	3	4	1
s298	4	8	4	2
s386	4	4	9	1
s8	4	8	5	1
sand	13	5	30	2
shifreg	1	3	0	1
sse	9	4	9	1
styr	10	5	20	2
tav	1	2	0	1
tbk	2	5	0	2
tma	22	16	42	1
train11	4	9	5	1
train4	1	2	0	1

7.4 Hardware Reduction for BRLC

Let us discus the case when the following condition takes place:

$$S_A > G + R. \tag{7.17}$$

In (7.17), the value S_A determines the number of address bits of EMB for given t_F. If (7.17) is true, then the following condition takes place:

$$S_0 = S_A - (G + R) > 0. \tag{7.18}$$

The value S_0 is equal to the number of "free" address inputs of EMB. These inputs are not connected with variables $T_r \in T$ and $p_g \in P$. We propose to use them for optimization of the circuit of LUTer.

Let us start from MP Mealy FSM. Let us represent the set X as $X^1 \cup X^2$ where the following condition takes place:

$$\begin{aligned} &X^1 \cap X^2 = \emptyset; \\ &X^1 \cup X^2 = X; \\ &|X^1| = S_0; \\ &|X^2| = L - S_0. \end{aligned} \tag{7.19}$$

Let us replace the logical conditions $x_l \in X^2$ by additional variables $p_g \in P$. It leads to M_0P Mealy FSM shown in Fig. 7.19.

In M_0P FSM, the LUTer executes the replacement of logical conditions $x_l \in X^2$. It implements the system

$$P = P(T, X^2). \tag{7.20}$$

The EMB implements functions

$$Y = Y(T, P, X^1); \tag{7.21}$$

$$\Phi = \Phi(T, P, X^1). \tag{7.22}$$

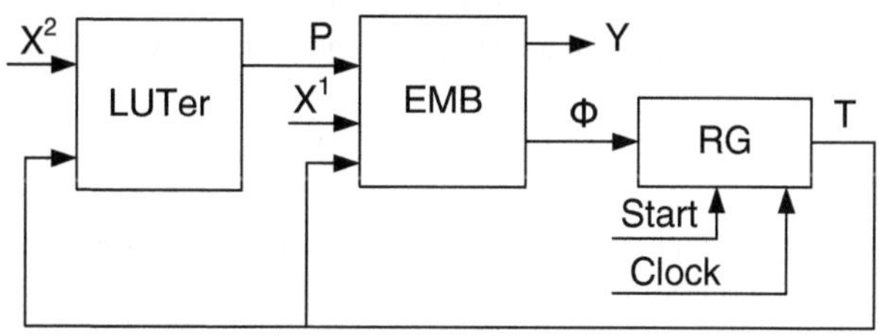

Fig. 7.19 Structural diagram of M_0P Mealy FSM

The following design method is proposed for M_0P Mealy FSM.

1. Constructing the set of states A.
2. State assignment.
3. Constructing the structure table of Mealy FSM.
4. Partitioning the set X and finding the sets X^1 and X^2.
5. Replacement of logical conditions $x_l \in X^2$.
6. Constructing the transformed structure table of Mealy FSM $M_0P(\Gamma_j)$.
7. Constructing the table of BIMF.
8. Implementing the logic circuit of FSM for a given FPGA chip.

Let us discus an example of design for Mealy FSM $M_0P(\Gamma_{11})$. Let us use the FPGA chip including the EMBs having the following configurations: $512 \times 1, 256 \times 2, 128 \times 4$ and 64×8 (bits). For the Mealy FSM $P(\Gamma_{11})$, there is $N + R_0 = 8$. Therefore, the configuration 64×8 can be used with $S_A = 6$. There is $R_0 = 2$, then it is possible to have $G + S_0 = 4$.

There is the structure table of Mealy FSM $P(\Gamma_{11})$ shown in Table 7.17. Let us find the sets $X(a_m)$ from this table. There are the following sets: $X(a_1) = \{x_1, x_2\}$, $X(a_2) = \{x_3, x_4\}, X(a_3) = \{x_5, x_6\}$. Let us represent each set $X(a_m) \subseteq X$ as $X(a_m)^1 \cup X(a_m)^2$ where $X(a_m)^1 \cap X(a_m)^2 = \emptyset$ $(m = \overline{1, M})$. Let us find the sets X^1 and X^2 using the rules:

$$X^1 = \bigcup_{m=1}^{M} X(a_m)^1; \tag{7.23}$$

$$X^2 = \bigcup_{m=1}^{M} X(a_m)^2. \tag{7.24}$$

Let us construct the following sets: $X(a_1)^1 = \{x_1\}$, $X(a_1)^2 = \{x_2\}$, $X(a_2)^1 = \{x_3\}$, $X(a_2)^2 = \{x_4\}$, $X(a_3)^1 = \{x_5\}$ and $X(a_3)^2 = \{x_6\}$. It leads to the sets $X^1 = \{x_1, x_3, x_5\}$ and $X^2 = \{x_2, x_4, x_6\}$. Obviously, there is $G = 1$ and $p_1 = A_1x_2 \vee A_2x_4 \vee A_3x_6$.

The transformed structure table of Mealy FSM $M_0P(\Gamma_j)$ includes the following columns: a_m, $K(a_m)$, a_s, $K(a_s)$, X_h^1, P_h, Y_h, Φ_h, h. In the discussed case, it is Table 7.24.

The table of BIMF includes the following columns: $K(a_m)$, X^1,P, Y, Φ, v, h. The first three columns create the address of some cell of EMB. This table includes $V_3(\Gamma_j)$ rows where:

$$Q_3(\Gamma_j) = 2^{S_0+G+R_0}. \tag{7.25}$$

It can be found that $Q_3(\Gamma_{11}) = 64$. There is the following expression determining the number of cells required to represent the transitions from the state $a_m \in A$:

$$H(a_m) = 2^{S_0+G}. \tag{7.26}$$

Table 7.24 Transformed structure table of Mealy FSM $M_0P(\Gamma_{11})$

a_m	$K(a_m)$	a_s	$K(a_s)$	X_h^1	P_h	Y_h	Φ_h	h
a_1	00	a_2	01	x_1	1	y_1y_2	D_2	1
		a_2	01	$\bar{x_1}$	p_1	y_2y_3	D_2	2
		a_2	01	$\bar{x_1}$	$\bar{p_1}$	$y_3y_4y_5$	D_2	3
a_2	01	a_2	01	x_3	1	y_1y_2	D_2	4
		a_3	10	$\bar{x_3}$	p_1	y_3y_6	D_1	5
		a_2	11	$\bar{x_3}$	$\bar{p_1}$	$y_3y_4y_5$	D_2	6
a_3	10	a_4	11	x_5	1	y_4y_6	D_1D_2	7
		a_4	11	$\bar{x_5}$	p_1	y_4y_6	D_1D_2	8
		a_1	00	$\bar{x_5}$	$\bar{p_1}$	y_1y_2	–	9
a_4	11	a_1	00	1	1	y_1y_2	–	10

Table 7.25 Part of table of BIMF of Mealy FSM $M_0P(\Gamma_{11})$

$K(a_m)$	X^1	P	Y	Φ	v	h
T_1T_2	$x_1x_3x_5$	p_1	$y_1y_2y_3y_4y_5y_6$	D_1D_2		
00	000	0	001110	01	1	3
00	000	1	011000	01	2	2
00	001	0	001110	01	3	3
00	001	1	011000	01	4	2
00	010	0	001110	01	5	3
00	010	1	011000	01	6	2
00	011	0	001110	01	7	3
00	011	1	011000	01	8	2
00	100	0	110000	01	9	1
00	100	1	110000	01	10	1
00	101	0	110000	01	11	1
00	101	1	110000	01	12	1
00	110	0	110000	01	13	1
00	110	1	110000	01	14	1
00	111	0	110000	01	15	1
00	111	1	110000	01	16	1

There is $H(a_m) = 16$ in the discussed case.

There is a part of the table of BIMF for FSM $M_0P(\Gamma_{11})$ represented by Table 7.25. This table represents the transitions from the state $a_1 \in A$.

There is the logic circuit of FSM $M_0P(\Gamma_{11})$ shown in Fig. 7.20. We use LUTs with $S = 5$ for implementing the circuit of LUTer. It is enough a single LUT to implement the LUTer. There are two LUTs used for implementing the RG.

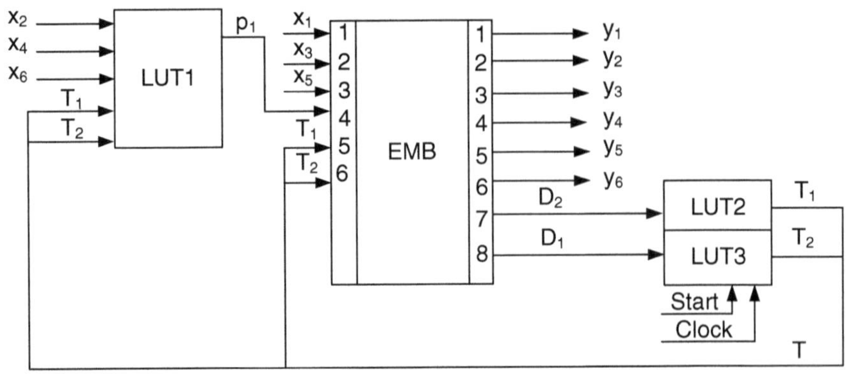

Fig. 7.20 Logic circuit of Mealy FSM $M_0P(\Gamma_{11})$

We investigated this approach for the benchmarks [18]. The results of investigations are represented by Table 7.26.

As follows from Table 7.26, this approach cannot be used for some benchmarks. It is connected with the equality $S_0 = 0$ for these cases. But it is quite possible that the model of M_0P FSM can be used for other FPGA chips.

Now let us discuss the ways for hardware reduction of LUTer for *MPY* Mealy FSMs. Three approaches can be used [17]. The first approach is based on the splitting the set X. The second is connected with the special state assignment. The third approach is based on the encoding of the logical conditions [17].

The splitting the set X by subsets X^1, X^2 is executed in the same manner as for M_0P Mealy FSM. It leads to M_0PY Mealy FSM shown in Fig. 7.21.

The LUTer implements the system $P = P(T, X^2)$, the block EMB1 the system $\Phi = \Phi(T, X^1, P)$ and the system of additional variables

$$Z = Z(T, P, X^1). \tag{7.27}$$

The block EMB2 implements the microoperations $Y = Y(Z)$.

There are the following conditions determining the possibility for using the M_0PY FSM:

$$2^{S_0+G+R_0} \cdot (R_0 + R_Z) \leq V_0; \tag{7.28}$$

$$2^{R_Z} \cdot N \leq V_0. \tag{7.29}$$

The proposed design method for Mealy FSM $M_0PY(\Gamma_j)$ includes the following steps:

1. Constructing the set of states A.
2. State assignment.
3. Constructing the structure table of FSM $P(\Gamma_j)$.

Table 7.26 Results of investigations for M_0P Mealy FSMs

Name	Slice	Slice FF	LUT	BRAM
bbara	3	4	2	1
bbsse	4	4	3	1
bbtas	2	3	1	1
bbcount	2	3	1	1
cse	20	15	52	0
dk14	2	3	1	1
dk15	1	2	0	1
dk16	3	5	1	1
dk17	1	3	0	1
dk27	– – M_0P cannot be used – –			
dk512	– – M_0P cannot be used – –			
donfile	3	5	1	1
ex1	6	5	8	12
ex2	12	11	26	0
ex3	3	10	6	0
ex4	3	4	2	1
ex5	3	9	5	0
ex6	2	3	1	1
ex7	4	9	5	0
keyb	30	16	73	0
kirkman	28	21	59	0
lion	1	2	0	1
lion9	3	9	5	0
mark1	3	4	2	1
mc	2	2	1	1
modulo12	– – M_0P cannot be used – –			
opus	13	13	30	0
planet	9	6	14	7
planet1	9	6	14	7
pma	4	5	5	1
s1	7	5	11	3
s1488	5	6	7	7
s1494	5	6	7	7
s1a	7	5	11	3
s208	10	12	15	0
s27	2	3	2	1
s298	3	8	2	1
s386	3	4	3	1
s8	4	7	8	0
sand	5	5	6	28

(continued)

Table 7.26 (continued)

Name	Slice	Slice FF	LUT	BRAM
shifreg	– – M_0P cannot be used – –			
sse	4	4	3	1
styr	6	5	10	8
tav	1	2	0	1
tbk	2	5	0	1
tma	28	16	70	0
train11	5	9	5	0
train4	1	2	0	1

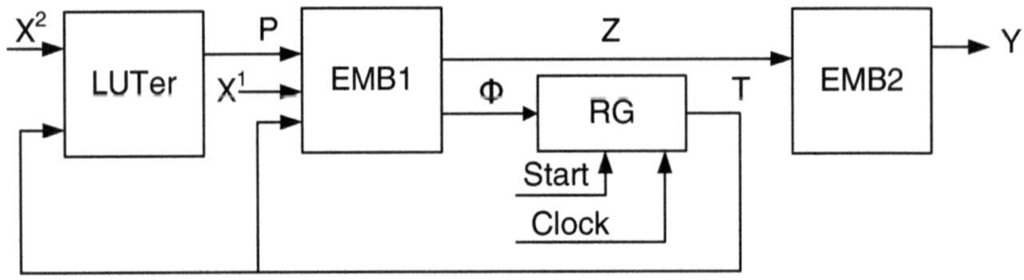

Fig. 7.21 Structure diagram of M_0PY Mealy FSM

4. Partitioning the set X by classes X^1 and X^2.
5. Replacement of logical conditions $x_l \in X^2$.
6. Encoding of the collections of microoperations.
7. Constructing the transformed structure table.
8. Constructing the the table of BIMF.
9. Constructing the table of BMO.
10. Implementing the FSM logic circuit.

Let us discuss an example of design for Mealy FSM $M_0PY(\Gamma_{12})$. There is the GSA Γ_{12} shown in Fig. 7.22.

It is possible to find the following sets and their parameters: $X = \{x_1, \ldots, x_5\}$, $L = 5$, $Y = \{y_1, \ldots, y_{10}\}$, $N = 10$, $A = \{a_1, \ldots, a_6\}$, $M = 6$, $R_0 = 3$, $T = \{T_1, T_2, T_3\}$, $\Phi = \{D_1, D_2, D_3\}$. Let us execute the state assignment in the trivial way: $K(a_1) = 000, \ldots, K(a_6) = 101$.

It allows constructing ST of Mealy FSM $P(\Gamma_{12})$ This table includes $H_0 = 12$ rows (Table 7.27).

Let the FPGA chip in use include EMBs with the following configurations: 512 × 1, 256 × 2, 128 × 4, 64 × 8, 32 × 16 (bits). To implement the circuit of FSM $P(\Gamma_{12})$, it is necessary $V_0 = 2^8 \times 13 = 3328$ (bits). So, the replacement of logical conditions should be used.

There are $T_0 = 7$ different collections of microoperations in the vertices of GSA Γ_{12}. They are the following: $Y_1 = \emptyset$, $Y_2 = \{y_1, y_2, y_8\}$, $Y_3 = \{y_3, y_9\}$, $Y_4 = \{y_2, y_4\}$,

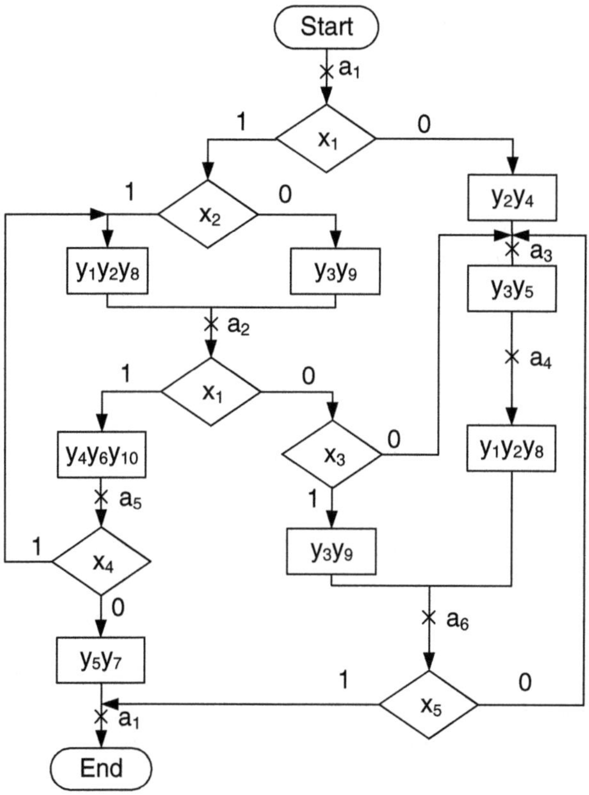

Fig. 7.22 Initial GSA Γ_{12}

$Y_5 = \{y_3, y_5\}$, $Y_6 = \{y_4, y_6, y_{10}\}$, $Y_7 = \{y_5, y_7\}$. It is enough $R_Z = 3$ variables $z_r \in Z$ for encoding of these collections.

Because of $R_0 + R_Z = 6$, the configuration 64×8 should be chosen with $S_A = 6$. It gives $G = 1$, $S_0 = 2$; therefore, there is $|X^1| = 2$, $|X^2| = 3$.

Let us represent the set X as $X = X^1 \cup X^2$ where $X^1 = \{x_2, x_3\}$ and $X^2 = \{x_1, x_4, x_5\}$. The table of replacement of logical conditions for FSM $M_0PY(\Gamma_{12})$ is represented by Table 7.28.

The following equation can be derived from Table 7.28

$$p_1 = \bar{T}_1\bar{T}_2x_1 \vee T_1\bar{T}_2\bar{T}_3x_4 \vee T_1\bar{T}_2T_3x_5. \quad (7.30)$$

There are six literals in the equation (7.30). It means that it is enough a single LUT with $S = 6$ to implement the circuit of LUTer for Mealy FSM $M_0PY(\Gamma_{12})$.

Let us encode the collections of microoperations $Y_t \subseteq Y$ in the trivial way: $K(Y_1) = 000, \ldots, K(Y_7) = 110$. Now, the transformed structure table of Mealy FSM $M_0PY(\Gamma_j)$ can by constructed. The table includes the columns a_m, $K(a_m)$, a_s,

Table 7.27 Structure table of Mealy FSM $P_1(\Gamma_{12})$

a_m	$K(a_m)$	a_s	$K(a_s)$	X_h	Y_h	Φ_h	h
a_1	000	a_2	001	x_1x_2	$y_1y_2y_8$	D_3	1
		a_2	001	$x_1\bar{x}_2$	y_3y_9	D_3	2
		a_3	010	$\bar{x}_1$	y_2y_4	D_2	3
a_2	001	a_5	100	x_1	$y_4y_6y_{10}$	D_1	4
		a_6	101	$\bar{x}_1x_3$	y_3y_9	D_1D_3	5
		a_4	011	$\bar{x}_1\bar{x}_3$	y_3y_5	D_2D_3	6
a_3	010	a_4	011	1	y_3y_5	D_2D_3	7
a_4	011	a_6	101	1	$y_1y_2y_8$	D_1D_3	8
a_5	100	a_2	001	x_4	$y_1y_2y_8$	D_3	9
		a_1	000	$\bar{x}_4$	y_5y_7	–	10
a_6	101	a_1	000	x_5	–	–	11
		a_4	011	$\bar{x}_5$	y_3y_5	D_2D_3	12

Table 7.28 Replacement of logical conditions for Mealy FSM $M_0PY(\Gamma_{12})$

a_m	a_1	a_2	a_3	a_4	a_5	a_6
p_1	x_1	x_1	–	–	x_4	x_5

Table 7.29 Transformed table of Mealy FSM $M_0PY(\Gamma_{12})$

a_m	$K(a_m)$	a_s	$K(a_s)$	X_h^1	P_h	Z_h	Φ_h	h
a_1	000	a_2	001	x_2	p_1	z_3	D_3	1
		a_2	001	$\bar{x}_2$	p_1	z_2	D_3	2
		a_3	010	1	$\bar{p_1}$	z_2z_3	D_2	3
a_2	001	a_5	100	1	p_1	z_1z_3	D_1	4
		a_6	101	x_3	$\bar{p_1}$	z_2	D_1D_3	5
		a_4	011	$\bar{x}_3$	$\bar{p_1}$	z_1	D_2D_3	6
a_3	010	a_4	011	1	1	z_1	D_2D_3	7
a_4	011	a_6	101	1	1	z_3	D_1D_3	8
a_5	100	a_2	001	1	p_1	z_3	D_3	9
		a_1	000	1	$\bar{p_1}$	z_1z_2	–	10
a_6	101	a_1	000	1	p_1	–	–	11
		a_4	011	1	$\bar{p_1}$	z_1	D_2D_3	12

$K(a_s), X_h^1, P_h, Z_h, \Phi_h, h$. The column Z_h contains additional variables $z_r \in Z$ which are equal to 1 in the code $K(Y_t)$ written in the h-th row of the table. The transformed ST of Mealy FSM $M_0PY(\Gamma_{12})$ is represented by Table 7.29.

This table is a base for constructing the table of BIMF. The table of BIMF includes the following columns: $K(a_m), P, X^1, Z, \Phi, q$. In the discussed case, transitions from each state $a_m \in A$ are represented by 8 rows of the table of BIMF. There is the part of

Table 7.30 Part of table of BIMF for Mealy FSM $M_0PY(\Gamma_{12})$

$K(a_m)$	P	X^1	Z	Φ	q	h
$T_1T_2T_3$	p_1	x_2x_3	$x_1x_2x_3$	$D_1D_2D_3$		
000	0	00	011	101	1	3
000	0	01	011	101	2	3
000	0	10	011	011	3	3
000	0	11	011	011	4	3
000	1	00	010	001	5	2
000	1	01	010	001	6	2
000	1	10	001	001	7	1
000	1	11	001	001	8	1

Table 7.31 Table of BMO for Mealy FSM $M_0PY(\Gamma_{12})$

$K(Y_t)$	Y										t
$z_1z_2z_3$	y_1	y_2	y_3	y_4	y_5	y_6	y_7	y_8	y_9	y_{10}	
000	0	0	0	0	0	0	0	0	0	0	1
001	1	1	0	0	0	0	0	1	0	0	2
010	0	0	1	0	0	0	0	0	1	0	3
011	0	1	0	1	0	0	0	0	0	0	4
100	0	0	1	0	1	0	0	0	0	0	5
101	0	0	0	1	0	1	0	0	0	1	6
110	0	0	0	0	1	0	1	0	0	0	7
111	0	0	0	0	0	0	0	0	0	0	8

this table for the discussed case (Table 7.30). It represents the transitions from state $a_1 \in A$.

The table of BMO includes the columns $K(Y_t)$, Y, t. In the discussed case, this table is represented by Table 7.31.

Let the FPGA chip in use include LUTs having $S = 4$. It means that the expression (7.30) should be transformed in the following way:

$$p_1 = \bar{T}_1\bar{T}_2x_1 \vee T_1(\bar{T}_2\bar{T}_3x_4 \vee \bar{T}_2T_3x_5) = A \vee T_1B. \tag{7.31}$$

There is the circuit of BRLC corresponding to the Eq. (7.31) shown in Fig. 7.23a. It includes three of LUTs.

The number of LUTs in the circuit of BRLC can be decreased due to the special state assignment [17]. In this case set A is represented as $A^1 \cup A^2$. The set A^1 includes states $a_m \in A$ with conditional transitions, as well as the initial state $a_1 \in A$. The set A^2 includes states $a_m \in A$ with the unconditional transitions. The state assignment

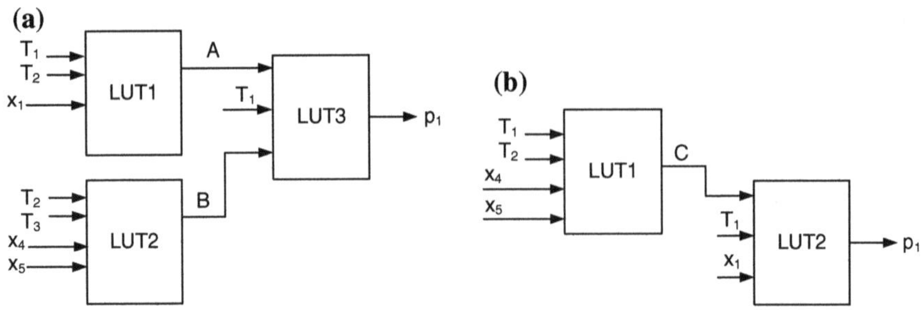

Fig. 7.23 Logic circuit of BRLC for Mealy FSM $M_0PY(\Gamma_{12})$ (**a**) and $M_EPY(\Gamma_{12})$ (**b**)

Fig. 7.24 Outcome of special state assignment for Mealy FSM $P(\Gamma_{12})$

T_1 \ T_2T_3	00	01	11	10
0	a_1	a_2	a_6	a_5
1	a_3	a_4	*	*

starts from the states $a_m \in A^1$. The codes $K(a_m)$ for states $a_m \in A^1$ correspond to decimal numbers from 0 to $M_1 - 1$, where $|A^1| = M_1$.

It is enough R_E variables $T_r \in T$ for encoding of the states $a_m \in A^1$, where

$$R_E = \lceil \log_2 M_1 \rceil . \tag{7.32}$$

In the discussed case, there are the following sets: $A^1 = \{a_1, a_2, a_5, a_6\}$ and $A^2 = \{a_3, a_4\}$. There is $R_E = 2$, so, the states $a_m \in A^1$ can be determined using only state variables T_2 and T_3 (Fig. 7.24).

This approach leads to M_EPY Mealy FSM (Fig. 7.25). There are R_E variables in the set $T' \subseteq T$. The only difference between the models of M_0PY and M_EPY FSMs is reduced to the different forms of the system P. In the later case, it is represented as

$$P = P(T', X^2). \tag{7.33}$$

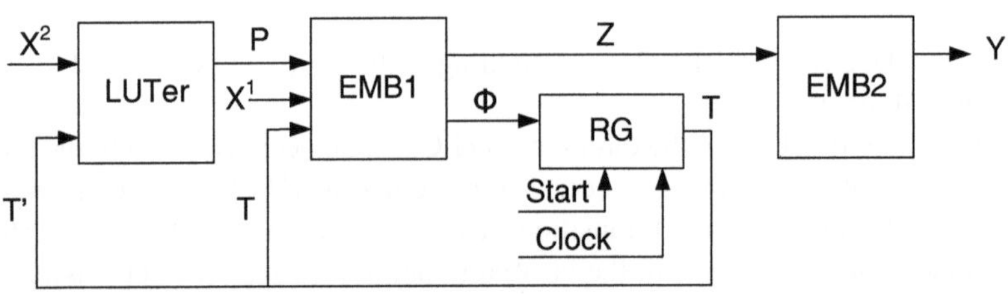

Fig. 7.25 Structural diagram of M_EPY Mealy FSM

In the discussed case, the following equation can be found for the function $p_1 \in P$:

$$p_1 = \bar{T}_2 x_1 \vee T_2 \bar{T}_3 x_4 \vee T_2 T_3 x_5 = \bar{T}_2 x_1 \vee C. \quad (7.34)$$

There is the logic circuit of BRLC Fig. 7.23b corresponding to (7.34). It requires 1,5 less amount of LUTs than its counterpart shown in Fig. 7.23a.

The only difference between design methods for M_0PY and M_EPY FSMs is reduced to the different state assignments. For M_EPY Mealy FSM, the special state assignment should be executed.

If $G = 1$, then the approach of encoding of logical conditions can be applied. Let the symbol $X(P)$ stand for the set of logical conditions replaced by the variable $p_g \in P$. It is enough R_L variables for encoding of the logical conditions $x_l \in X(P)$:

$$R_L = \lceil \log_2 |X(P)| \rceil . \quad (7.35)$$

Let us use the variables $b_r \in B$ for encoding of logical conditions. This approach leads to M_CPY Mealy FSM (Fig. 7.26).

There are the following functions that differ this model from others. The LUTer implements the function

$$P = P(B, X^2). \quad (7.36)$$

The EMB1 implements the functions B used for the encoding of logical conditions. It is the following system:

$$B = B(T, P, X^1). \quad (7.37)$$

The design method for Mealy FSM $M_CPY(\Gamma_j)$ includes all steps of the method for M_0PY FSM. But the encoding of logical conditions is executed before the step of their replacement. Let us discuss an example of design for Mealy FSM $M_CPY(\Gamma_{12})$.

The steps from 1 to 4 are executed before. There is the set $X(P) = X^2 = \{x_1, x_4, x_5\}$ with $L_P = 3$. These logical conditions can be encoded using $R_L = 2$ variables. It gives the set $B = \{b_1, b_2\}$. Let us encode the logical conditions in the following manner: $K(x_1) = 00$, $K(x_4) = 01$ and $K(x_5) = 10$. It gives the following equation for the block BRLC:

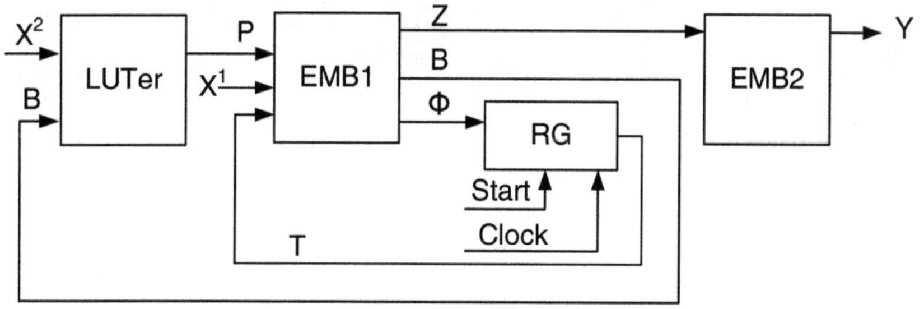

Fig. 7.26 Structural diagram of M_CPY Mealy FSM

Table 7.32 Structure table of Mealy FSM $M_CPY(\Gamma_{12})$

a_m	$K(a_m)$	a_s	$K(a_s)$	X_h^1	P_h	Z_h	B_h	Φ_h	h
a_1	000	a_2	001	x_2	p_1	z_3	–	D_3	1
		a_2	001	$\bar{x}_2$	p_1	z_2	–	D_3	2
		a_3	010	1	$\bar{p_1}$	z_2z_3	–	D_2	3
a_2	001	a_5	100	1	p_1	z_1z_3	–	D_1	4
		a_6	101	x_3	$\bar{p_1}$	z_2	–	D_1D_3	5
		a_4	011	$\bar{x}_3$	$\bar{p_1}$	z_1	–	D_2D_3	6
a_3	010	a_4	011	1	1	z_1	–	D_2D_3	7
a_4	011	a_6	101	1	1	z_3	–	D_1D_3	8
a_5	100	a_2	001	1	p_1	z_3	b_2	D_3	9
		a_1	000	1	$\bar{p_1}$	z_1z_2	b_2	–	10
a_6	101	a_1	000	1	p_1	–	b_1	–	11
		a_4	011	1	$\bar{p_1}$	z_1	b_1	D_2D_3	12

$$p_1 = \bar{b_1}\bar{b_2}x_1 \vee b_2x_4 \vee b_1x_5 = A \vee C. \tag{7.38}$$

Let us encode the collections of microoperations as it is done for the $M_0PY(\Gamma_7)$.

The transformed structure table of M_CPY Mealy FSM includes all columns of its counterpart for M_0PY Mealy FSM. Also, it includes the column B_h with the variables $b_r \in B$ equal to 1 in the code $K(x_l)$ from the h-th row of the table.

Table 7.32 represents the transformed ST of Mealy FSM $M_CPY(\Gamma_{12})$. There is the additional column B_h in the table of BIMF. It contains the code $K(x_l)$. There are the same tables of BMO for equivalent M_0PY and M_CPY Mealy FSMs. There is the logic circuit of Mealy FSM $M_CPY(\Gamma_{12})$ shown in Fig. 7.27.

This approach can be used if the following condition takes place:

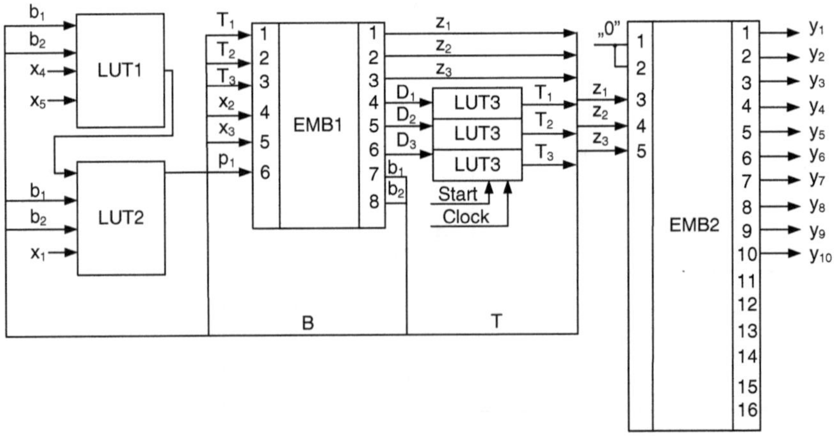

Fig. 7.27 Logic circuit of Mealy FSM $M_CPY(\Gamma_{12})$

$$2^{R_0+S_0+G} \cdot (R_0 + R_Z + R_L) \leq V_0. \tag{7.39}$$

It there is $G > 1$, then logical conditions from different sets $X(p_g)$ should be encoded using different variables $b_r \in B$. We do not discus this approach in this Chapter. Let us point out that the symbol R_L in (7.39) stands for the number of bits required for encoding the variables $b_r \in B$.

There are different numbers of LUTs in different models of Mealy FSMs. We show them in the final table for the benchmarks [18]. It is Table 7.33.

As we can see, there is diminishing for the number of LUTs for some benchmarks. For example, it is necessary less than 50 % of LUTs for *ex*1 and *planet*. Sometimes, there is increasing for this value. So, it is necessary to use a lot of different models for

Table 7.33 Results of investigations for different models of Mealy FSMs

Module	Number of LUT							
Name	P	MP	PY	MPY	PH	MPH	M_{OP}	M_{OPY}
bbara	0	4	0	4	0	4	2	2
bbsse	0	8	0	8	0	8	3	3
bbtas	0	2	0	2	0	2	1	1
bbcount	0	3	0	3	0	3	1	5
cse	40	63	0	8	19	27	52	37
dk14	0	3	0	3	0	3	1	1
dk15	0	0	0	0	0	0	0	0
dk17	0	0	3	3	0	0	0	3
dk27	0	1	0	1	0	1	Not use	Not use
dk512	0	1	0	1	0	1	Not use	Not use
donfile	0	2	1	3	0	2	1	1
ex1	0	18	0	18	0	18	8	8
ex2	12	12	27	15	11	13	26	27
ex3	6	6	9	9	6	8	6	6
ex4	0	4	0	4	0	4	2	2
ex5	5	5	7	7	0	2	5	5
ex6	0	2	0	2	0	2	1	1
ex7	5	5	1	3	5	7	5	5
keyb	74	99	2	25	44	44	73	60
kirkman	54	188	77	99	88	409	59	106
lion	0	0	1	1	0	0	0	0
lion9	5	5	5	5	5	5	5	5
mark1	20	4	13	4	0	4	2	2
mc	0	2	0	2	0	2	1	1
modulo12	0	1	0	0	0	1	Not use	Not use
opus	21	6	17	6	10	6	30	19

(continued)

Table 7.33 (continued)

Module	Number of LUT							
Name	P	MP	PY	MPY	PH	MPH	M_{OP}	M_{OPY}
planet	0	29	0	29	0	29	14	14
planet1	0	29	0	29	0	29	14	14
pma	89	15	85	15	89	15	5	5
s1	0	24	0	24	0	24	11	11
s1488	0	27	0	27	0	27	7	7
s1494	0	23	0	23	0	23	7	7
s1a	0	24	0	0	0	24	11	0
s208	13	19	0	9	28	37	15	15
s27	0	4	1	5	0	4	2	2
s298	0	4	5	9	0	4	2	7
s386	0	9	0	9	0	9	3	3
s8	8	11	0	0	5	5	8	9
sand	0	30	0	30	0	30	6	6
shifreg	0	0	0	0	0	0	Not use	Not use
sse	0	9	0	9	0	9	3	3
styr	0	20	0	20	0	20	10	10
tav	0	0	0	0	0	0	0	0
tbk	0	0	3	3	0	0	0	3
tma	52	54	59	49	38	42	70	60
train11	5	5	7	7	5	5	5	5
train4	0	0	1	1	0	0	0	0

the same GSA Γ. It means that an expert system is necessary in CAD tools used to design FSM circuits. This system should make choosing the best model for a given GSA Γ and a given FPGA chip.

References

1. I. Grout, *Digital Systems Design with FPGAs and CPLDs* (Elsevier Science, Oxford, 2008)
2. C. Maxfield, *The Design Warrior's Guide to FPGAs* (Academic Press Inc, Orlando, 2004)
3. Xilinx, http://www.xilinx.com. Accessed Jan 2015
4. Altera, http://www.altera.com. Accessed Jan 2015
5. T. Kim, T. Vella, R. Brayton, A. Sangiovanni-Vincentalli, *Synthesis of Finite State Machines: Functional Optimization* (Kluwer Academic Publishers, Boston, 1997)
6. M. Nowicka, T. Łuba, M. Rawski, FPGA-based decomposition of Boolean functions: algorithms and implementation. Adv Comput Syst, 502–509 (1999)
7. C. Scholl, *Functional Decomposition with Application to FPGA Synthesis* (Kluwer Academic Publishers, Boston, 2001)

8. G. Sutter, E. Todorovich, S. López-Buedo, E. Boemo, Low-power FSMs in FPGA: encoding alternatives, in *Integrated Circuit Design. Power and Timing Modeling, Optimization and Simulation* (Springer, 2002), pp. 363–370
9. X. Wu, M. Pedram, L. Wang, Multi-code state assignment for low-power design. IEEE Proc. Circuits Devices Syst. **147**, 271–275 (2000)
10. J. Cong, K. Yan, Synthesis for FPGAs with embedded memory blocks, in *Proceedings of the 2000 ACM/SIGDA 8th International Symposium on FPGAs* (2000), pp. 75–82
11. M. Rawski, H. Selvaraj, T. Łuba, An application of functional decomposition in ROM-based FSM implementation in FPGA devices. J. Syst. Archit. **51**(6–7), 423–434 (2005)
12. M. Rawski, P. Tomaszewicz, G. Borowski, T. Łuba, Logic Synthesis Method of Digital Circuits Designed for Implementation with Embedded Memory Blocks on FPGAs, in *Design of Digital Systems and Devices*, LNEE 79, ed. by M. Adamski, A. Barkalov, M. Węgrzyn (Springerg, Berlin, 2011), pp. 121–144
13. V. Sklyarov, Synthesis and implementation of RAM-based finite state machines in FPGAs, in *Proceedings of Field-Programmable Logic and Applications: The Roadmap to Reconfigurable Computing* (Springer, Villach, 2000), pp. 718–728
14. A. Tiwari, K. Tomko, Saving power by mapping finite-state machines into Embedded Memory Blocks in FPGAs, in *Proceedings of the conference on Design, Automation and Test in Europe*, vol. 2 (IEEE Computer Society, 2004), pp. 916–921
15. L. Garcia-Vargas, R. Senhadji-Navarro, A. Civit-Balcells, P. Guerra-Gutierrezz, ROM-based finite state machine implementation in low cost FPGAs, in *IEEE International Simposium on Industrial Electronics* (Vigo, 2007), pp. 2342–2347
16. S. Baranov, *Logic Synthesis of Control Automata* (Kluwer Academic Publishers, 1994)
17. V. Sklyarov, I. Skliarova, A. Barkalov, L. Titarenko, *Synthesis and Optimization of FPGA-based Systems*, vol. 294, Lecture notes in electrical engineering (Springer, Berlin, 2014)
18. LGSynth93, International Workshop on logic synthesis benchmark suite (LGSynth93). TAR, Benchmarks test, http://www.cbl.ncsu.edu:16080/benchmarks/LGSynth93/LGSynth93.tar (1993)
19. Xilinx, Virtex-5 Family Overview. PDF, Xilinx Corporation, http://www.xilinx.com/support/documentation/data_sheets/ds100.pdf, 2009

Chapter 8
Design of EMB-Based Moore FSMs

8.1 Trivial Implementing Moore FSMs

A Moore FSM can be implemented using single EMB if the following condition takes place:

$$2^{L+R_1} \cdot (R_1 + N) \leq V_0. \tag{8.1}$$

In this case a single EMB represents both blocks BIMF and BMO (Fig. 8.1). It is a P Moore FSM.

There are the following steps in the design method of P Moore FSM:

1. Constructing the set of states A.
2. State assignment.
3. Constructing the structure table.
4. Constructing the table of EMB.
5. Implementing the FSM logic circuit with EMBs and LUTs of given FPGA chip.

Let us discuss an example of design for the Moore FSM $P(\Gamma_{13})$. The GSA Γ_{13} is shown in Fig. 8.2. It is marked by states of Moore FSM using the rules [2]. It is possible to derive the following sets and their parameters from GSA Γ_{13}: $A = \{a_1, \ldots, a_5\}$, $M_1 = 5$, $X = \{x_1\}$, $L = 1$, $Y = \{y_1, \ldots, y_4\}$, $N = 4$, $R_1 = 3$, $T = \{T_1, T_2, T_3\}$, and $\Phi = \{D_1, D_2, D_3\}$.

Let us encode the states $a_m \in A$ in the trivial way: $K(a_1) = 000, \ldots, K(a_5) = 100$. Now, the ST can be constructed for the FSM $P(\Gamma_{13})$. It is Table 8.1. The column a_m includes both the current state and the collection of MOs $Y(a_m) \subseteq Y$.

There are the following columns in the table of EMB: $K(a_m)$, X, Φ, Y, q. In the discussed case, it includes 16 rows, but only 10 of them contain the useful data (the input memory functions and microoperations). These rows are shown in Table 8.2. We add columns m (to show the current state a_m) and h (to shown the correspondence among the rows of ST and table of EMB).

A. Barkalov et al., *Logic Synthesis for FPGA-Based Finite State Machines*,
Studies in Systems, Decision and Control 38, DOI 10.1007/978-3-319-24202-6_8

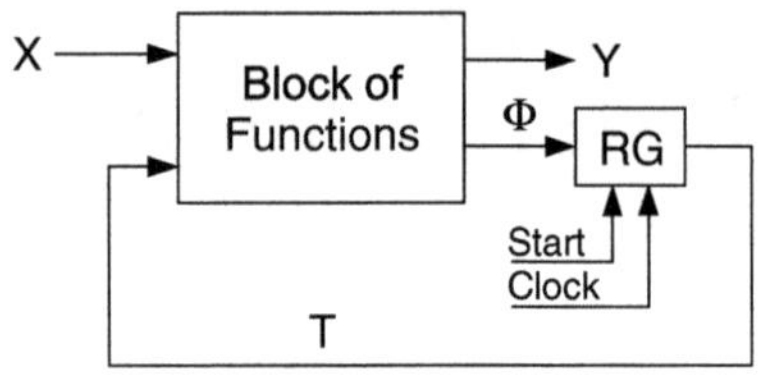

Fig. 8.1 Structural diagram of single EMB-based P Moore FSM

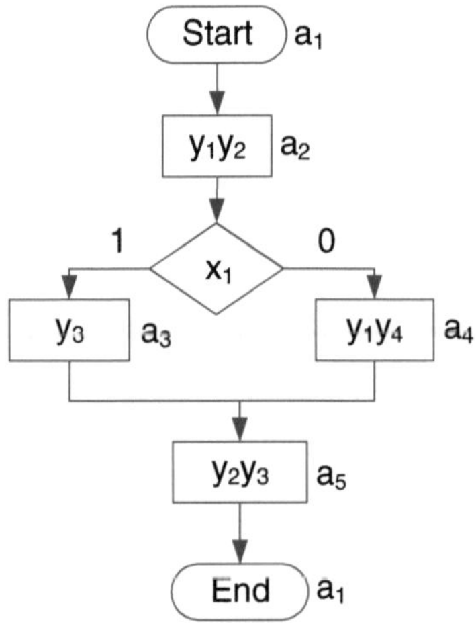

Fig. 8.2 Graph-scheme of algorithm Γ_{13}

Table 8.1 Structure table of Moore FSM $P(\Gamma_{13})$

a_m	$K(a_m)$	a_s	$K(a_s)$	X_h	Φ_h	h
a_1	000	a_2	001	1	D_3	1
$a_2(y_1y_2)$	001	a_3	010	x_1	D_2	2
		a_4	011	$\bar{x}_1$	D_2D_3	3
$a_3(y_3)$	010	a_5	100	1	D_1	4
$a_4(y_1y_4)$	011	a_5	100	1	D_1	5
$a_5(y_2y_3)$	100	a_1	000	1	–	6

Analysis of benchmarks [9] shows that the condition (8.1) takes place for approximately 20 % of all examples. There is more probability that the following conditions take places:

$$2^{L+R_1} \cdot (R_1 + N) > V_0; \tag{8.2}$$

$$R_1 \cdot 2^{L+R_1} \leq V_0; \tag{8.3}$$

$$N \cdot 2^{R_1} \leq V_0. \tag{8.4}$$

Table 8.2 Table of EMB for Moore FSM $P(\Gamma_{13})$

$K(a_m)$	X	Φ	Y	q	m	h
$T_1T_2T_3$	x_1	$D_1D_2D_3$	$y_1y_2y_3y_4y_5$			
000	0	001	00000	1	1	1
000	1	001	00000	2	1	1
001	0	011	11000	3	2	3
001	1	010	11000	4	2	2
010	0	100	00100	5	3	4
010	1	100	00100	6	3	4
011	0	100	10010	7	4	5
011	1	100	10010	8	4	5
100	0	000	01100	9	5	6
100	1	000	01100	10	5	6

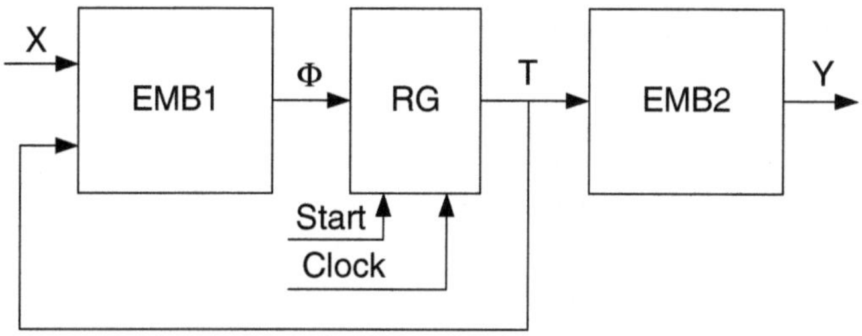

Fig. 8.3 Structural diagram of P Moore FSM

The condition (8.2) shows that it is impossible to use the model shown in Fig. 8.1. The condition (8.3) shows that it is enough a single EMB to implement the system (1.4). The condition (8.4) shows that it is enough a single EMB to im3plement the system (1.5). In this case, it is enough two of EMBs to implement the circuit of P Moore FSM (Fig. 8.3). Let us name this model as P Moore FSM, whereas the P_t Moore FSM describes the model shown in Fig. 8.1.

In P FSM, the EMB1 implements the system of input memory functions Φ, whereas the EMB2 the system of MOs Y. There are the following steps in the design method of P Moore FSM:

1. Constructing the set of states A.
2. State assignment.
3. Constructing the structure table.
4. Transformation of the structure table. Constructing the table of EMB1.
5. Constructing the table of microoperations (Table of EMB2).
6. Implementing the FSM logic circuit with EMBs and LUTs of given FPGA chip.

Table 8.3 Transformed structure table of Moore FSM $P(\Gamma_{13})$

$K(a_m)$	X	Φ	q	h
$T_1T_2T_3$	x_1	$D_1D_2D_3$		
000	0	001	1	1
000	1	001	2	1
001	0	011	3	3
001	1	010	4	2
010	0	100	5	4
010	1	100	6	4
011	0	100	7	5
011	1	100	8	5
100	0	000	9	6
100	1	000	10	6

Table 8.4 Table of microoperations of Moore FSM $P(\Gamma_{13})$

$K(a_m)$	Y	m
$T_1T_2T_3$	$y_1y_2y_3y_4$	
000	0000	1
001	1100	1
010	0010	3
011	1001	2
100	0110	4

Let us discuss an example when the GSA Γ_{13} is used for implementing P Moore FSM. We execute steps 1–3 before. There is the transformed ST of Moore FSM $P(\Gamma_{13})$ represented by Table 8.3. Table 8.4 represents the table of BMO for the discussed case.

Comparison of Tables 8.2, 8.3 and 8.4 shows that Table 8.3 includes three first columns of Table 8.2. There are only 5 rows in the table of BMO in contrast to 10 rows of Table 8.2. It is connected with the fact that the system Y does not depend on logical conditions $x_e \in X$.

Let an FPGA chip in use include EMB having configurations 16×4 and 8×8. The first of them is used for implementing the transformed structure table. Both configurations can be used for implementing the table of microoperations. Let us choose the configuration 8×8 for implementing the system Y. There is the logic circuit of Moore FSM $P(\Gamma_{13})$ shown in Fig. 8.4.

In this circuit, LUT1–LUT3 are used for implementing the register RG. Let us point out that both EMB1 and EMB2 have unused resources. If there are special inputs of synchronization in EMBs, then there are no LUTs used in EMB-based Moore FSMs.

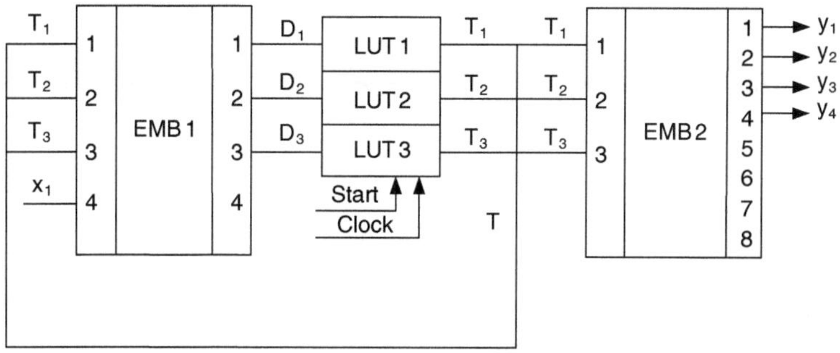

Fig. 8.4 Logic circuit of Moore FSM $P(\Gamma_{13})$

If the conditions (8.3)–(8.4) are violated, then it is necessary to use different methods of structural decomposition [2, 4, 5, 12] to diminish the hardware amount in the EMB-based circuits of Moore FSMs.

The structural decomposition is reduced to increasing the number of structural levels in an FSM circuit. There are the following methods of structural decomposition [2, 4, 12]: (1) replacement of logical conditions; (2) encoding of collections of microoperations; (3) encoding of the fields of compatible microoperations; (4) encoding of the rows of structure table. Some of these methods were discussed in this book. Let us continue the discussion.

8.2 Structural Decomposition for Moore FSMs

Let $X(a_m)$ be a set of logical conditions determining transitions from the state $a_m \in A$ and let us define the following variable:

$$G = \max(|X(a_1)|, \ldots, |X(a_m)|). \tag{8.5}$$

If the following condition takes place

$$G \ll L, \tag{8.6}$$

then the method of replacement of logical conditions [2] can be applied. Let $P = \{p_1, \ldots, p_G\}$ be a set of additional variables used for the replacement of logical conditions. To execute the replacement, a special table of replacement of logical conditions should be constructed. In this table, the columns are marked by variables $p_g \in P$, whereas the rows by states $a_m \in A$. So, the table includes G columns and M rows. If a variable $p_g \in P$ replaces a logical condition $x_l \in X$ in a state $a_m \in A$, then the symbol x_l should be written on the intersection of the row a_m and the column p_g of the table. To minimize the hardware amount for a logic circuit used for the

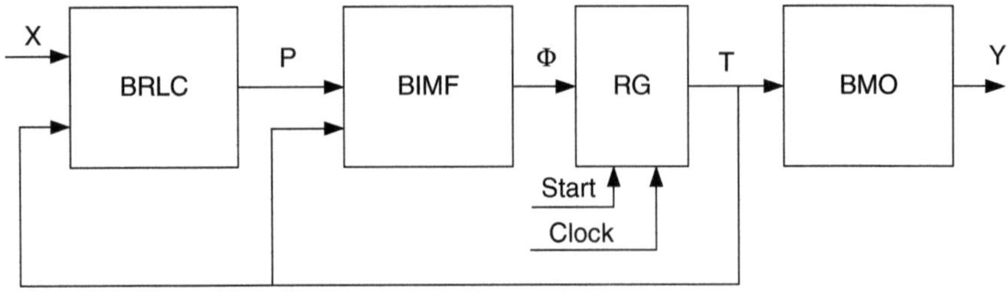

Fig. 8.5 Structural diagram of MP Moore FSM

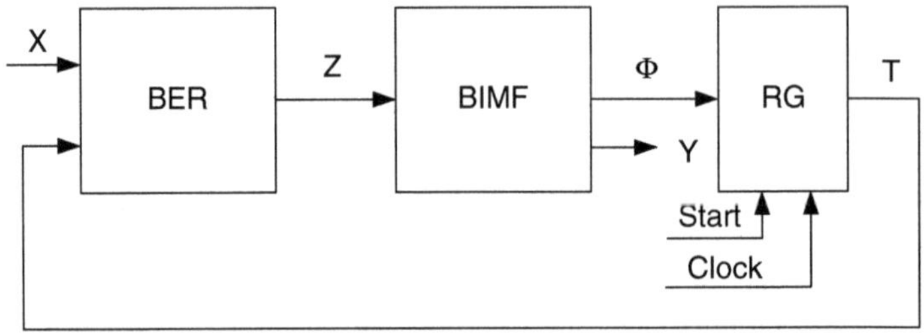

Fig. 8.6 Structural diagram of PH Moore FSM

replacement, the distribution of logical conditions is executed in such a manner that each variable $x_l \in X$ is always placed in the same column of the table. Of course, such a distribution is not always possible.

The following system can be derived from the table of replacement:

$$P = P(T, X). \tag{8.7}$$

This system is implemented by a LUTer corresponding to the BRLC. It leads to MP Moore FSM shown in Fig. 8.5.

We discussed the methods of encoding of CMOs and fields of compatible MOs in previous Chapters. In the case of Moore FSM, they are used only with the methods of OCT. But it is possible to use the encoding of the rows of ST in case of Moore FSM. It leads to the PH Moore FSM (Fig. 8.6).

In PH FSM, the BER implements the functions (2.30). The number of variables $z_r \in Z$ is determined as

$$R_{H1} = \lceil log_2 H_1 \rceil . \tag{8.8}$$

If EMBs are used for implementing the PH Moore FSM, then blocks BER and BIMF are represented by the blocks EMB1 and EMB2, respectively (Fig. 8.7).

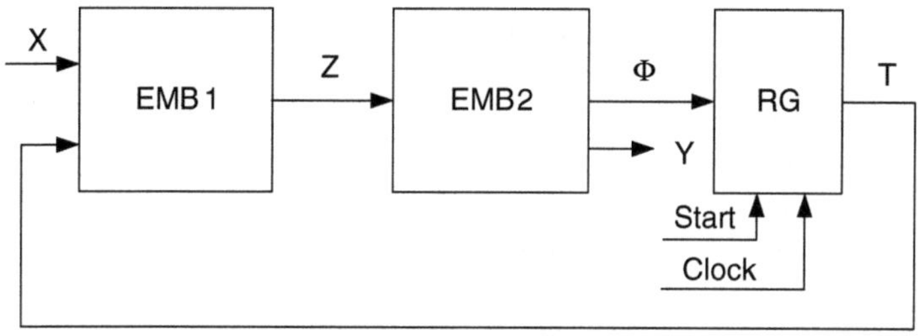

Fig. 8.7 Structural diagram of EMB-based *PH* Moore FSM

There are the following steps in the design of *PH* Moore FSM:

1. Constructing the set of states A.
2. State assignment.
3. Constructing the structure table of Moore FSM $P(\Gamma_j)$.
4. Encoding of the rows of structure table.
5. Constructing the transformed structure table.
6. Constructing the table of BER.
7. Constructing the table of BIMF.
8. Implementing the logic circuit of FSM with a particular FPGA chip.

Let us discuss an example of design for Moore FSM $PH(\Gamma_{13})$. We executed the first three steps of this method before. There is $H_1 = 6$ (it follows from Table 8.1). Using (8.8), the value $R_{H1} = 3$ can be found, as well as the set $Z = \{z_1, z_2, z_3\}$.

Let us encode the terms F_h in the trivial way: $K(F_1) = 000, \ldots, K(F_6) = 101$. There are the following columns in the transformed ST: a_m, $K(a_m)$, X_h, $K(F_h)$, Z_h, h. There are the functions $z_r \in Z$ shown in the columns Z_h if the corresponding bits are equal to 1 in the row h of the initial ST ($h = \overline{1, H_1}$). There is the transformed ST of Moore FSM $PH(\Gamma_{13})$ represented by the Table 8.5.

The table of BER is constructed on the base of the transformed ST. It includes the following columns: $K(a_m)$, X, Z, q. There are 16 rows in this table but only 10 of them contains the useful data (Table 8.6).

Table 8.5 Transformed table of Moore FSM $PH(\Gamma_{13})$

a_m	$K(a_m)$	X_h	$K(F_h)$	Z_h	h
a_1	000	1	000	–	1
a_2	001	x_1	001	z_3	2
		x_1	010	z_2	3
a_3	010	1	011	z_2z_3	4
a_4	011	1	100	z_1	5
a_5	100	1	101	z_1z_3	6

Table 8.6 Table of BER of Moore FSM $PH(\Gamma_{13})$

$K(a_m)$	X	Z	q	h
$T_1T_2T_3$	x_1	$z_1z_2z_3$		
000	0	000	1	1
000	1	000	2	1
001	0	010	3	3
001	1	001	4	2
010	0	011	5	4
010	1	011	6	4
011	0	100	7	5
011	1	100	8	5
100	0	101	9	6
100	1	101	10	6

Table 8.7 Table of BIMF of Moore FSM $PH(\Gamma_{13})$

$K(F_h)$	Φ	X	h
$z_1z_2z_3$	$D_1D_2D_3$	$y_1y_2y_3$	
000	001	0000	1
001	011	1100	2
010	011	1100	3
011	100	0010	4
100	100	1001	5
101	000	0110	6

This table is constructed if BER is implemented using EMBs. If LUTs are used for implementing the circuit of BER, then the functions $Z(T, X)$ should be derived from the transformed ST. There are R_{H1} of such equations. Each equation determines a logic circuit of a particular LUTerz_r $(r = \overline{1, R_{H1}})$.

The table of BIMF is constructed on the base of the initial ST. It includes the following columns: $K(F_h)$, Φ, Y, h. In the discussed case, it is Table 8.7.

We use the codes $K(a_s)$ from the h-th row of the initial ST to fill the column Φ of the table of BIMF. The CMOs $Y_t \subseteq Y$ are repeated for the rows corresponding to the same current state $a_m \in A$. Because of it, there is the same CMO in the rows 2 and 3 of Table 8.7.

The PH model can be used if the following conditions take places:

$$2^{L+R_1} \cdot R_{H1} \leq V_0; \tag{8.9}$$

$$2^{R_{H1}} \cdot (N + R_1) \leq V_0. \tag{8.10}$$

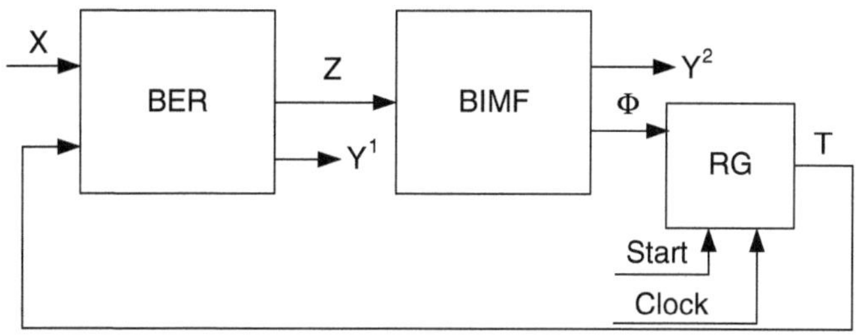

Fig. 8.8 Structural diagram of $PH1$ Moore FSM

Let it be violated the condition (8.10). Let us denote as $t_1 (t_2)$ the number of outputs for EMB1 (EMB2). Let it be Δ_t free outputs in EMB1, where

$$\Delta_t = t_1 = R_{H1}. \tag{8.11}$$

In this case, it is possible to implement up to Δ_t functions $\Phi \cup Y$ using the free outputs of EMB1. Now, the following condition should be checked:

$$\left\lceil \frac{N + R_1 - \Delta_t}{t_2} \right\rceil = 1. \tag{8.12}$$

In the condition (8.12) takes place, then the BIMF is implemented as a single EMB. It leads to $PH1$ Moore FSM (Fig. 8.8).

As follows from Fig. 8.8, the set Y is divided by two disjoint sets Y^1 and Y^2. There are up to Δ_t microoperations in the set Y^1.

Let us discuss the following examples. Let it be the following configurations of EMB for a particular FPGA chip: 16×4 and 8×4, bits. Let us check the conditions (8.9)–(8.10) for the Moore FSM $P(\Gamma_{13})$. There is $V_0 = 64$ (for EMB1) and $V_0 = 48$ (for EMB2). So the following relations can be obtained:

$$16 \times 4 \le 64; \tag{8.13}$$

$$8 \times 7 = 56 > 48; \tag{8.14}$$

$$\Delta_t = 1; \tag{8.15}$$

$$8 \times 6 = 48. \tag{8.16}$$

The following conclusion can be done from analysis of (8.13)–(8.16). It is enough a single EMB to implement the circuit of BER. It is necessary more than one EMB to implement the BIMF. It is possible to implement one additional function by BER. If one MO is deleted from Y, then it is possible to implement the circuit of BIMF by a single EMB. There is the logic circuit of Moore FSM $PH1(\Gamma_{13})$ shown in Fig. 8.9.

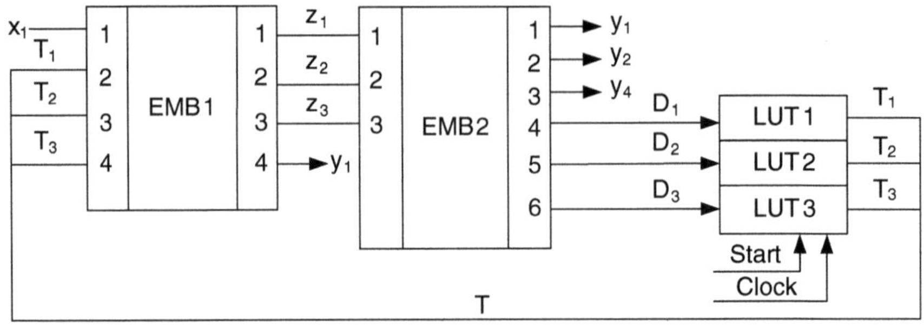

Fig. 8.9 Logic circuit of Moore FSM $PH1(\Gamma_{13})$

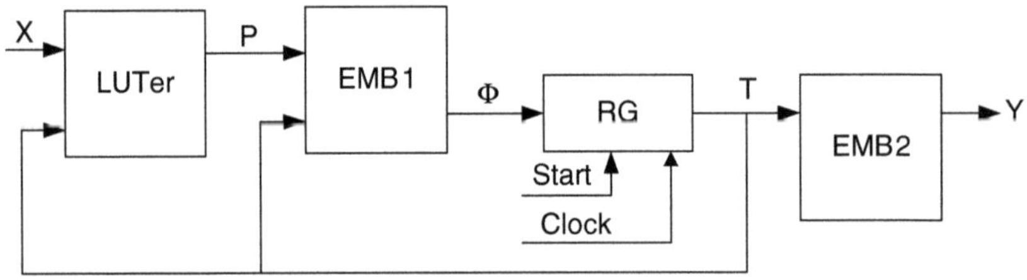

Fig. 8.10 Structural diagram of MP Moore FSM

Let us name this approach as the distribution of microoperations (DMO). Let us point out that this approach can be used in other models of FSMs (including Mealy FSMs).

The RLC can be used for optimization of P Moore FSM. It leads to MP Moore FSM (Fig. 8.10). A block LUTer represents BRLC, a block EMB1 represents BIMF, and a block EMB2 represents BMO.

As in the case of MP Mealy FSM, the LUTer implements the system (1.47). The EMB1 implements functions (1.49), whereas the EMB2 the system (1.5). This model can be applied if the following conditions take places:

$$2^{G+R_1} \cdot R_1 \leq V_0; \tag{8.17}$$

$$2^R \cdot N \leq V_0. \tag{8.18}$$

There are the following steps in the design method of MP Moore FSM:

1. Constructing the set of states A.
2. State assignment.
3. Constructing the structure table of FSM $P(\Gamma_j)$.
4. Replacement of logical conditions.
5. Constructing the system (1.47).
6. Constructing the transformed structure table.
7. Constructing the table of BIMF.

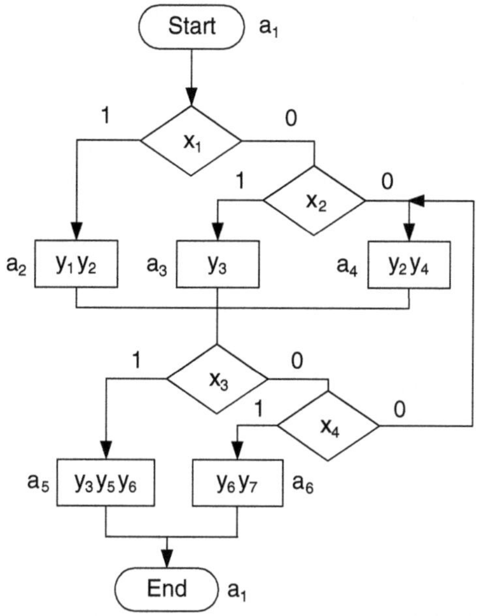

Fig. 8.11 Initial GSA Γ_{14}

8. Constructing the table of BMO.
9. Implementing the FSM logic circuit with EMBs and LUTs of a particular FPGA chip.

Let us discuss an example of design for Moore FSM $MP(\Gamma_{14})$. The GSA Γ_{14} is shown in Fig. 8.11.

This GSA is marked by states of Moore FSM using the rules from [10]. The following sets and their characteristics can be found for Moore FSM $P(\Gamma_{14})$: $A = \{a_1, \ldots, a_6\}$, $M_1 = 6$, $X = \{x_1, \ldots, x_4\}$, $L = 4$, $Y = \{y_1, \ldots, y_7\}$, $N = 7$, $R = 3$, $T = \{T_1, T_2, T_3\}$ and $\Phi = \{D_1, D_2, D_3\}$.

Let the FPGA chip in use have $V_0 = 128$ and let the following configurations of EMBs exist: 128×1, 64×2, 32×4 and 16×8 (bits). For the FSM $P(\Gamma_{14})$ the following relation takes place: $2^{L+R_1} \cdot R_1 = 128 \times 3 > 128$. Therefore, this model cannot be used in the discussed case.

Let us encode the states $a_m \in A$ in the trivial way: $K(a_1) = 000, \ldots, K(a_6) = 101$. Now, the ST can be constructed (Table 8.8) for the Moore FSM $P(\Gamma_{14})$.

As follows from this table, there are the following sets $X(a_m) \subseteq X$: $X(a_1) = \{x_1, x_2\}$, $X(a_2) = X(a_3) = X(a_4) = \{x_3, x_4\}$, $X(a_5) = X(a_6) = \emptyset$. Obviously, there is $G = 2$ and $P = \{p_1, p_2\}$.

Let us check the conditions (8.17)–(8.18). For the discussed case there are:

$$2^5 \cdot 3 = 96 < 128;$$
$$2^3 \cdot 7 = 56 < 128.$$

Table 8.8 Structure table of FSM $P(\Gamma_{14})$

a_m	$K(a_m)$	a_s	$K(a_s)$	X_h	Φ_h	h
$a_1(-)$	000	a_2	001	x_1	D_3	1
		a_3	010	$\bar{x}_1 x_2$	D_2	2
		a_4	011	$\bar{x}_1 \bar{x}_2$	$D_2 D_3$	3
$a_2(y_1 y_2)$	001	a_5	100	x_3	D_1	4
		a_6	101	$\bar{x}_3 x_4$	$D_1 D_3$	5
		a_4	011	$\bar{x}_3 \bar{x}_4$	$D_2 D_3$	6
$a_3(y_3)$	010	a_5	100	x_3	D_1	7
		a_6	101	$\bar{x}_3 x_4$	$D_1 D_3$	8
		a_4	011	$\bar{x}_3 \bar{x}_4$	$D_2 D_3$	9
$a_4(y_2 y_4)$	011	a_5	100	x_3	D_1	10
		a_6	101	$\bar{x}_3 x_4$	$D_1 D_3$	11
		a_4	011	$\bar{x}_3 \bar{x}_4$	$D_2 D_3$	12
$a_5(y_3 y_5 y_6)$	100	a_1	000	1	–	13
$a_6(y_6 y_7)$	101	a_1	000	1	–	14

Table 8.9 Table of replacement of logical conditions for Moore FSM $MP(\Gamma_{14})$

a_m	a_1	a_2	a_3	a_4	a_5	a_6
p_1	x_1	x_3	x_3	x_3	–	–
p_2	x_2	x_4	x_4	x_4	–	–

It means that the model $MP(\Gamma_{14})$ can be used. The table of replacement of logical conditions is represented by Table 8.9.

The following system of equations can be derived from Table 8.9:

$$\begin{aligned} p_1 &= A_1 x_1 \vee (A_2 \vee A_3 \vee A_4) x_3; \\ p_2 &= A_1 x_2 \vee (A_2 \vee A_3 \vee A_4) x_4. \end{aligned} \tag{8.19}$$

If variables $A_m \in A$ are replaced by corresponding conjunctions, the system (8.19) represents the system (1.47) for the given example.

The transformed structure table of Moore FSM $MP(\Gamma_j)$ is constructed in the same way as its counterpart for Mealy FSM $MP(\Gamma_j)$. In the discussed example, it is represented by Table 8.10.

This table is a base for constructing the table of BIMF containing the following columns: $K(a_m)$, P, Φ, q. In the discussed example, this table includes 32 rows. It is necessary 4 rows to represent transitions from any state $a_m \in A$. Table 8.11 represents the part of table BIMF for the state $a_3 \in A$.

Table 8.10 Transformed structure table of Moore FSM $MP(\Gamma_{14})$

a_m	$K(a_m)$	a_s	$K(a_s)$	P_h	Φ_h	h
$a_1(-)$	000	a_2	001	p_1	D_3	1
		a_3	010	$\bar{p_1}p_2$	D_2	2
		a_4	011	$\bar{p_1}\bar{p_2}$	D_2D_3	3
$a_2(y_1y_2)$	001	a_5	100	p_1	D_1	4
		a_6	101	$\bar{p_1}p_2$	D_1D_3	5
		a_4	011	$\bar{p_1}\bar{p_2}$	D_2D_3	6
$a_3(y_3)$	010	a_5	100	p_1	D_1	7
		a_6	101	$\bar{p_1}p_2$	D_1D_3	8
		a_4	011	$\bar{p_1}\bar{p_2}$	D_2D_3	9
$a_4(y_2y_4)$	011	a_5	100	p_1	D_1	10
		a_6	101	$\bar{p_1}p_2$	D_1D_3	11
		a_4	011	$\bar{p_1}\bar{p_2}$	D_2D_3	12
$a_5(y_3y_5y_6)$	100	a_1	000	1	–	13
$a_6(y_6y_7)$	101	a_1	000	1	–	14

Table 8.11 Part of table of BIMF for FSM $MP(\Gamma_{14})$

$K(a_m)$	P	Φ	v	h
$T_1T_2T_3$	p_1p_2	$D_1D_2D_3$		
010	00	011	9	9
010	01	101	10	8
010	10	100	11	7
010	11	100	12	7

8.3 Optimization of BIMF Based on Pseudoequivalent States

One of the specific features of Moore FSM is existence of classes of pseudoequivalent states [4]. The states $a_m, a_s \in A$ are pseudoequivalent states if outputs of corresponding operator vertices are connected with the input of the same vertex of GSA Γ. Lets us find the partition Π_A of the set A by classes of pseudoequivalent states $B_1, \ldots, B_I$. Let us repeat the main issues concerning the methods based on PES. Let us point out that the corresponding methods where proposed for FPGA-based Moore FSM in [12].

Two approaches can be used for optimizing the BIMF of Moore FSM. The first of them is the optimal state assignment. In this case, the states are encoded in such a way that each class $B_i \in \Pi_A$ is represented by minimum possible amount of generalized intervals of R_1-dimensional Boolean space. The second approach is connected with encoding of the classes $B_i \in \Pi_A$. Let us discuss these approaches and corresponding

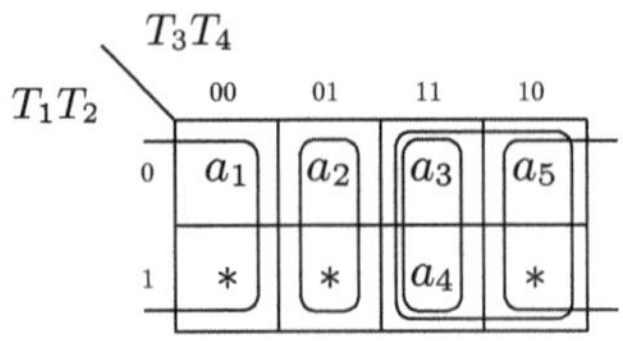

Fig. 8.12 Optimal state codes for Moore FSM $P(\Gamma_{13})$

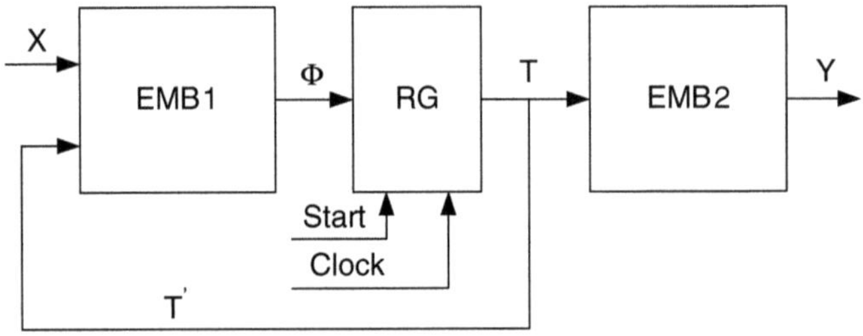

Fig. 8.13 Structural diagram of P_0 Moore FSM

model of EMB-based Moore FSMs. Let us use the GSA Γ_{13} for illustrating these approaches.

There is the following partition $\Pi_A = \{B_1, \ldots, B_4\}$ for the GSA Γ_{13} (Fig. 8.2). There are the following classes $B_i \in \Pi_A$: $B_1 = \{a_1\}$, $B_2 = \{a_2\}$, $B_3 = \{a_3, a_4\}$ and $B_4 = \{a_5\}$. So there is $I = 4$. Let us encode the states $a_m \in A$ in the optimal way (Fig. 8.12).

The transitions from the state a_5 do not present in the structure table because they are executed automatically (using only pulse Clock). Because of it, the code of state a_5 can be treated as “don’t care” and can be included into the cubes for one of other classes $B_i \in \Pi_A$.

Taking it into account, the following codes can be obtained for classes $B_i \in \Pi_A$ in the discussed case: $K(B_1) = **0$, $K(B_2) = *01$, $K(B_3) = *11$. So, the value of T_1 is not significant to determine the classes $B_i \in \Pi_A$. In the general case this approach leads to EMB-based Moore FSM $P_0(\Gamma_j)$. Its structural diagram is shown in Fig. 8.13.

In the P_0 FSM, the BIMF is represented by the block EMB1. It implements the system

$$\Phi = \Phi(T', X). \tag{8.20}$$

The block EMB2 implements the system (1.5). The following condition should take place for implementing this model:

$$R_1 \cdot 2^{L+R_E} \le V_0. \tag{8.21}$$

The value of R_E is determined by the capital number of the set $T' \subseteq T$. The proposed design method for P_0 Moore FSM is the following one:

1. Constructing the set of states A.
2. Optimal state assignment.
3. Constructing the transformed structure table.
4. Constructing the table of BIMF.
5. Constructing the table of microoperations.
6. Implementation of the FSM logic circuit.

Let us discuss an example of design for $P_0(\Gamma_{13})$. Let it be the EMBs with the following configurations: 32×1, 16×2, 8×4, bits. Because there is $R_1 = 3$, the configuration 8×4 should be chosen. But because of $R_1 + L = 4$, the number of cells should be equal to 16 for $t_F = 4$. It is $t_F = 2$ for $V = 16$. So, the model $P(\Gamma_{13})$ cannot be used.

Let us use the state codes from Fig. 8.12. Analysis of the state codes shows that there is $R_E = 2$. Now, there is $2^{1+2} \times 3 = 24 < 32$. It means that the condition (8.21) is satisfied and it is possible to use the model $P_0(\Gamma_{13})$.

To construct the transformed structure table of P_0 Moore FSM, it is necessary to construct the system of generalized formulae of transitions for classes $B_i \in \Pi_A$. This system does not include the class $B_4 \in \Pi_A$ because the state $a_5 \in B_4$ is connected only with state $a_1 \in A$. The following system can be derived from GSA Γ_{13}:

$$\begin{aligned} B_1 &\rightarrow a_2; \\ B_2 &\rightarrow x_1 a_3 \vee \bar{x}_1 a_4; \\ B_3 &\rightarrow a_5. \end{aligned} \tag{8.22}$$

The system (8.22) is the base for constructing the transformed ST of Moore FSM $P_0(\Gamma_{13})$ shown by the Table 8.12.

Using this table, it is possible to construct the table of BIMF. It contains the columns $K(B_i)$, X, Φ, q. It includes only 8 rows for discussed case (Table 8.13).

We hope there is the transparent connection between Tables 8.12 and 8.13. For example, four rows of Table 8.13 (namely, the rows 1, 2, 5 and 6) correspond to the row 1 of the Table 8.12. Next, two rows of Table 8.13 (rows 7, 8) correspond to the row 4 of Table 8.12. And so on.

Table 8.12 Transformed structure table of Moore FSM $P_0(\Gamma_{13})$

B_i	$K(B_i)$	a_s	$K(a_s)$	X_h	Φ_h	h
B_1	**0	a_2	001	1	D_3	1
B_2	*01	a_3	011	x_1	D_2D_3	2
		a_4	111	$\bar{x}_1$	$D_1D_2D_3$	3
B_3	*11	a_5	010	1	D_2	4

Table 8.13 Table of BIM of Moore FSM $P_0(\Gamma_{13})$

$K(B_i)$	X	Φ	q	h
T_2T_3	x_1	$D_1D_2D_3$		
00	0	001	1	1
00	1	001	2	1
01	0	111	3	3
01	1	011	4	2
10	0	001	5	1
10	1	001	6	1
11	0	010	7	4
11	1	010	8	4

Table 8.14 Table of microoperations of Moore FSM $P_0(\Gamma_{13})$

$K(a_m)$	Y	m
$T_1T_2T_3$	$y_1y_2y_3y_4$	
000	0000	1
001	1100	2
010	0110	3
011	0010	4
100	0000	5
101	0000	6
110	0000	7
111	1001	8

There is the same approach used for constructing the table of MOs as, for example, for P Moore FSM. In the discussed case, the table of MOs is represented by Table 8.14.

There is the logic circuit of FSM $P_0(\Gamma_{13})$ shown in Fig. 8.14. We use EMBs with the configuration 8×4 to implement the circuits of BIMF and BMO.

It is quite possible the situation when $T' = T$ after the execution of optimal state assignment [4]. In this case the following approach can be used.

Let us encode each class $B_i \in \Pi_A$ by a binary code $K(B_i)$ having R_A bits. The value of R_A is determined as (4.17).

Let us use the variables $\tau_r \in \tau$ for the encoding, where $|\tau| = R_A$. Let the following condition take place:

$$\begin{aligned} &2^{R_A+L} \cdot R_1 \leq V_0; \\ &2^{R} \cdot (N + R_A) \leq V_0. \end{aligned} \tag{8.23}$$

In this case, we propose to use the P_C Moore FSM. Its structural diagram is shown in Fig. 8.15.

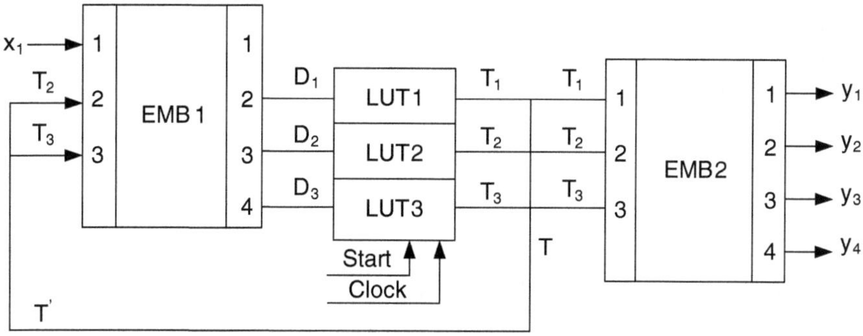

Fig. 8.14 Logic circuit of Moore FSM $P_0(\Gamma_{13})$

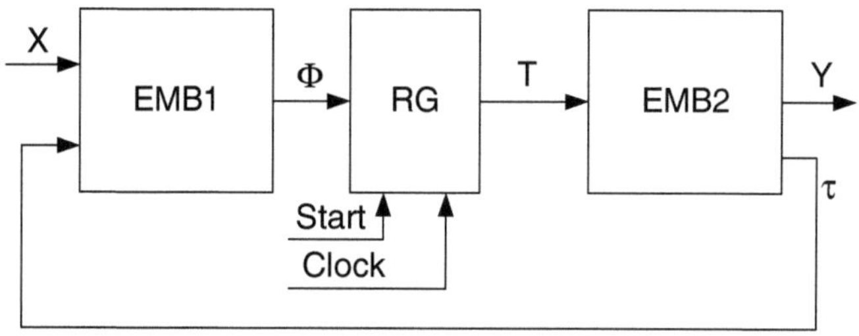

Fig. 8.15 Structural diagram of P_C Moore FSM

In P_C Moore FSM, the block EMB1 corresponds to BIMF. It implements the system of input memory functions

$$\Phi = \Phi(\tau, X). \tag{8.24}$$

The block EMB2 implements the circuit of BMO. It generates the functions Y and the system of additional variables

$$\tau = \tau(T). \tag{8.25}$$

The proposed design method for P_C Moore FSM includes the following steps:

1. State assignment.
2. Finding the partition $\Pi_A = \{B_1, \ldots, B_I\}$.
3. Encoding of the classes $B_i \in \Pi_A$.
4. Constructing the transformed structure table.
5. Constructing the table of BIMF.
6. Constructing the table of BMO.
7. Implementation of the FSM logic circuit.

Table 8.15 Transformed structure table of Moore FSM $P_C(\Gamma_{13})$

B_i	$K(B_i)$	a_s	$K(a_s)$	X_h	Φ_h	h
B_1	00	a_2	001	1	D_3	1
B_2	01	a_3	010	x_1	D_2	2
		a_4	011	$\bar{x}_1$	D_2D_3	3
B_3	10	a_5	100	1	D_1	4

Table 8.16 Table of BIMF of Moore FSM $P_C(\Gamma_{13})$

$K(B_i)$	X	Φ	v	h
$\tau_1\tau_2$	x_1	$D_1D_2D_3$		
00	0	001	1	1
00	1	001	2	1
01	0	011	3	3
01	1	010	4	2
10	0	100	5	4
10	1	100	6	4
11	0	000	7	0
11	1	000	8	0

Let us discuss an example of design for Moore FSM $P_C(\Gamma_{13})$. The set A includes $M_1 = 5$ elements and there is $R_1 = 3$. Let us encode the states $a_m \in A$ in the trivial way: $K(a_1) = 000, \ldots, K(a_5) = 100$.

There is the partition $\Pi_A = \{B_1, \ldots, B_4\}$ with $I_A = 4$. It gives $R_A = 2$. Let us encode the classes $B_i \in \Pi_A$ in the trivial way: $K(B_1) = 00, \ldots, K(B_4) = 11$.

To construct the transformed structure table, the system of generalized formulae of transitions should be derived from a GSA Γ_j. In the discussed case, this system is represented by (8.22). The transformed structure table of P_C Moore FSM includes the same columns as its counterpart for P_0 Moore FSM (Table 8.15).

The table of BIMF is constructed on the base of the transformed structure table. In the discussed case, it is represented by Table 8.16.

The table of BMO includes the additional column τ (Table 8.17). If there is $a_m \in B_i$, then the row corresponding to the state a_m includes the code $K(B_i)$.

There is the logic circuit of Moore FSM $P_C(\Gamma_{13})$ shown in Fig. 8.16. There is an EMB with configuration 8×6 required for implementing the circuit of BMO.

It is possible to decrease the requirements to EMBs implementing the BIMF. But using PES makes impossible the applying DMO approach. There are not all variables $T_r \in T$ entering the BIMF for P_0 Moore FSM. There are no variables $T_r \in T$ entering the BIMF for P_C Moore FSM.

Table 8.17 Table of BMO of Moore FSM $P_C(\Gamma_{13})$

$K(a_m)$	Y	τ	m
$T_1T_2T_3$	$y_1y_2y_3y_4$	$\tau_1\tau_2$	
000	0000	00	1
001	1100	01	2
010	0010	10	3
011	1001	10	4
100	0110	11	5
101	0000	00	6
110	0000	00	7
111	1001	00	8

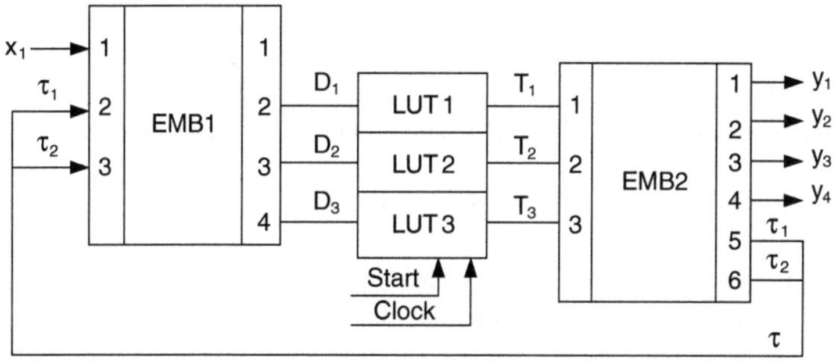

Fig. 8.16 Logic circuit of FSM of $P_C(\Gamma_{13})$

8.4 Optimizing LUTer in Replacement of Logical Conditions

There are eight different approaches targeting the hardware reduction for LUT-based BRLC. There are the following approaches for Moore FSMs [6, 12]:

1. Optimal state assignment (M_EP_E FSMs).
2. Transforming the codes of PES into the codes of classes of PES (M_CP_C FSMs).
3. Splitting logical conditions (M_0P FSMs).
4. Special state assignment (M_SP FSMs).
5. Special class assignment (M_SP_C FSMs).
6. Splitting logical conditions and optimal state assignment ($M_{0E}P_E$ FSMs).
7. Splitting logical conditions and transformation of PES ($M_{0C}P_C$ FSMs).
8. Encoding of logical conditions (M_KP FSMs).

Let us discuss these approaches. The approaches 1, 2, 5, 6, 7 are based on the partition $\Pi_A = \{B_1, \ldots, B_I\}$.

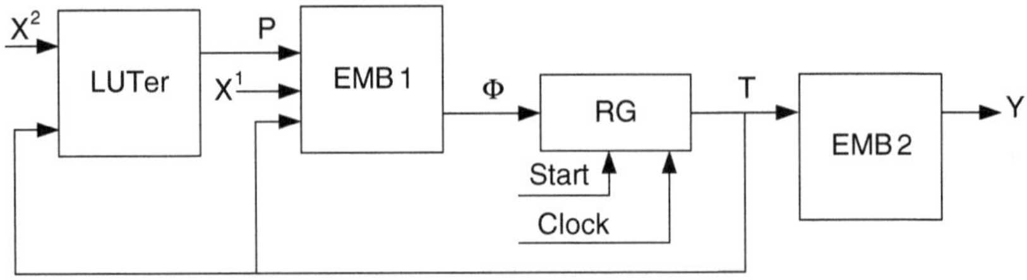

Fig. 8.17 Structural diagram of FSM of $P_C(\Gamma_{13})$

Let us start from the splitting logical conditions. Let the following condition take place:

$$S_A > G + R_1. \tag{8.26}$$

It means that there are S_0 "free" inputs of EMB implementing the circuit of BIMF. As it is for the case of Mealy FSMs, we propose to use these inputs for optimizing the LUTer representing the BRLC [8]. The number S_0 is determined as:

$$S_0 = S_A - (G + R_1). \tag{8.27}$$

As it is in the case of M_0P Mealy FSMs, let us split the set X and represent it as $X^1 \cup X^2$. Let the following condition take place:

$$S_A = G + R_1 + |X^1|. \tag{8.28}$$

It means that only logical conditions $x_l \in X^2$ should be transformed by the LUTer. The logical conditions $x_l \in X^1$ enter the BIMF. It leads to M_0P Moore FSM (Fig. 8.17).

The LUTer implements the system (7.20), the EMB1 implements the system (7.21). As in the previous case, the EMB2 implements the system (1.5). This model can be used if the following condition takes place:

$$2^{G+S_C+R_1} \cdot R_1 \leq V_0. \tag{8.29}$$

There are the following steps in the design method for M_0P Moore FSM:

1. Constructing the set of states A.
2. State assignment.
3. Constructing the structure table of FSM $P(\Gamma_j)$.
4. Partitioning the set X by subsets X^1 and X^2.
5. Replacement of logical conditions $x_l \in X^2$.
6. Constructing the transformed structure table.
7. Constructing the table of BIMF.
8. Constructing the table of BMO.
9. Implementing the FSM logic circuit.

Let us discus an example of design for Moore FSM $M_0P(\Gamma_{14})$. In the previous sections, we found all sets and their parameters for the FSM $P(\Gamma_{14})$. Let us encode the states $a_m \in A$ in the trivial way: $K(a_1) = 000, \ldots, K(a_6) = 101$. The ST of $P(\Gamma_{14})$ is represented by Table 8.8.

Let an FPGA chip in use include EMBs having the following configurations: 256×1, 128×2, 64×4, 32×8 (bits). In this case the model $P(\Gamma_{14})$ cannot be applied. For Moore FSM $MP(\Gamma_{14})$, there is $G = 2$. So, there is $2^{G+R_1} \cdot R_1 = 32 \times 3 = 96 < 256$. It means that the model $MP(\Gamma_{14})$ can be used. In this case, there are 160 bits of EMB which are not used. Let us try to use them due to applying the model $M_0P(\Gamma_{14})$.

There is $R_1 = 3$, then the configuration 64×4 can be chosen. In this case $S_A = 6$ and $S_A - R_1 = 3$. So, two possibilities can be used for the replacement of logical conditions: (1) $G = 2$ and $|X^1| = 1$ and (2) $G = 1$ and $|X^1| = 2$. There are the following sets $X(a_m) \subseteq X$: $X(a_1) = \{x_1, x_2\}$, $X(a_2) = X(a_3) = X(a_4) = \{x_3, x_4\}$ and $X(a_5) = X(a_6) = \emptyset$. Let us divide the set X by following subsets: $X^1 = \{x_2, x_4\}$ and $X^2 = \{x_1, x_3\}$. It gives $P = \{p_1\}$ and the following equation can be found:

$$p_1 = A_1x_1 \vee A_2x_3 \vee A_3x_3 \vee A_4x_3. \tag{8.30}$$

There is the ST of Moore FSM $M_0P(\Gamma_{14})$ shown in Table 8.18. It is constructed as a transformation of the initial ST (Table 8.8).

There are $Q = 64$ rows in the table of BIMF in the discussed case. It is determined by the formula

$$Q = 2^{G+R_1+L_1}. \tag{8.31}$$

Table 8.18 Transformed structure table of Moore FSM $M_0P(\Gamma_{14})$

a_m	$K(a_m)$	a_s	$K(a_s)$	X_h^1	P_h	Φ_h	h
$a_1(-)$	000	a_2	001	1	p_1	D_3	1
		a_3	010	x_2	$\bar{p_1}$	D_2	2
		a_4	011	$\bar{x_2}$	$\bar{p_1}$	D_2D_3	3
$a_2(y_1y_2)$	001	a_5	100	1	p_1	D_1	4
		a_6	101	x_4	$\bar{p_1}$	D_1D_3	5
		a_4	011	$\bar{x_4}$	$\bar{p_1}$	D_2D_3	6
$a_3(y_3)$	010	a_5	100	1	p_1	D_1	7
		a_6	101	x_4	$\bar{p_1}$	D_1D_3	8
		a_4	011	$\bar{x_4}$	$\bar{p_1}$	D_2D_3	9
$a_4(y_2y_4)$	011	a_5	100	1	p_1	D_1	10
		a_6	101	x_4	$\bar{p_1}$	D_1D_3	11
		a_4	011	$\bar{x_4}$	$\bar{p_1}$	D_2D_3	12
$a_5(y_3y_5y_6)$	100	a_1	000	1	1	–	13
$a_6(y_6y_7)$	101	a_1	000	1	1	–	14

Table 8.19 Part of table of BIMF of Moore FSM $M_0P(\Gamma_{14})$

$K(a_m)$	X^1	P	Φ	v	h
$T_1T_2T_3$	x_2x_4	p_1	$D_1D_2D_3$		
000	00	0	011	1	3
000	00	1	001	2	1
000	01	0	011	3	3
000	01	1	001	4	1
000	10	0	010	5	2
000	10	1	001	6	1
000	11	0	010	7	2
000	11	1	001	8	1

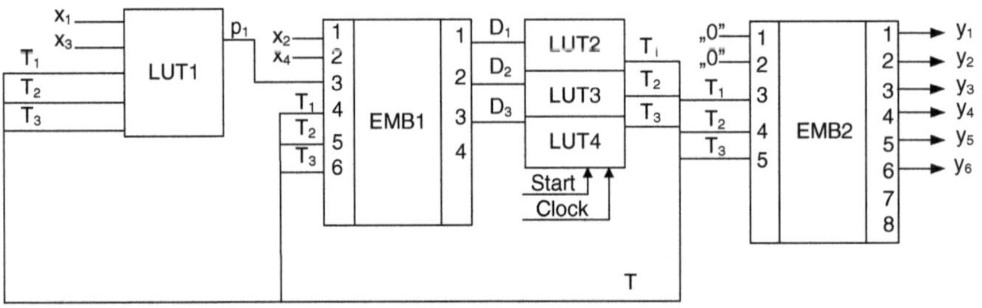

Fig. 8.18 Logic circuit of Moore FSM $M_0P(\Gamma_{14})$

There is $L_1 = |X^1|$ in the formula (8.31). Table 8.19 represents a part of the table of BIMF for $M_0P(\Gamma_{14})$. There are the transitions from the state $a_1 \in A$ shown in Table 8.19.

The table of BMO is the same as for $P(\Gamma_{14})$. There is a logic circuit of Moore FSM $M_0P(\Gamma_{14})$ shown in Fig. 8.18. It is enough a single LUT having $S = 5$ to implement the LUTer. To implement the circuit of BIMF, they use an EMB with the configuration 64×4. There is an EMB with the configuration 32×8 used for implementing the circuit of BMO. Let us point out that both EMBs have free outputs. There is one free output for EMB1 and two free outputs for EMB2.

Let us use the state codes $K(a_m)$ and represent the Eq. (8.30) as the following one:

$$p_1 = \bar{T}_1\bar{T}_2\bar{T}_3x_1 \vee \bar{T}_1\bar{T}_2T_3x_3 \vee \bar{T}_1T_2\bar{T}_3x_3 \vee T_1\bar{T}_2\bar{T}_3x_3. \tag{8.32}$$

We used LUT with $S = 5$ to implement this equation.

Now let us discuss the case when there is $S = 4$.

If there is $S = 4$, then the expression (8.32) should be transformed using the rules of functional decomposition [1, 7]. The transformed equation is the following:

$$p_1 = \bar{T}_1(\bar{T}_2\bar{T}_3x_1 \vee \bar{T}_2T_3x_3 \vee T_2\bar{T}_3x_3) \vee T_1(\bar{T}_2\bar{T}_3x_3). \tag{8.33}$$

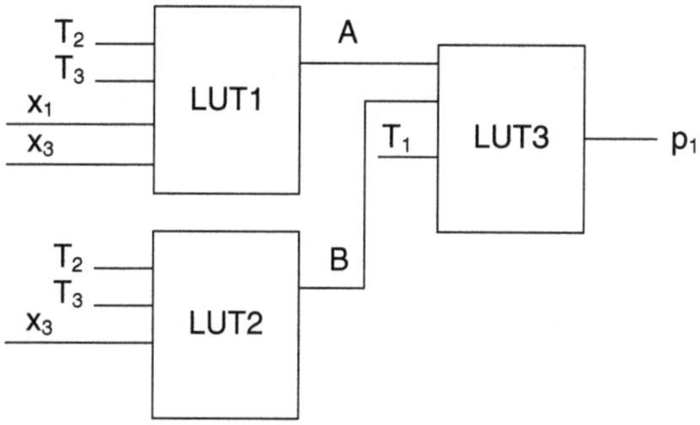

Fig. 8.19 Logic circuit of LUTer for $S = 4$

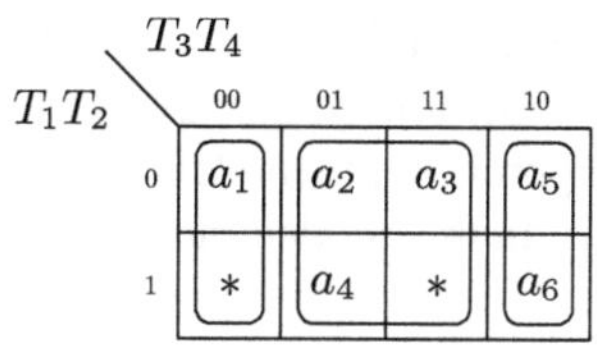

Fig. 8.20 Optimal state codes for Moore FSM $P(\Gamma_{14})$

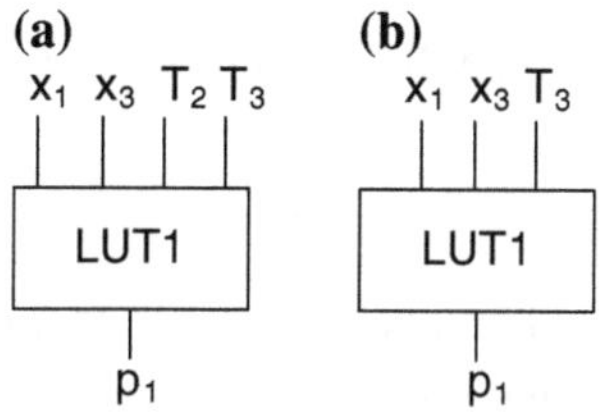

Fig. 8.21 Optimal circuit of LUTer

The equation (8.33) corresponds to the logic circuit of LUTer shown in Fig. 8.19. In Fig. 8.19, there are $A = \bar{T}_2\bar{T}_3x_1 \vee \bar{T}_2T_3x_3 \vee T_2\bar{T}_3x_3$ and $B = \bar{T}_2\bar{T}_3x_3$. This circuit has 2 levels and uses 3 LUTs with $S = 4$. Let us try to improve this circuit using pseudoequivalent states of Moore FSM [11, 13].

There is the partition $\Pi_A = \{B_1, B_2, B_3\}$ in the case of Moore FSM $P(\Gamma_{14})$. There are the following classes $B_i \in \Pi_A$: $B_1 = \{a_1\}$, $B_2 = \{a_1, a_2, a_3\}$ and $B_3 = \{a_5, a_6\}$. Let us encode the states $a_m \in A$ as it is shown in Fig. 8.20.

Using this codes, we can find that the Eq. (8.32) is transformed into the following one:

$$p_1 = \bar{T}_2\bar{T}_3x_1 \vee T_3x_3. \tag{8.34}$$

This equation corresponds to the single-level logic circuit shown in Fig. 8.21a.

Because $X(a_5) = X(a_6) = \emptyset$, their codes can be treated as "don't care" for function p_1. It allows obtaining the following equation:

$$p_1 = \bar{T}_3x_1 \vee T_3x_3. \tag{8.35}$$

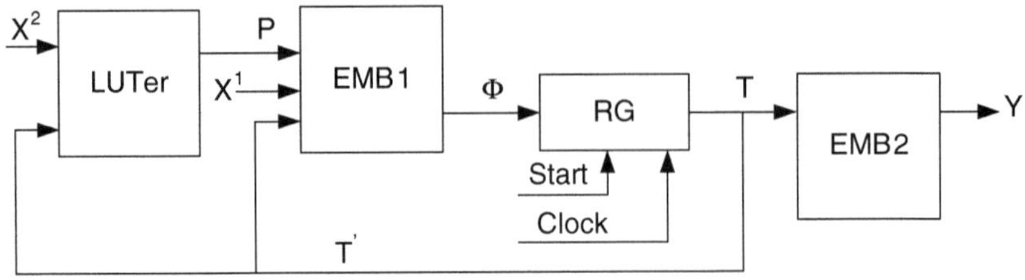

Fig. 8.22 Structural diagram of Moore FSM $M_{0E}P_E(\Gamma_j)$

It is enough a single LUT having $S = 3$ to implement the circuit of LUTer corresponding to (8.35). This circuit is shown in Fig. 8.21.

Obviously, this approach is based on the optimal state assignment [3]. Using this approach, the model of $M_{0E}P_E$ Moore FSM can be proposed (Fig. 8.22).

In this model, the LUTer implements the system

$$P = P(T', X^2). \tag{8.36}$$

The EMB1 implements the system

$$\Phi = \Phi(T', P, X^2). \tag{8.37}$$

Let it be $R_E = |T'|$. In this case the proposed model can be used if the following condition takes place:

$$2^{R_E+S_0+G} \cdot R_1 \leq V_0. \tag{8.38}$$

There are the following codes $K(B_i)$ derived from Fig. 8.20: $K(B_1) = *00$, $K(B_2) = **1$ and $K(B_3) = *10$. It gives $T' = \{T_2, T_3\}$ and $R_E = 2$.

There are the following steps in the design method for $M_{0E}P_E$ Moore FSM:

1. Constructing the set of states A.
2. Constructing the partition $\Pi_A = \{B_1, \ldots, B_I\}$.
3. Optimal state assignment.
4. Constructing the structure table of Moore FSM $P_E(\Gamma_j)$.
5. Partitioning the set X by subsets X^1 and X^2.
6. Replacement of logical conditions $x_l \in X^2$.
7. Constructing the transformed structure table.
8. Constructing the table of BIMF.
9. Constructing the table of BMO.
10. Implementing the FSM logic circuit.

Let us discuss an example of design of Moore FSM $M_{0E}P_E(\Gamma_{14})$. There are the sets $A = \{a_1, \ldots, a_6\}$ and $\Pi_A = \{B_1, B_2, B_3\}$ found before. Let us encode the states $a_m \in A$ in the optimal way as it is shown in Fig. 8.20.

Table 8.20 Structure table of $P_E(\Gamma_{14})$

B_i	$K(B_i)$	a_s	$K(a_s)$	X_h	Φ_h	h
B_1	*00	a_2	001	x_1	D_3	1
		a_3	011	$\bar{x_1}x_2$	D_2D_3	2
		a_4	101	$\bar{x_1}\bar{x_2}$	D_1D_3	3
B_2	**1	a_5	010	x_3	D_2	4
		a_6	110	$\bar{x_3}x_4$	D_1D_2	5
		a_4	101	$\bar{x_3}\bar{x_4}$	D_1D_3	6
B_3	*10	a_1	000	1	–	7

To construct the structure table of $P_E(\Gamma_{14})$, let us find the system of generalized formulae of transitions [4]. In the discussed case, it is the following system:

$$\begin{aligned} B_1 &\rightarrow x_1a_2 \vee \bar{x_1}x_2a_3 \vee \bar{x_1}\bar{x_2}a_4; \\ B_2 &\rightarrow x_3a_5 \vee \bar{x_3}x_4a_6 \vee \bar{x_3}\bar{x_4}a_4; \\ B_3 &\rightarrow a_1. \end{aligned} \tag{8.39}$$

This system is used for constructing the transformed ST of Moore FSM $P_E(\Gamma_{14})$. It is Table 8.20. This table includes the following columns: B_i, $K(B_i)$, a_s, $K(a_s)$, X_h, Φ_h, h. The codes $K(B_i)$ of classes $B_i \in \Pi_A$ are taken from Fig. 8.20, as well as the codes of states $a_m \in A$.

Let the FPGA chip in use have EMBs with the following configurations: 128×1, 64×2, 32×4, 16×8 (bits). Because there is $R_1 = 3$, we should choose the configuration 32×4 for implementing the circuit of BIMF. There is $T' = \{T_2, T_3\}$ and, therefore, $R_E = 2$. For given configuration, there is $S_A = 5$. It means that three inputs can be used for logical conditions $x_l \in X^1$ and additional variables $p_g \in P$. Let us make the following partition of the set X: $X^1 = \{x_2, x_4\}$ and $X^2 = \{x_1, x_3\}$. It gives $P = \{p_1\}$.

It is possible to derive the following equation from Table 8.20:

$$p_1 = B_1x_1 \vee B_2x_3 = \bar{T_2}\bar{T_3}x_1 \vee T_3x_3. \tag{8.40}$$

Let $X(B_i)$ be a set of logical conditions determining transitions from states $a_m \in B_i$ $(i = \overline{1, I})$. Because $X(B_3) = \emptyset$, the codes of states $a_5, a_6 \in B_3$ can be treated as "don't cares". It gives the final form of the system (1.47) for the given example:

$$p_1 = \overline{T_3}x_1 \vee T_3x_3. \tag{8.41}$$

There are the following columns in the transformed ST of $M_{0E}P_E$ Moore FSM: B_i, $K(B_i)$, a_s, $K(a_s)$, P_h, X_h^1, Φ_h, h. In the discussed case it is represented by Table 8.21.

Table 8.21 Transformed structure table of Moore FSM $M_{0E}P_E(\Gamma_{14})$

B_i	$K(B_i)$	a_s	$K(a_s)$	P_h	X_h^1	Φ_h	h
B_1	*00	a_2	001	p_1	1	D_3	1
		a_3	011	$\bar{p_1}$	x_2	D_2D_3	2
		a_4	101	$\bar{p_1}$	$\bar{x_2}$	D_1D_3	3
B_2	**1	a_5	010	p_1	1	D_2	4
		a_6	110	$\bar{p_1}$	x_4	D_1D_2	5
		a_4	101	$\bar{p_1}$	$\bar{x_4}$	D_1D_3	6
B_3	*10	a_1	000	1	1	–	7

Table 8.22 Part of the table of BIMF for Moore FSM $M_{0E}P_E(\Gamma_{14})$

$K(B_i)$	P	X^1	Φ	v	h
T_2T_3	p_1	x_2x_4	$D_1D_2D_3$		
00	0	00	101	1	3
00	0	01	101	2	3
00	0	10	011	3	2
00	0	11	011	4	2
00	1	00	001	5	1
00	1	01	001	6	1
00	1	10	001	7	1
00	1	11	001	8	1

The table of BIMF of $M_{0E}P_E$ Moore FSM includes the following columns: $K(B_i), P, X^1, \Phi, h$. The columns $K(B_i), P, X^1$ create the address of a cell inside the EMB. In the discussed case, the transitions from each class $B_i \in \Pi_A$ are represented by 8 rows of this table. The transitions from the class $B_1 \in \Pi_A$ are represented by Table 8.22.

The table of BMO is always the same for given GSA. It includes the columns $K(a_m), Y, m$. The address of a cell is determined by the state code $K(a_m)$. We do not show this table for $M_{0E}P_E$ Moore FSM.

Let us use the following logic elements for implementing the logic circuit of $M_{0E}P_E(\Gamma_{14})$: a LUT having $S = 3$, EMBs with the configurations 32×4 and 16×8. In this case, all parts of the FSM circuit are implemented as single elements. The only difference is that RG required R_1 of LUTs. There is the logic circuit of $M_{0E}P_E(\Gamma_{14})$ shown in Fig. 8.23.

Due to the optimal state encoding, the circuit of BRLC is implemented using only one LUT with $S = 3$. To implement the Eq. (8.32) with LUTs having three inputs, it should be decomposed: $p_1 = T_1(T_2\bar{T_3}x_3) \vee \bar{T_1}(T_2(\bar{T_3}x_3) \vee \bar{T_2}(\bar{T_3}x_1 \vee T_3x_3)) = T_1A \vee \bar{T_1}(T_2B \vee \bar{T_2}C) = T_1A \vee \bar{T_1}D$.

This equation corresponds to the circuit having three layers formed by 5 LUTs. This circuit is shown in Fig. 8.24.

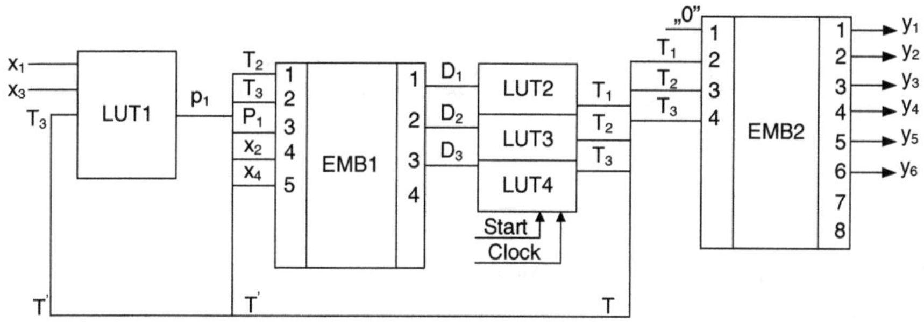

Fig. 8.23 Logic circuit of Moore FSM $M_{0E}P_E(\Gamma_{14})$

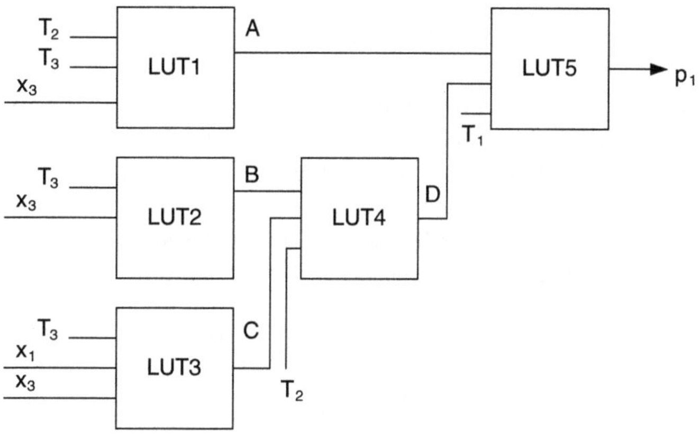

Fig. 8.24 Logic circuit of BRLC based on trivial state codes

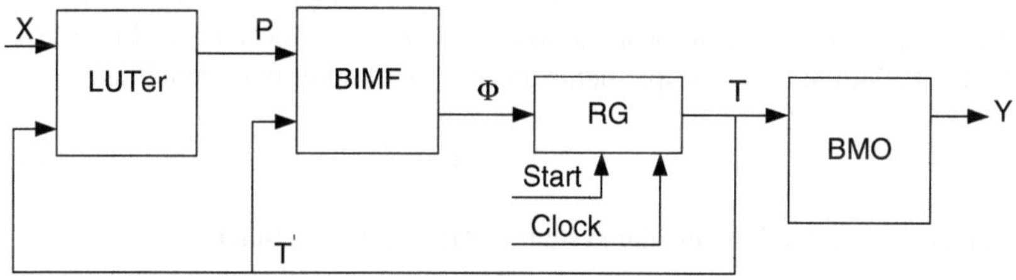

Fig. 8.25 Structural diagram of $M_E P_E$ Moore FSM

So, the proposed approach allows 5 times reduction for the hardware of BRLC, as well as 3 times acceleration for the propagation time. Of course, it is true only for the given example.

There are two approaches combined in $M_{0E}P_E$ Moore FSM: (1) the splitting logical conditions and (2) the optimal state assignment. Let us point out that the LUTer can be improved without the splitting logical conditions. It is possible if $S_0 = 0$. In this case, the Moore FSM is represented by $M_E P_E$ Moore FSM (Fig. 8.25).

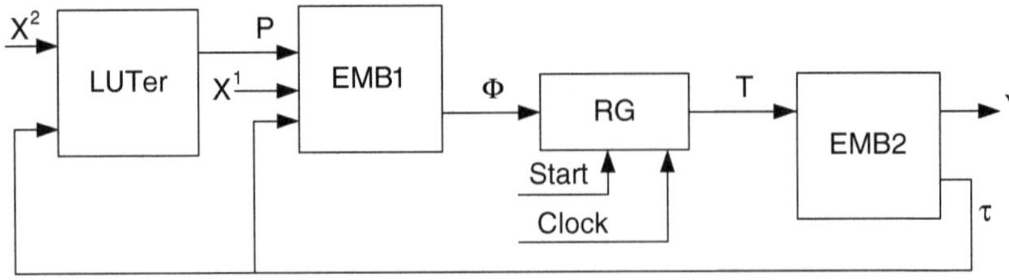

Fig. 8.26 Structure diagram of $M_{OC}P_C$ Moore FSM

There is the following system P implemented by LUTer:

$$P = P(T', X). \tag{8.42}$$

There are the same steps in design methods of MP and $M_E P_E$ FSMs. The only difference is reduced to the method of state assignment.

If there is no hardware reduction due to the optimal state assignment, then they can use the transformation of state codes $a_m \in B_i$ into the class codes $B_i \in \Pi_A$. There are R_A bits in the codes $K(B_i)$. This value is determined by (4.3).

Let the following conditions take places:

$$S_0 = S_A - (R_A + G) > 0; \tag{8.43}$$

$$2^{R_A+S_0+G} \cdot R_1 \leq V_0. \tag{8.44}$$

The condition (8.43) shows that it is possible to use the splitting logical conditions. The condition (8.44) shows that only single EMB is enough to implement the circuit of BIMF. In this case, we propose the model of $M_{OC}P_C$ Moore FSM (Fig. 8.26).

In this FSM, the LUTer implements the system of additional variables

$$P = P(\tau, X^2). \tag{8.45}$$

The block EMB1 implements the system of input memory functions

$$\Phi = \Phi(\tau, X^1, P). \tag{8.46}$$

The block EMB2 implements microoperations $y_n \in Y$ and the system (4.4).

There are the following steps in the design method for $M_{OC}P_C$ Moore FSM:

1. State assignment.
2. Encoding of the classes $B_i \in \Pi_A$.
3. Constructing the structure table of $P_C Y$ Moore FSM.
4. Partitioning the set X by subsets X^1 and X^2.
5. Replacement of logical conditions $x_l \in X^2$.
6. Constructing the transformed structure table.

Fig. 8.27 Initial GSA Γ_{15}

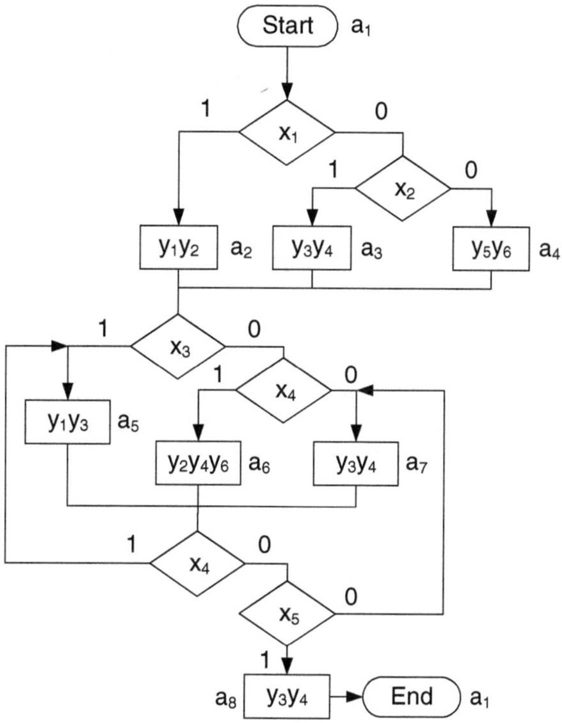

7. Constructing the table of BIMF.
8. Constructing the table of BMO.
9. Implementing the FSM logic circuit.

Let us discuss an example of design for FSM $M_{0C}P_C(\Gamma_{15})$. There is GSA Γ_{15} shown in Fig. 8.27.

Let us analyse the GSA Γ_{15}. It is possible to find the following sets and their parameters from this analysis: $A = \{a_1, \ldots, a_8\}$, $M_1 = 8$, $X = \{x_1, \ldots, x_5\}$, $L = 5$, $Y = \{y_1, \ldots, y_6\}$, $N = 6$, $R_1 = 3$, $T = \{T_1, T_2, T_3\}$, $\Phi = \{D_1, D_2, D_3\}$.

The following partition Π_A can be found for Moore FSM: $\Pi_A = \{B_1, \ldots, B_4\}$, where $B_1 = \{a_1\}$, $B_2 = \{a_2, a_3, a_4\}$, $B_3 = \{a_5, a_6, a_7\}$, $B_4 = \{a_8\}$. So, there is $I_A = 4$ and $R_B = 2$. A state assignment is treated as optimal if any class $B_i \in \Pi_A$ is represented by a single generalized interval of R_1-dimensional Boolean space [3]. It is impossible to encode the states $a_m \in A$ in this way for the discussed example. Because of it, let us encode the states in the trivial way: $K(a_1) = 000, \ldots, K(a_8) = 111$.

Because of $R_A = 2$, there is $\tau = \{\tau_1, \tau_2\}$. Let us encode the classes $B_i \in \Pi_A$ in the trivial way: $K(B_1) = 00, \ldots, K(B_4) = 11$. Let us point out that there is no need in the representing the transitions from state $a_8 \in B_4$ by the structure table. In the case of D flip-flops such transitions are executed automatically (using only pulse Clock). Therefore, the code 11 can be treated as "don't care" input assignment.

To construct the structure table of P_C Moore FSM, it is necessary to construct the system of generalized formulae of transitions. In the discussed case, this system is the following one:

$$\begin{aligned} B_1 &\rightarrow x_1a_2 \vee \bar{x}_1x_2a_3 \vee \bar{x}_1\bar{x}_2a_4;\\ B_2 &\rightarrow x_3a_5 \vee \bar{x}_3x_4a_6 \vee \bar{x}_3\bar{x}_4a_7;\\ B_3 &\rightarrow x_4a_5 \vee \bar{x}_4x_5a_8 \vee \bar{x}_4\bar{x}_5a_7. \end{aligned} \tag{8.47}$$

Let us point out that there is no formula for the class $B_4 \in \Pi_A$ in the system (8.47). It is connected with the fact that this class is treated as "don't care".

There are the following columns in the ST of P_C Moore FSM: B_i, $K(B_i)$, a_s, $K(a_s)$, X_h, Φ_h, h. This table is constructed using the system of GFT. In the discussed case, they use the system (8.47) to construct the Table 8.23.

Let us use the FPGA chip including EMBs with the following configurations: 256×1, 128×2, 64×4, 32×8 (bits). It is necessary to use EMBs with $V_0 = 2^{3+8} \cdot 3 = 6144$ bits on the case of Moore FSM $MP(\Gamma_{15})$. Obviously, the replacement of logical conditions should be used in the discussed case.

In the discussed case, there is $R_1 = 3$. It means that the configuration 64×4 must be used. Because there is $R_A = 2$, the EMB has $S_A - R_A = 6 - 2 = 4$ free inputs. It gives $|X^1| = 3$, $|X^2| = 2$ and $G = 1$. Let us represent the set X as $X^1 \cup X^2$ where $X^1 = \{x_2, x_3, x_5\}$ and $X^2 = \{x_1, x_4\}$.

Now, let us replace the logical conditions $x_l \in X^2$. The RLC is shown in Table 8.24.

Table 8.23 Structure table of Moore FSM $P_C(\Gamma_{15})$

B_i	$K(B_i)$	a_s	$K(a_s)$	X_h	Φ_h	h
B_1	00	a_2	001	x_1	D_3	1
		a_3	010	$\bar{x}_1x_2$	D_2	2
		a_4	011	$\bar{x}_1\bar{x}_2$	D_2D_3	3
B_2	01	a_5	100	x_3	D_1	4
		a_6	101	$\bar{x}_3x_4$	D_1D_3	5
		a_7	110	$\bar{x}_3\bar{x}_4$	D_1D_2	6
B_3	10	a_5	100	x_4	D_1	7
		a_8	111	$\bar{x}_4x_5$	$D_1D_2D_3$	8
		a_7	110	$\bar{x}_4\bar{x}_5$	D_1D_2	9

Table 8.24 Table of replacement of logical conditions for Moore FSM $M_0P_C(\Gamma_{15})$

B_i	B_1	B_2	B_3	B_4
p_1	x_1	x_4	x_4	–

Table 8.25 Transformed structure table of Moore FSM $M_{0C}P_C(\Gamma_{15})$

B_i	$K(B_i)$	a_s	$K(a_s)$	X_h^1	P_h	Φ_h	h
B_1	00	a_2	001	1	p_1	D_3	1
		a_3	010	x_2	$\bar{p_1}$	D_2	2
		a_4	011	$\bar{x_2}$	$\bar{p_1}$	D_2D_3	3
B_2	01	a_5	100	x_3	1	D_1	4
		a_6	101	$\bar{x_3}$	p_1	D_1D_3	5
		a_7	110	$\bar{x_3}$	$\bar{p_1}$	D_1D_2	6
B_3	10	a_5	100	1	p_1	D_1	7
		a_8	111	x_5	p_1	$D_1D_2D_3$	8
		a_7	110	$\bar{x_5}$	p_1	D_1D_2	9

The following equation can be derived from Table 8.24: $p_1 = B_1x_1 \vee B_2x_4 \vee B_3x_4 = \bar{\tau_1}\bar{\tau_2}x_1 \vee \bar{\tau_1}\tau_2x_4 \vee \tau_1\bar{\tau_2}x_4$. This formula can be implemented as a single-level circuit using LUTs with $S \geq 4$. Let us point out that this equation can be simplified due to appropriate encoding of the classes $B_i \in \Pi_A$. For example, if there is $K(B_1) = 00$, $K(B_2) = 10$, $K(B_3) = 11$, $K(B_4) = 01$, then there is the following equation: $p_1 = \bar{\tau_1}x_1 \vee \tau_1x_4$. It can be implemented using only one LUT having $S = 3$.

There are the following columns in the transformed ST of $M_{0C}P_C$ Moore FSM: B_i, $K(B_i)$, a_s, $K(a_s)$, X_h^1, P_h, Φ_h, h. It is Table 8.25 in discussed case.

The table of BIMF includes the same columns as it is for the case of FSM $M_{0E}P_E$. In the discussed case, the transitions from each class $B_i \in \Pi_A$ are represented by 16 rows of the table. The table of BMO includes the following columns: $K(a_m)$, Y, $K(B_i)$, m. The column $K(B_i)$ includes the code of class $B_i \in \Pi_A$ such that $a_m \in B_i$ (for the row number m of the table). In the case of Moore FSM $M_{0C}P_C(\Gamma_{15})$ this table is represented by Table 8.26.

Table 8.26 Table of BMO for Moore FSM $M_{0C}P_C(\Gamma_{15})$

$K(a_m)$	Y	$K(B_i)$	m
$T_1T_2T_3$	$y_1y_2y_3y_4y_5y_6$	$\tau_1\tau_2$	
000	000000	00	1
001	110000	01	2
010	001100	01	3
011	000011	01	4
100	101000	10	5
101	010101	10	6
110	001100	10	7
111	101000	11	8

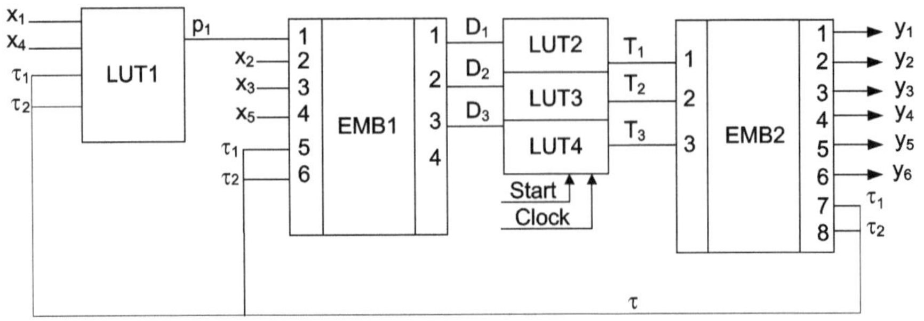

Fig. 8.28 Logical circuit of Moore FSM $M_{OC}P_C(\Gamma_{15})$

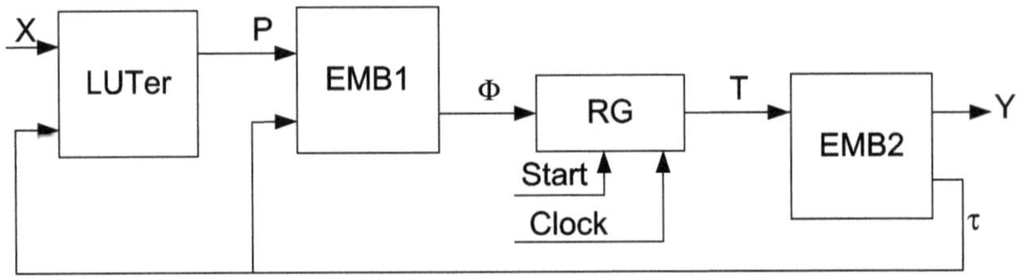

Fig. 8.29 Structural diagram of $M_C P_C$ Moore FSM

There is the logic circuit of Moore FSM $M_{OC}P_C(\Gamma_{15})$ shown in Fig. 8.28. There is a single LUT with $S = 4$ used for implementing the circuit of LUTer. To implement the circuit of BIMF, it is enough an EMB having the configuration 64×4 (one output is free). They need the configuration 8×8 to implement the circuit of BMO.

It is possible to use the transformation of PES without the splitting logical conditions. It results in $M_C P_C$ Moore FSM (Fig. 8.29).

Now, let us discuss the models based on the special state assignments. It is reduced to representing the set A as two disjoint sets A_C and A_U. If there is $X(a_m) \neq \emptyset$, then the state $a_m \in A$ is placed into the set A_U.

Let us discuss the following example. There is an FSM S_1 with $A = \{a_1, \ldots, a_{16}\}$ and $\Pi_A = \{B_1, \ldots, B_{10}\}$. Let it be the following classes $B_i \in \Pi_A$: $B_1 = \{a_1\}$, $B_2 = \{a_2, a_3\}$, $B_3 = \{a_4, a_5, a_6\}$, $B_4 = \{a_7, a_8\}$, $B_5 = \{a_9\}$, $B_6 = \{a_{10}, a_{11}, a_{12}\}$, $B_7 = \{a_{13}\}$, $B_8 = \{a_{14}\}$, $B_9 = \{a_{15}\}$, $B_{10} = \{a_{16}\}$. Let the logical conditions $x_l \in X$ be represented by the additional variables $p_g \in P$ as it is show in Table 8.27.

Table 8.27 Table of RLC for Moore FSM S_1

B_i	B_1	B_2		B_3			B_4		B_5	B_6			B_7	B_8	B_9	B_{10}
a_m	a_1	a_2	a_3	a_4	a_5	a_6	a_7	a_8	a_9	a_{10}	a_{11}	a_{12}	a_{13}	a_{14}	a_{15}	a_{16}
p_1	x_1	x_2	x_2	–	–	–	–	–	x_1	x_6	x_6	x_6	–	–	–	–
p_2	–	x_3	x_3	–	–	–	–	–	x_5	x_4	x_4	x_4	–	–	–	–

Fig. 8.30 Optimal state codes for Moore FSM S_1

T_3T_4 \ T_1T_2	00	01	11	10
00	a_1	a_2	a_8^*	a_{10}
01	a_9	a_3	a_{13}^*	a_{11}
11	a_{14}^*	a_{15}^*	a_{16}^*	a_{12}
10	a_4^*	a_5^*	a_7^*	a_6^*

It is possible to derive the following equations from the Table 8.27:

$$p_1 = A_1x_1 \vee (A_2 \vee A_3)x_2 \vee A_9x_1 \vee (A_{10} \vee A_{11} \vee A_{12})x6; \tag{8.48}$$

$$p_2 = (A_2 \vee A_3)x_3 \vee A_9x_5 \vee (A_{10} \vee A_{11} \vee A_{12})x_4. \tag{8.49}$$

Let us encode the states in the trivial way: $K(a_1) = 0000, \ldots, K(a_{16}) = 1111$. Let us discuss the expression (8.48).

There is the following expression for p_1 corresponding the this style of state assignment:

$$p_1 = \bar{T}_1\bar{T}_2x_1 \vee T_2\bar{T}_3x_2 \vee T_1\bar{T}_3x_2 \vee T_2T_3\bar{T}_4x_6 \vee T_1T_3\bar{T}_4x_6. \tag{8.50}$$

Let us use the LUTs having $S = 4$ for implementing the LUTer. It is possible to find that there are 6 LUTs in the circuit corresponding to (8.50). This circuit has 4 layers.

Let us divide the set A by following subsets: $A_C = \{a_1, a_2, a_3, a_9, \ldots, a_{12}\}$, $A_U = \{a_4, \ldots, a_8, a_{13}, \ldots, a_{16}\}$. To optimize the expressions for $p_g \in P$, we can treat as "don't care" the state codes for $a_m \in A_U$. Let us encode the states $a_m \in A$ in the optimal way: if $x_l \in X$ is replaced by p_g for states $a_m, a_s \in A_C$, then the codes $K(a_m)$ and $K(a_s)$ should be placed into the same generalized interval of R_1-dimensional Boolean space. There is the outcome of this approach shown in Fig. 8.30.

There is the sign "*" near the states $a_m \in A_U$. These states are treated as "don't care". The following expression can be derived from Fig. 8.30:

$$p_1 = \bar{T}_1\bar{T}_2x_1 \vee T_2x_2 \vee T_1x_6. \tag{8.51}$$

There are 2 of LUTs with $S = 4$ in the circuit corresponding to (8.51). Analysis of (8.51) shows that there is a set $T' \subseteq T$ whose elements are used in this equation. It is the set $T' = \{T_1, T_2\}$.

This approach leads to M_SP Moore FSM (Fig. 8.31). In this FSM, the LUTer implements the system

$$P = P(T', X). \tag{8.52}$$

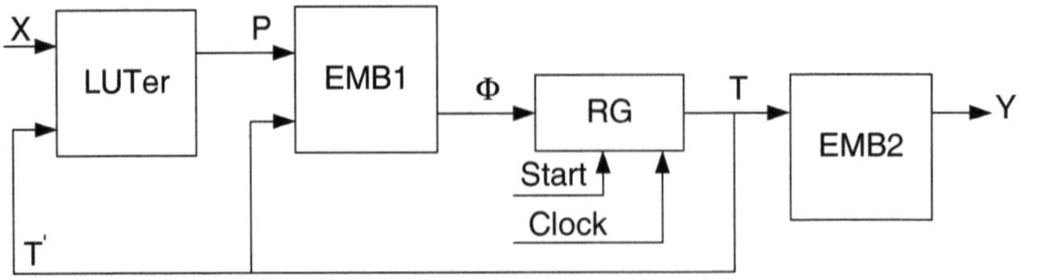

Fig. 8.31 Structural diagram of $M_S P$ Moore FSM

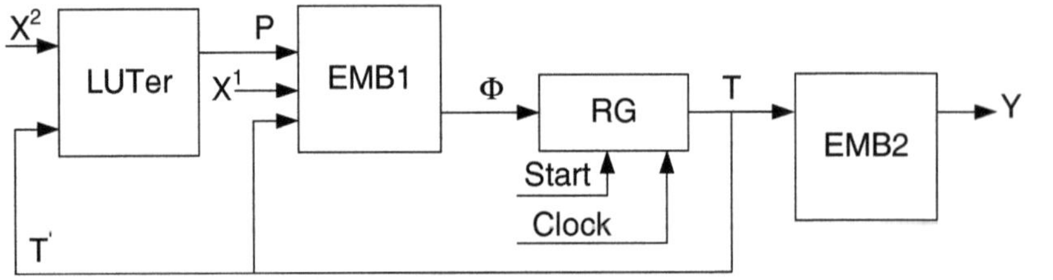

Fig. 8.32 Structural diagram of $M_{0S} P$ Moore FSM

It is possible to combine this approach with the splitting logical conditions. It leads to $M_{0S} P$ Moore FSM shown in Fig. 8.32.

In this model, the LUTer implements the system

$$P = P(T', X^2). \tag{8.53}$$

The same approach can be used for $M_C P_C$ Moore FSM. Let us discuss the FSM S_1. As follows from Table 8.27, there is $I_A = 9$. It determines $R_A = 4$ and $\tau = \{\tau_1, \ldots, \tau_4\}$. Let us encode the classes $B_i \in \Pi_A$ in the trivial way: $K(B_1) = 0000, \ldots, K(B_{10}) = 1001$. In this case, the following expression can be constructed for the function p_1:

$$p_1 = \bar{\tau}_1 \bar{\tau}_2 x_1 \vee \tau_2 \bar{\tau}_4 x_2 \vee \tau_2 \tau_3 x_6. \tag{8.54}$$

This equation can be implemented using four of LUTs having $S = 4$. The circuit has two layers of logic.

Now, let us divide the set Π_A by two disjoint subsets Π_{AC} and Π_{AU}. If there are conditional transitions from states $a_m \in B_i$, then $B_i \in \Pi_{AC}$. Otherwise there is the following relation: $B_i \in \Pi_{AU}$. There are the following sets in the discussed case: $\Pi_{AC} = \{B_1, B_2, B_5, B_6\}$ and $\Pi_{AU} = \{B_3, B_4, B_7, \ldots, B_{10}\}$.

Let us encode the classes $B_i \in \Pi_{AC}$ in the optimal way: if the logical condition $x_l \in X$ is replaced by variable $p_g \in P$ for the states $a_m \in B_i \cup B_j$, then the these classes should have codes from the same generalized interval of R_A-dimensional Boolean space. There is one of the possible outcomes of optimal class assignment shown in Fig. 8.33.

$\tau_3\tau_4$ \ $\tau_1\tau_2$	00	01	11	10
00	B_1	B_2	$*$	B_3
01	B_5	B_4^*	$*$	B_6^*
11	$*$	$*$	$*$	$*$
10	B_7^*	B_8^*	B_{10}^*	B_9

Fig. 8.33 Optimal class codes for Moore FSM S_1

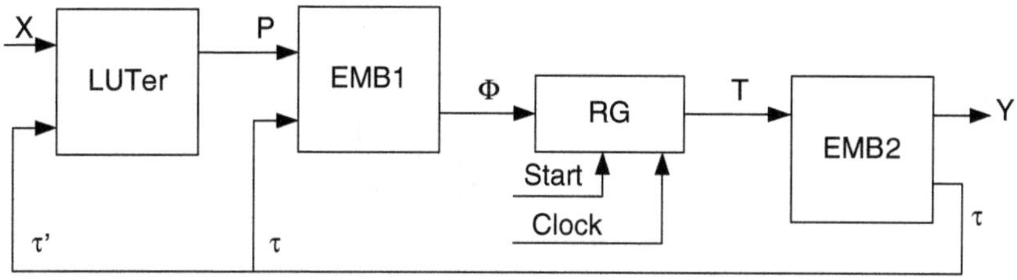

Fig. 8.34 Structural diagram of $M_{CS}P_C$ Moore FSM

There is the sign "*" near the classes $B_i \in \Pi_{AU}$. It means that their codes are treated as insignificant input assignments. Using this rule, it is possible to get the following equation:

$$p_1 = \bar{\tau}_1\bar{\tau}_2 x_1 \vee \tau_2 x_2 \vee \tau_1 x_6. \quad (8.55)$$

The comparison of (8.54) and (8.55) shows that the later is much simpler. It corresponds to the logic circuit with only two LUTs having $S = 4$. Also let us point out that there is only a subset τ' of the set τ in equation (8.55). There is $\tau' = \{\tau_1, \tau_2\}$. Let us denote the Moore FSM based on this approach as $M_{CS}P_C$ Moore FSM. Its structural diagram is shown in Fig. 8.34.

In this FSM, the LUTer implements the system

$$P = P(\tau', X). \quad (8.56)$$

Both EMB1 and EMB2 execute the same function as it is for $M_C P_C$ Moore FSM.

This approach can be combined with the splitting logical conditions. It leads to $M_{0CS}P_C$ Moore FSM (Fig. 8.35).

In this model, the LUTer generates the functions

$$P = P(\tau', X^2). \quad (8.57)$$

All discussed models can be modified using the method of encoding of logical conditions [6, 12]. We discussed it before in this book. For example, there are four pairs of logical conditions in the Table 8.27: $\langle x_1, -\rangle$, $\langle x_2, x_3\rangle$, $\langle x_1, x_5\rangle$ and $\langle x_4, x_6\rangle$.

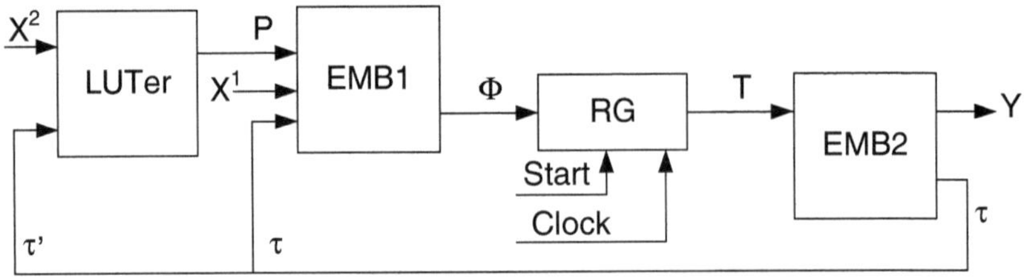

Fig. 8.35 Structural diagram of $M_{OCS}P_C$ Moore FSM

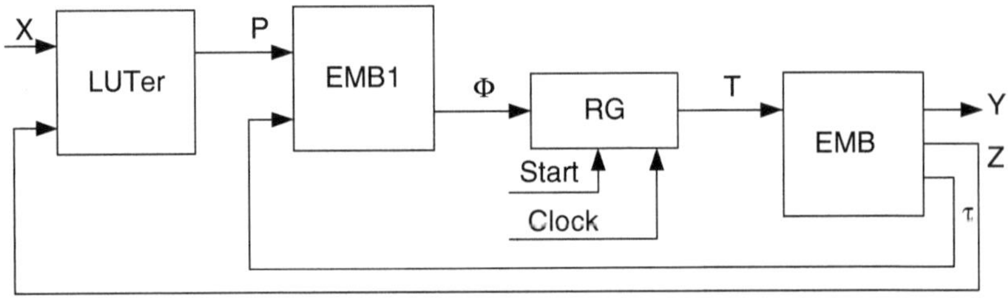

Fig. 8.36 Structural diagram of $M_Z P_C$ Moore FSM

So, there is $n_P = 4$. Using (2.20), it is possible to find the value of $R_P = 2$. Let us use the variables $z_r \in Z$ for encoding the pairs $\langle x_l, x_m \rangle$. Let us use the following codes: $K(x_1, -) = 00$, $K(x_2, x_3) = 01$, $K(x_1, x_5) = 10$ and $K(x_4, x_6) = 11$. Now, it is possible to construct the following system of equations:

$$\begin{aligned} p_1 &= \bar{z_2}x_1 \vee \bar{z_1}z_2x_2 \vee z_1z_2x_6; \\ p_2 &= z_2x_3 \vee z_1x_5 \vee z_1z_2x_4. \end{aligned} \tag{8.58}$$

Each of equations (8.58) corresponds to the logic circuit having 2 LUTs with $S = 4$. Let us point out that the variables $z_r \in Z$ should be generated by BMO.

For example, there is a structural diagram of $M_Z P_C$ Moore FSM shown in Fig. 8.36. It is based on the encoding of logical conditions and transformation of PES.

In this model, the LUTer implements the system

$$P = P(Z, X). \tag{8.59}$$

The EMB2 implements the systems (1.5), (4.1) and (4.4).

There are the characteristics of different models of Moore FSMs with replacement of logical conditions shown in Table 8.28.

The row 17 of Table 8.28 represents the MP Moore FSM where everything is encoded in the trivial way. The rows 1–16 show different approach used for hardware reducing in the blocks BRLC or/and BIMF. We have discussed some of these models. It is possible to work out design methods for each of models from Table 8.28. It is

Table 8.28 Characteristics of Moore FSM with RLC

№	FSM	BRLC	BIMF	BMO
1.	M_0P	$P = P(T, X^2)$	$\Phi = \Phi(T, P, X^1)$	$Y = Y(T)$
2.	$M_{0E}P_E$	$P = P(T', X^2)$	$\Phi = \Phi(T', P, X^1)$	$Y = Y(T)$
3.	$M_{0C}P_C$	$P = P(\tau, X^2)$	$\Phi = \Phi(\tau, P, X^1)$	$Y = Y(T)$ $\tau = \tau(T)$
4.	M_CP_C	$P = P(\tau, X)$	$\Phi = \Phi(\tau, X)$	$Y = Y(T)$ $\tau = \tau(T)$
5.	M_sP	$P = P(T', X)$	$\Phi = \Phi(T'X)$	$Y = Y(T)$
6.	$M_{0S}P$	$P = P(T', X^2)$	$\Phi = \Phi(T', X^1, P)$	$Y = Y(T)$
7.	$M_{CS}P_C$	$P = P(\tau', X)$	$\Phi = \Phi(\tau, P)$	$Y = Y(T)$ $\tau = \tau(T)$
8.	$M_{0CS}P_C$	$P = P(\tau', X^2)$	$\Phi = \Phi(\tau, P, X^1)$	$Y = Y(T)$ $\tau = \tau(T)$
9.	$M_{0Z}P$	$P = P(Z, X^2)$	$\Phi = \Phi(T, P, X^1)$	$Y = Y(T)$ $Z = Z(T)$
10.	$M_{0EZ}P_E$	$P = P(Z, X^2)$	$\Phi = \Phi(T', P, X^1)$	$Y = Y(T)$ $Z = Z(T)$
11.	$M_{0CZ}P_C$	$P = P(Z, X^2)$	$\Phi = \Phi(\tau, P, X^1)$	$Y = Y(T)$ $\tau = \tau(T)$ $Z = Z(T)$
12.	$M_{CZ}P_C$	$P = P(Z, X)$	$\Phi = \Phi(\tau, X)$	$Y = Y(T)$ $\tau = \tau(T)$ $Z = Z(T)$
13.	$M_{SZ}P$	$P = P(Z, X)$	$\Phi = \Phi(T', X)$	$Y = Y(T)$ $Z = Z(T)$
14.	$M_{0SZ}P$	$P = P(Z, X^2)$	$\Phi = \Phi(T', X^1, P)$	$Y = Y(T)$ $Z = Z(T)$
15.	$M_{CSZ}P_C$	$P = P(Z, X)$	$\Phi = \Phi(\tau, P)$	$Y = Y(T)$ $\tau = \tau(T)$ $Z = Z(T)$
16.	$M_{0CS}P_C$	$P = P(Z, X^2)$	$\Phi = \Phi(\tau, P, X^1)$	$Y = Y(T)$ $\tau = \tau(T)$ $Z = Z(T)$
17.	MP_E	$P = P(X, T)$	$\Phi = \Phi(T, X)$	$Y = Y(T)$

necessary to have an expert system assisting in the choice of the approach leading to the circuit with minimum amount of LUTs in BRLC.

If input memory functions depend on variables $T_r \in T$, then it is possible to use the method of distribution of microoperations. It is necessary to optimize the hardware amount in BMO. This method can be applied for MP, M_0P, and $M_{0Z}P$ Moore FSM. We do not discuss these models in our book.

References

1. Altera, http://www.altera.com. Accessed Jan 2015
2. S. Baranov, *Logic Synthesis of Control Automata* (Kluwer Academic Publishers, 1994)
3. A. Barkalov, Principles of logic optimization for Moore microprogram automaton. Cybern. Syst. Anal. **34**(1), 54–60 (1998)
4. A. Barkalov, L. Titarenko, *Logic Synthesis for FSM-Based Control Units* (Springer, Berlin, 2009)
5. A. Barkalov, L. Titarenko, A. Barkalov, Structural decomposition as a tool for the optimization of an FPGA-based implementation of a Mealy FSM. Cybern. Syst. Anal. **48**(2), 313–323 (2012)
6. A. Barkalov, I. Zelenjova, Optimization of replacement of logical conditions for an automaton with bidirectional transitions. Autom. Control Comput. Sci. **34**(5), 48–53, Allerton Press Inc
7. I. Grout, *Digital Systems Design with FPGAs and CPLDs* (Elsevier Science, Oxford, 2008)
8. M. Kołopieńczyk, A. Barkalov, L. Titarenko, Hardware reduction for RAM-based Moore FSMs, in *7th International Conference on Human System Interactions—HSI 2014*, Lisbon, Portugal (2014), pp. 255–260 [B.m.]
9. LGSynth93, International Workshop on logic synthesis benchmark suite (LGSynth93). TAR, Benchmarks test, http://www.cbl.ncsu.edu:16080/benchmarks/LGSynth93/LGSynth93.tar (1993)
10. C. Maxfield, *The Design Warrior's Guide to FPGAs* (Academic Press Inc, Orlando, 2004)
11. C. Scholl, *Functional Decomposition with Application to FPGA Synthesis* (Kluwer Academic Publishers, Boston, 2001)
12. V. Sklyarov, I. Skliarova, A. Barkalov, L. Titarenko, *Synthesis and Optimization of FPGA-based Systems*, vol. 294, Lecture notes in electrical engineering (Springer, Berlin, 2014)
13. Xilinx, http://www.xilinx.com. Accessed Jan 2015

Conclusion

Now we are witnesses of the intensive development of design methods targeting FPGA-based circuits and systems. The complexity of digital systems to de designed increases drastically, as well as the complexity of FPGA chips used for the design. The up-to-day FPGAs include up to seven billions of transistors and it is not a limit.

Development of digital systems with such complex logic elements is impossible without application of hardware description languages, computer-aided design tools and design libraries. But even the application of all these tools does not guarantee that some competitive product will be designed for appropriate time-to-market. To solve this problem, a designer should know not only CAD tools, but the design and optimization methods, too. It is especially important in case of such irregular devices as control units. Because of irregularity, their logic circuits are implemented without using of the standard library cells; only LUTs and EMBs of a particular FPGA chip can be used in FSM logic circuit design. In this case, the knowledge and experience of a designer become a crucial factor of the success. Many experiments conducted with use of standard industrial packages show that outcomes of their operation are, especially in case of complex control units design, too far from optimal. Thus, it is necessary to develop own program tools oriented on FSM optimization and use them together with industrial packages. This problem cannot be solved without fundamental knowledge in the area of logic synthesis. Besides, to be able to develop new design and optimization methods, a designer should know the existed methods.

We think that new FSM models and design methods proposed in our book will help in solution of this very important problem. We hope that our book will be useful for the designers of digital systems and scholars developing synthesis and optimization methods targeting implementation of logic circuits of FPGA-based finite state machines.

A. Barkalov et al., *Logic Synthesis for FPGA-Based Finite State Machines*,
Studies in Systems, Decision and Control 38, DOI 10.1007/978-3-319-24202-6

Index

A
Addressing of microinstructions
 combined, 10
 compulsory, 10, 13
 natural, 16

B
Block of
 code transformer, 66
 functions, 4
 identifiers, 100
 input memory functions, 7, 213
 microoperations, 7
 state codes transformer, 97
Boolean
 function, 193
 space, 49, 130, 135, 176, 251, 267, 271, 272
 variable, 3, 6, 8, 24, 25, 40
BRAM, 197, 200, 211

C
Circuit
 LUT-based, 39, 41
Class of
 pseudoequivalent states, 9, 46
CLB, 33, 34
Collection of microoperations, 2, 207
Combinational circuit, 28, 193
Complex programmable logic device (CPLD), 21
Compositional microprogram control unit (CMCU), 3, 15, 16, 18
Computer aided design (CAD) system, 7
Consumed power, 37, 42, 50

D
Decomposition, 27
 functional, 193, 260
 structural, 26, 197, 216, 243
Don't care input assignment, 140, 268, 273

E
Embedded memory block (EMB), 193

F
Field-programmable gate array (FPGA), 21, 33
Field-programmable logic device (FPLD), 21
Fields of compatible microoperations, 21, 22, 66, 100, 112, 143, 243
Finite state-machine (FSM), 1
 Mealy, 2
 Moore, 2
Flip-flop, 2

G
Generalized formula of transitions, 253, 256, 263, 268
Generalized interval of Boolean space, 267, 271, 272
Generic array logic (GAL), 27
Graph-schemes of algorithm
 block, 2, 3, 10

H
Hardware
 amount, 13, 45, 57, 60, 101, 120, 148, 186, 197, 210, 243, 275

A. Barkalov et al., *Logic Synthesis for FPGA-Based Finite State Machines*,
Studies in Systems, Decision and Control 38, DOI 10.1007/978-3-319-24202-6

reduction, 21, 50, 65, 103, 143, 158, 161, 211, 226, 257, 266
Hardware description language, 37

I
Interconnections, 33, 37, 43, 193

J
JEDI, 45, 49, 50

K
Karnaugh map, 20, 105, 109, 161, 166, 172

L
Logical condition, 2

M
Macrocell, 27
Microoperation, 2
Model of FSM, 2, 22
Multiplexer, 14, 27, 34, 134

O
Object codes transformation (OCT), 65, 97
Object of FSM
heterogeneous, 65
Operational linear chain (OLC), 15

P
PALer, 129
Performance, 38, 45, 46, 50, 133, 195
Product term, 5, 24, 27, 28, 41, 129
Programmable array logic (PAL), 21
Programmable logic array (PLA), 21
Programmable logic sequencer (PLS), 27
Programmable read-only memory (PROM), 21
Pseudoequivalent states, 9, 46, 60, 251, 261

R
Read-only memory (ROM), 14

S
State
code, 1
internal, 1
variables, 2
State assignment
energy-saving, 46
Structural diagram, 4, 7, 11, 16, 26–28
Sum-of-products (SOP), 6
Synchronization, 19, 242
Synthesis, 3, 4, 14, 18, 37, 38, 43, 68, 71, 77, 135
System of
Boolean functions, 25, 57
generalized formulae of transitions, 253, 256, 263, 268

T
Table of
BCT, 66, 77, 78
BIMF, 18, 19, 69, 72, 75, 77, 82, 84, 124, 135, 145, 150, 155, 162, 168, 175, 176, 210, 216, 224, 230, 234, 246, 250, 253, 256, 259, 264, 269
BMO, 19, 70, 72, 78, 82, 85, 89, 157, 165, 175, 183, 190, 204, 208, 231, 242, 256, 260, 264, 269
BSCT, 152, 156, 175
EMBer, 43, 45, 74, 78, 83, 105, 106, 110, 114, 115, 119, 140, 152, 175
LUTer, 51, 59, 60, 79, 83, 86, 106, 116–118, 123, 135, 138
MX, 175
Transformation of
initial GSA, 12
object codes, 60, 65, 97
state codes, 60, 97, 101, 129, 135, 154, 266
Transformed
structure table, 202, 209, 224, 229, 234, 242, 250, 253, 256
Truth table, 8

V
Vertex
conditional, 3
end, 2
operator, 2
start, 2
Verticalization of GSA, 125